Handbook of
3D Machine Vision

Optical Metrology and Imaging

SERIES IN OPTICS AND OPTOELECTRONICS

Series Editors: **E Roy Pike**, Kings College, London, UK

Robert G W Brown, University of California, Irvine, USA

Recent titles in the series

Handbook of
3D Machine Vision

Optical Metrology and Imaging

Edited by
Song Zhang

CRC Press
Taylor & Francis Group
Boca Raton London New York

CRC Press is an imprint of the
Taylor & Francis Group, an **informa** business

A TAYLOR & FRANCIS BOOK

CRC Press
Taylor & Francis Group
6000 Broken Sound Parkway NW, Suite 300
Boca Raton, FL 33487-2742

First issued in paperback 2017

ISBN-13: 978-1-4398-7219-2 (hbk)
ISBN-13: 978-1-138-19957-6 (pbk)

Library of Congress Cataloging-in-Publication Data

Handbook of 3D machine vision : optical metrology and imaging / author/editor Song Zhang.
  pages cm. -- (Series in optics and optoelectronics ; 16)
  Includes bibliographical references and index.
  ISBN 978-1-4398-7219-2 (alk. paper)
  1. Imaging systems. 2. Three-dimensional imaging. 3. Computer vision. I. Zhang, Song.

TK8315.H36 2013
006.3'7--dc23   2012037858

Visit the Taylor & Francis Web site at
http://www.taylorandfrancis.com

and the CRC Press Web site at
http://www.crcpress.com

Contents

Preface

In August 2010, John Navas and I were sitting in a restaurant, across the road from the San Diego Convention Center, brainstorming the idea of a book on the recent hot topic of three-dimensional (3D) machine vision. We found that the success of any 3D vision technique is founded upon the capabilities of its real-world measurement strategy, and thus a new book covering a wide variety of existing 3D measurement approaches might generate significant interest.

With the release of *Avatar* and other 3D movies and the emergence of 3D TVs and monitors, 3D imaging technologies have started penetrating into our daily lives. The popularity of Microsoft Kinect now enables the general audience easily to capture and access 3D measurements of the real world. However, the Kinect has many limitations, and therefore it becomes increasingly important to know how to choose an optimal 3D imaging technique for a particular application. Yet, with numerous 3D vision methods and no single comprehensive resource for understanding them, the process of choosing a method becomes very frustrating.

This book strives to fill this gap by providing readers with the most popular 3D imaging techniques in depth and breadth. The book focuses on optical methods (optical metrology and imaging), which are the most popular due to their noninvasive and noncontact nature. We start the book (Chapter 1) with stereo vision, as it is a well-studied method. Because of the fundamental difficulty of stereo matching, adding random speckle patterns or space–time varying patterns substantially improves the results of stereo vision; this approach is covered in Chapters 2 and 3. Stereo particle image velocimetry (PIV) will also be discussed in this book (Chapter 4) as a major experimental means in fluid dynamics. Chapter 5 presents the structured-light technique, an extremely popular method in the field of computer science due to its ease of implementation and robustness. However, structured light is usually restricted to macroscale measurement.

Chapter 6 will discuss digital holography, which is capable of performing micro- to nanoscale measurement. Precisely measuring dynamically deformable natural objects (i.e., without surface treatment) is vital to understanding kinematics, dynamics, mechanics, etc. Grating, interferometry, and fringe projection techniques have been successfully used for these types of applications; Chapters 7–9 will cover these methods. All these aforementioned techniques require triangulation to recover a 3D shape; yet, some techniques do not. Chapters 10 and 11 present two representative methods. Moreover, Chapters 12–14 cover 3D measurement techniques that are not restricted to surface capture: 3D ultrasound (Chapter 12), optical coherence tomography (OCT, Chapter 13), and 3D endoscopy (Chapter 14).

Finally, the ultimate goal behind the development of novel 3D imaging techniques is to conquer the challenges we face daily and improve our quality of life. For this reason, the book cannot be complete without dedicating one chapter to a representative application. Therefore, Chapter 15 covers the promising field of biometrics, which may prove essential to security and public safety.

I am deeply indebted to the inventors and/or the major players of each individual technology for contributing the chapter on that particular method. I thank the authors for their time and willingness to work with me, and I genuinely believe that their efforts will be greatly appreciated by readers. I am extremely grateful to John Navas for his tremendous help in originating and navigating the book to its current form and Jennifer Ahringer and Rachel Holt for their incredible effort in publishing this book in a timely fashion. I sincerely thank my professional mentor, Dr. James Oliver, at Iowa State University, who has provided constant encouragement and support over the past few years, which ultimately made me believe that I could edit this book during my early career! I also thank my students, especially Laura Ekstrand, for spending many hours helping me organize this book. Last, but not least, I wholeheartedly thank my wife, Xiaomei Hao, for allowing me to occupy myself many nights and weekends on this book without any complaint. Without all of these wonderful people, this book would not be possible.

I hope that this book is a handy *handbook* for students, engineers, and scientists as a reference for learning the essence of 3D imaging techniques and a time-saver for choosing the optimal method to study or implement. Most of the images are currently printed in black and white, but you can download the color figures from the CRC website for better clarity.

Song Zhang
Iowa State University

The Editor

Dr. Song Zhang is an assistant professor of mechanical engineering at Iowa State University. His research interests include investigation of the fundamental physics of optical metrology, new mathematical and computational tools for 3D shape analysis, and the utilization of those insights for designing superfast 3D imaging and sensing techniques. He has often been credited for developing the first high-resolution, real-time 3D imaging system. Recently, he has developed a 3D imaging system that could achieve kilohertz with hundreds of thousands of measuring points per frame by inventing the defocusing technology.

Dr. Zhang has published over 40 peer-reviewed journal articles, authored four book chapters, and filed five US patents (granted or pending). Among all the journal papers he has published, seven were featured on the cover pages, and one was highlighted as "Image of the Week" by Optics InfoBase. Numerous media have reported his work, and rock band Radiohead has utilized his technology to produce a music video House of Cards. He serves as a reviewer for over 20 international journals, as a committee member for numerous conferences, and as cochair for a few conferences. He won the NSF CAREER award in 2012.

Contributors

Anand Asundi
Nanyang Technological University
of Singapore
Singapore

Seung-Hae Baek
Graduate School of Electrical
Engineering and Computer
Science
Kyungpook National University
Daegu, Korea

Hujun Bao
State Key Lab of CAD & CG
Zhejiang University
Hangzhou, China

Jeff Bax
Imaging Research Laboratories
Robarts Research Institute
Graduate Program in Biomedical
Engineering
London, Ontario, Canada

Wenjing Chen
Opto-Electronics Department
Sichuan University
Chengdu, China

Bernard Chiu
City University of Hong Kong
Hong Kong, China

Chee Oi Choo
Nanyang Technological University
of Singapore
Singapore

Brian Curless
University of Washington
Seattle, Washington

Maria De Marsico
Department of Computer Science
University of Rome
Rome, Italy

Kapil Dev
Nanyang Technological University
of Singapore
Singapore

Laura Ekstrand
Department of Mechanical
Engineering
Iowa State University
Ames, Iowa

Aaron Fenster
Imaging Research Laboratories
Robarts Research Institute,
Department of Medical Imaging
Graduate Program in Biomedical
Engineering, Department of
Medical Biophysics
London, Ontario, Canada

Sergio Fernandez
Institute of Informatics and
Applications
University of Girona
Girona, Spain

Jason Geng
IEEE Intelligent Transportation
System Society
Rockville, Maryland

Yan Hao
Nanyang Technological University
of Singapore
Singapore

Hui Hu
Department of Aerospace
   Engineering
Iowa State University
Ames, Iowa

Yuan Hao Huang
Singapore–MIT Alliance for
   Research and Technology
   (SMART) Center
Singapore

Nikolaus Karpinsky
Department of Mechanical
   Engineering
Iowa State University
Ames, Iowa

Shoji Kawahito
Shizuoka University
Shizuoka, Japan

Michael K. K. Leung
Department of Electrical and
   Computer Engineering
Ryerson University
Toronto, Canada

Yusheng Liu
Aerospace Image Measurement
   and Vision Navigation Research
   Center
National University of Defense
   Technology
Changsha, China

Michele Nappi
Biometric and Image Processing
   Laboratory (BIPLab)
University of Salerno
Fisciano (SA), Italy

Yukitoshi Otani
Center for Optical Research and
   Education
Utsunomiya University
Utsunomiya, Japan

Soon-Yong Park
School of Computer Science and
   Engineering
Kyungpook National University
Daegu, Korea

Grace Parraga
Imaging Research Laboratories
Robarts Research Institute,
   Department of Medical Imaging
Graduate Program in Biomedical
   Engineering, Department of
   Medical Biophysics
London, Ontario, Canada

Daniel Riccio
Biometric and Image Processing
   Laboratory (BIPLab)
University of Salerno
Fisciano (SA), Italy

Joaquim Salvi
Institute of Informatics and
   Applications
University of Girona
Girona, Spain

Steven M. Seitz
University of Washington
Seattle, Washington

Yang Shang
Aerospace Image Measurement
   and Vision Navigation Research
   Center
National University of Defense
   Technology
Changsha, China

Noah Snavely
University of Wisconsin
Madison, Wisconsin

Beau A. Standish
Department of Electrical and
    Computer Engineering
Ryerson University
Toronto, Canada

Xianyu Su
Opto-Electronics Department
Sichuan University
Chengdu, China

Catherine E. Towers
School of Mechanical Engineering
University of Leeds
Leeds, United Kingdom

David P. Towers
School of Mechanical Engineering
University of Leeds
Leeds, United Kingdom

Yajun Wang
Department of Mechanical
    Engineering
Iowa State University
Ames, Iowa

Qu Weijuan
Nanyang Technological University
    of Singapore
Singapore

Li Zhang
University of Wisconsin
Madison, Wisconsin

Qican Zhang
Opto-Electronics Department
Sichuan University
Chengdu, China

Song Zhang
Department of Mechanical
    Engineering
Iowa State University
Ames, Iowa

1

Stereo Vision

Soon-Yong Park and Seung-Hae Baek

CONTENTS

Stereo vision has been of interest to the computer and robot vision community. This chapter introduces fundamental theories of stereo vision such as stereo calibration, rectification, and matching. Practical examples of stereo calibration and rectification are also presented. In addition, several stereo matching techniques are briefly introduced and their performance is compared by using a couple of reference stereo images.

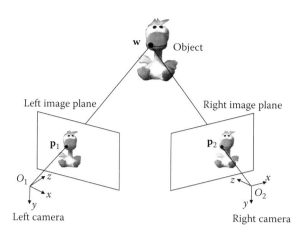

FIGURE 1.1
Stereo correspondence problem.

1.1 Introduction

Stereo vision is motivated from the human vision system, which can perceive depth properties of a scene. The human vision system obtains two different images of a scene from the slightly different views of the eyes and interprets the images for depth perception. Because depth perception is one of the powerful abilities of the human vision system, stereo vision has many application areas, such as three-dimensional (3D) scene reconstruction, object recognition, entertainment, robot navigation, etc. To obtain the depth information of a scene from a pair of images, it needs to solve one of the inherent problems in stereo vision, called the *correspondence problem*. Because two different image planes are defined in stereo vision, the projections of an object into the planes are represented with respect to different image coordinates. Therefore, the correspondence problem can be defined as determining the coordinate difference between the images of the object. Solving the stereo correspondence problem is called *stereo matching*. The result of stereo matching is commonly represented by a disparity map whose intensity represents the coordinate difference between corresponding image points, called *disparity*.

In general, a stereo vision system consists of two identical vision cameras, which capture the left and right images of an object. A conventional configuration of a stereo vision system is shown in Figure 1.1. A 3D point $\mathbf{W}$ on the surface of a real object is projected into the image planes of the cameras. Thus, two two-dimensional (2D) points, $\mathbf{p}_1$ and $\mathbf{p}_2$, on the image planes are the projections of point $\mathbf{W}$. As mentioned in the previous paragraph, the correspondence problem is to find points $\mathbf{p}_1$ and $\mathbf{p}_2$ between stereo images. Finding stereo correspondences needs a huge number of computations because the disparity of every image point should be determined to obtain a 2D disparity or depth map. For example, if the resolution of each image is $N \times M$, a brute force approach for finding all correspondences needs $(N \times M)^2$ computations.

To reduce the computation complexity of the stereo matching problem, the 3D geometry of a stereo vision system should be considered so that stereo matching is restricted to certain image areas. To know the stereo geometry, calibration of a stereo vision with respect to a reference coordinate system is needed. A direct calibration method is presented to obtain a geometric relationship between the two cameras, which is represented by 3D Euclidean transformations, rotation, and translation. From the stereo calibration, we know that stereo correspondences are related by a 2D point-to-line projection, called *epipolar geometry*.

The epipolar geometry between the stereo images provides a very strong constraint to find stereo correspondences. The epipolar constraint is essential in stereo matching because an image point in one image has its conjugate point on an epipolar line in the other image, while the epipolar line is derived by the original point. Therefore, the computation complexity and time of stereo correspondences can be greatly reduced. If the epipolar lines are parallel to the horizontal image axis, stereo matching can be done in an effective and fast way since the correspondences between stereo images lie only along the same horizontal line. For this reason, it is better to convert nonparallel epipolar lines to parallel ones. In terms of stereo configuration, this is the same as converting a general stereo configuration to the parallel stereo configuration, which is called *stereo rectification*. Rectification of stereo images transforms all epipolar lines in the stereo images parallel to the horizontal image axis. Therefore, in the rectified stereo images, corresponding image points are always in the same horizontal lines. This geometric property also greatly reduces computation time for stereo matching.

For many years, various stereo matching techniques have been introduced. Most stereo matching techniques are categorized into two types. In terms of matching cost and energy aggregation, they are categorized as either local or global stereo matching techniques. Local stereo matching techniques use image templates defined in both stereo images to measure their correlation [14,15]. Common techniques are SAD (sum of absolute difference), SSD (sum of squared difference), and NCC (normalized cross correlation).

In the template-based method, a cost function is defined based on the similarity between two image templates of their left and right images. Suppose an image template is defined from the left image and many comparing templates are defined along an epipolar line in the right image. Then a matching template from the right image is determined in the sense of minimizing the matching cost. Local matching techniques are useful when only some parts in an image are of interest for obtaining the depth map of the area.

By the way, most recent investigations in stereo vision address global error minimization. In global matching methods, a cost function is defined in terms of image data and depth continuity [14]. The data term measures the similarity of matching templates, while the continuity term additionally measures the global matching cost based on the disparities of adjacent image pixels. In other words, the disparity of an image point is determined not only from its image properties but also from disparity continuities around it. The global approaches produce more accurate results than the local methods. However, high computation time is needed. Examples of global matching are belief propagation (BP) [3,10,12], semiglobal matching (SGM) [7,8], graph cut (GC) [2], and cooperative region (CR) [19], among others.

In terms of the density of a disparity map, stereo matching techniques are divided into two categories: dense and sparse matching methods. The dense matching method decides the disparity of every image pixel, thus producing a disparity map in which resolution is the same as that of stereo images. Obtaining the disparity of every pixel produces very dense depth information. However, the dense matching method needs high computation time. On the other hand, the sparse matching method computes the disparities of only some image pixels. Feature-based stereo matching techniques belong to this method. Image features are relevant subjects in an image such as corners or edges. Therefore, obtaining disparities of feature points is easier and more robust than obtaining featureless points.

This chapter presents fundamental theories and practices of stereo vision. In Section 1.2, the stereo geometry is presented to define the epipolar geometry and constraint. In Section 1.3, calibration of a stereo vision system, which describes the geometric relationship between two independent camera coordinate systems, is presented. In Section 1.4, rectification of stereo images, which aligns epipolar lines along the same horizontal lines, is presented. Section 1.5 shows some practical examples of stereo calibration and rectification. In Section 1.6, two simple methods to compute 3D coordinates from a pair of stereo correspondences are presented. In Section 1.7, basic theories of several stereo matching techniques are briefly introduced and their performances are compared using standard stereo test images.

1.2 Stereo Geometry

In this section, the epipolar geometry of a stereo vision system is presented in order to understand the *epipolar constraint*. In a pair of stereo images, two corresponding points in the left and right images are always on the epipolar lines, which are defined in the stereo image planes. Figure 1.2 shows the geometry of a stereo vision system. Here, O_1 and O_2 are the centers of the left and right cameras, respectively. In the two image planes Π_1 and Π_2, 2D image points p_1 and p_2 are the projections of a 3D point W to each image plane. Since W is projected toward the center of each camera, p_1 and p_2 lie on the projection lines from W to both centers.

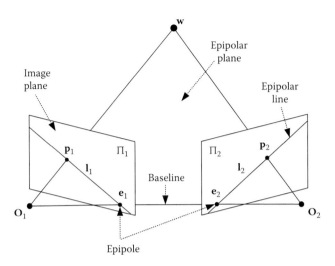

FIGURE 1.2
Stereo geometry.

In Figure 1.2, it is noted that three points, **W**, $\mathbf{O}_1$, and $\mathbf{O}_2$, form a 3D plane that intersects the two image planes. Here, the plane is called the epipolar plane and the intersection lines are called epipolar lines $\mathbf{l}_1$ and $\mathbf{l}_2$ in both image planes. As shown in the figure, two image points are on the epipolar lines, which is an important and strong constraint for finding matching points between the stereo images. For example, in the left image plane, $\mathbf{p}_1$ is the projection of **W** and its conjugate point in the right image plane is on the epipolar line $\mathbf{l}_2$. Therefore, it is possible to search the matching point only on the line $\mathbf{l}_2$, instead of searching all areas in the right image plane.

As shown in Figure 1.2, two image points $\mathbf{p}_1$ and $\mathbf{p}_2$ in the left and right image planes can be represented by either 2D vectors which are represented with respect to two different image coordinate systems or 3D vectors which are represented with respect to two different camera coordinate systems. In other words, $\mathbf{p}_1$ and $\mathbf{p}_2$ are 2D vectors in the image planes as well as 3D vectors in the camera coordinate systems. Now consider $\mathbf{p}_1$ and $\mathbf{p}_2$ as 3D vectors with respect to the left and right camera coordinate systems. In this case, the z component of $\mathbf{p}_1$ and $\mathbf{p}_2$ is the focal length of each camera. If the rotation and translation between two camera systems are **R** and **t**, respectively, $\mathbf{p}_2$ is transformed from $\mathbf{p}_1$ such that

$$\mathbf{p}_2 = \mathbf{R}\mathbf{p}_1 + \mathbf{t} \tag{1.1}$$

Here, the rotation and translation represent the transformations from the left to the right camera coordinate systems. Because **t** is the translation vector from the left to the right cameras, the cross product of $\mathbf{t}$ and $\mathbf{p}_2$ is orthogonal to $\mathbf{p}_2$. Let $\mathbf{Tp}_2$ be the cross product of $\mathbf{t}$ and $\mathbf{p}_2$, where

$$T = \begin{bmatrix} 0 & -t_z & t_y \\ t_z & 0 & -t_x \\ -t_y & t_x & 0 \end{bmatrix} \tag{1.2}$$

In this equation, T is a 3×3 matrix and t_x, t_y, t_z are the x, y, z components of $\mathbf{t}$. Therefore, $\mathbf{Tp}_2$ is a matrix representation of the cross product of two vectors.

Since three vectors, $\mathbf{p}_1$, $\mathbf{p}_2$, and $\mathbf{T}$, are in the same epipolar plane, the relations between the vectors are expressed by the following equations such that

$$\mathbf{p}_2 \cdot (T\mathbf{p}_2) = 0 \tag{1.3}$$

$$\mathbf{p}_2 \cdot (T(R\mathbf{p}_1 + \mathbf{t})) = 0 \tag{1.4}$$

$$\mathbf{p}_2 \cdot (T(R\mathbf{p}_1)) = 0 \tag{1.5}$$

and

$$\mathbf{p}_2^T E \mathbf{p}_1 = 0 \tag{1.6}$$

Here, a 3×3 matrix $\mathbf{E}$ is called *essential matrix* and is expressed as

$$E = TR \tag{1.7}$$

From Equation (1.6), it is found that

$$\mathbf{l}_2 = E\mathbf{p}_1 \tag{1.8}$$

which means that

$$\mathbf{p}_2^T \mathbf{l}_2 = 0 \tag{1.9}$$

because $\mathbf{p}_2$ is on the right epipolar line $\mathbf{l}_2$. This provides a very strong constraint in finding correspondences between stereo images. If an image point $\mathbf{p}_1$ is defined, its correspondence exists only in the right epipolar line $\mathbf{l}_2$. If an image point in the right image plane is not on the epipolar line, it is not possible for the point to be a correspondence. To decide if an image point $\mathbf{p}_2$ in the right image plane is on line $\mathbf{l}_2$, their dot product should be zero. However, in reality, the dot product will not be perfect zero because of the discrete property of the image coordinates. Therefore, if the dot product is very close to zero, the image point can be regarded as on the epipolar line.

Similarly, the left epipolar line is derived by

$$\mathbf{l}_1 = \mathbf{p}_2^T \mathbf{E} \tag{1.10}$$

and

$$\mathbf{l}_1 \mathbf{p}_1 = 0 \tag{1.11}$$

It is more convenient that any image point be represented with respect to the pixel coordinate system, rather than to the image coordinate system. To represent an image point in the pixel coordinates, it is necessary to convert the image coordinate system to the pixel coordinate system such that

$$\mathbf{p}_1' = \mathbf{K}_1 \mathbf{p}_1 \tag{1.12}$$

$$\mathbf{p}_2' = \mathbf{K}_2 \mathbf{p}_2 \tag{1.13}$$

In the preceding equations, $\mathbf{K}_1$ and $\mathbf{K}_2$ are the intrinsic matrices of the left and the right cameras; $\mathbf{p}_1'$ and $\mathbf{p}_2'$ are the coordinates of $\mathbf{p}_1$ and $\mathbf{p}_2$ in the pixel coordinate systems, respectively. Using Equations (1.12) and (1.13), Equation (1.6) is transformed as

$$\mathbf{p}_2'^T \mathbf{F} \mathbf{p}_1' = 0 \tag{1.14}$$

In this equation, $\mathbf{F}$ is called the *fundamental matrix* and it can be derived as

$$\mathbf{F} = \left\{ \mathbf{K}_2^{-1} \right\}^T \mathbf{E} \mathbf{K}_1^{-1} \tag{1.15}$$

In Equation (1.14), the image vectors are represented in the pixel coordinates, which is intuitive in obtaining epipolar lines also in the pixel coordinates. In contrast, the unit of vectors in Equation (1.6) follows the metric system. Therefore, to use the essential matrix, we need to convert the pixel coordinate system to the metric system by scaling and translating the center of the pixel coordinate system. For this reason, Equation (1.14) is used more often than Equation (1.6) in stereo vision.

To obtain the fundamental matrix using Equation (1.15), it is necessary to know the essential matrix in advance. If there are no geometric or calibration parameters between the left and the right cameras, the fundamental matrix cannot be derived. To overcome this inconvenience, some studies use the correspondences between stereo images to derive the fundamental matrix. Because the fundamental matrix is singular, it is possible to derive the matrix using only pixel coordinates of several matching points.

Let $\tilde{\mathbf{p}}'_1 = \begin{bmatrix} u & v & 1 \end{bmatrix}^T$ and $\tilde{\mathbf{p}}'_2 = \begin{bmatrix} u' & v' & 1 \end{bmatrix}^T$. Then, Equation (1.14) can be written with two image vectors and a square matrix such that

$$\begin{bmatrix} u' & v' & 1 \end{bmatrix} \begin{bmatrix} F_{11} & F_{12} & F_{13} \\ F_{21} & F_{22} & F_{23} \\ F_{31} & F_{32} & F_{33} \end{bmatrix} \begin{bmatrix} u \\ v \\ 1 \end{bmatrix} = 0 \tag{1.16}$$

This can be written as

$$\begin{bmatrix} uu' & uv' & u & u'v & vv' & v & u' & v' & 1 \end{bmatrix} \begin{bmatrix} F_{11} \\ F_{12} \\ F_{13} \\ F_{21} \\ F_{22} \\ F_{23} \\ F_{31} \\ F_{32} \\ F_{33} \end{bmatrix} = 0 \tag{1.17}$$

In this linear equation, F_{33} can be set 1 because the fundamental matrix is singular. Thus, there are only eight unknowns in this equation. If there are at least eight matched image pairs between the stereo images, we can write a linear equation to solve the elements of the fundamental matrix as follows. The solution of the equation can be easily obtained by using a linear algebra library. This method is called the *eight-point algorithm*, which is very useful not only in stereo vision but also in robot vision.

$$\begin{bmatrix} u_1u'_1 & u_1v'_1 & u_1 & v_1u'_1 & v_1 & u'_1 & v'_1 \\ u_2u'_2 & u_2v'_2 & u_2 & v_2u'_2 & v_2 & u'_2 & v'_2 \\ u_3u'_3 & u_3v'_3 & u_3 & v_3u'_3 & v_3 & u'_3 & v'_3 \\ u_4u'_4 & u_4v'_4 & u_4 & v_4u'_4 & v_4 & u'_4 & v'_4 \\ u_5u'_5 & u_5v'_5 & u_5 & v_5u'_5 & v_5 & u'_5 & v'_5 \\ u_6u'_6 & u_6v'_6 & u_6 & v_6u'_6 & v_6 & u'_6 & v'_6 \\ u_7u'_7 & u_7v'_7 & u_7 & v_7u'_7 & v_7 & u'_7 & v'_7 \\ u_8u'_8 & u_8v'_8 & u_8 & v_8u'_8 & v_8 & u'_8 & v'_8 \end{bmatrix} \begin{bmatrix} F_{11} \\ F_{12} \\ F_{13} \\ F_{21} \\ F_{22} \\ F_{23} \\ F_{31} \\ F_{32} \end{bmatrix} = - \begin{bmatrix} 1 \\ 1 \\ 1 \\ 1 \\ 1 \\ 1 \\ 1 \\ 1 \end{bmatrix} \tag{1.18}$$

1.3 Stereo Calibration

Calibration of a stereo vision system is to determine the 3D Euclidean transformations between the left and the right camera coordinate systems. After the geometrical relationship between two coordinate systems is determined, the relationship can be used for stereo rectification and 3D reconstruction from corresponding image points. In order to calibrate a stereo vision system, it is necessary to define a reference coordinate system, usually called the world coordinate system, so that the camera coordinate systems can be determined with respect to the reference coordinate system.

1.3.1 Pinhole Camera Model

In stereo vision, the pinhole camera model is commonly used. A pinhole camera is modeled by its optical center $\mathbf{O}$ and its image plane Π as shown in Figure 1.3. A 3D point $\mathbf{W}$ is projected into an image point $\mathbf{p}$, which is an intersection of Π with the line containing $\mathbf{O}$ and $\mathbf{W}$. Let $\mathbf{W} = [x\ y\ z]^T$ be the coordinates in the world coordinate system, $\mathbf{p} = [u\ v]^T$ in the image plane (CCD), and $\mathbf{p}' = [u'\ v']^T$ in the pixel plane. The mapping from 3D coordinates to 2D coordinates is the *perspective projection,* which is represented by a linear transformation in *homogeneous coordinates.* Let $\tilde{\mathbf{p}} = [u\,v1]^T$, $\tilde{\mathbf{p}}' = [u'\,v'1]^T$, and $\tilde{\mathbf{W}} = [x\,y\,z1]^T$ be the homogeneous coordinates of $\mathbf{p}$, $\mathbf{p}'$, and $\mathbf{W}$, respectively. Then the perspective transformation is given by the matrix $\tilde{\mathbf{M}}$:

$$\tilde{\mathbf{p}}' = \mathbf{K}\mathbf{p}' \cong \mathbf{K}\tilde{\mathbf{M}}\tilde{\mathbf{W}} \tag{1.19}$$

where "$\cong$" means equal up to a scale factor. The camera is therefore modeled by a scaling and translating matrix $\mathbf{K}$ and its perspective transformation matrix (PPM) $\tilde{\mathbf{M}}$, which can be decomposed, using the QR factorization, into the product

$$\tilde{\mathbf{M}} = \mathbf{A}\begin{bmatrix} \mathbf{R} | \mathbf{t} \end{bmatrix} \tag{1.20}$$

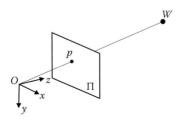

FIGURE 1.3
Pinhole camera model.

The matrices $\mathbf{K}$ and $\mathbf{A}$ depend on the intrinsic parameters only and have the following forms:

$$\mathbf{K} = \begin{bmatrix} k_u & 0 & 0 \\ 0 & k_v & 0 \\ 0 & 0 & 1 \end{bmatrix}, \mathbf{A} = \begin{bmatrix} f_u & \gamma & u_0 \\ 0 & f_v & v_0 \\ 0 & 0 & 1 \end{bmatrix} \tag{1.21}$$

where
f_u, f_v are the focal lengths in the horizontal and the vertical directions
k_u, k_v are the scaling factors from the image plane to the pixel plane
(u_0, v_0) are the coordinates of the principal point in the pixel plane
(u_i, v_i) are the coordinates of the offset of the principal point in the image plane
γ is a skew factor

The camera position and orientation (extrinsic parameters) are represented by a 3 × 3 rotation matrix $\mathbf{R}$ and a translation vector $\mathbf{t}$, representing a rigid transformation that brings the camera coordinate system onto the world coordinate system.

The PPM can be also written as

$$\tilde{\mathbf{M}} = \begin{bmatrix} \mathbf{q}_1^T & q_{14} \\ \mathbf{q}_2^T & q_{24} \\ \mathbf{q}_3^T & q_{34} \end{bmatrix} = \begin{bmatrix} \mathbf{Q} \mid \tilde{\mathbf{q}} \end{bmatrix} \tag{1.22}$$

The focal plane is the plane parallel to the image plane that contains the optical center $\mathbf{O}$, and the projection of $\mathbf{W}$ to the plane is 0. Therefore, the coordinate $\mathbf{O}$ is given by

$$\mathbf{O} = -\mathbf{Q}^{-1}\tilde{\mathbf{q}} \tag{1.23}$$

Therefore, $\tilde{\mathbf{M}}$ can be written as

$$\tilde{\mathbf{M}} = \begin{bmatrix} \mathbf{Q} \mid -\mathbf{Q}\mathbf{O} \end{bmatrix} \tag{1.24}$$

The optical ray associated to an image point $\mathbf{p}$ is the line $\mathbf{pO}$ that means the set of 3D points $\mathbf{W} : \tilde{\mathbf{p}} \cong \tilde{\mathbf{M}}\tilde{\mathbf{W}}$. In parametric form:

$$\mathbf{W} = \mathbf{O} + \lambda \mathbf{Q}^{-1}\tilde{\mathbf{p}}, \ \lambda \in \mathbb{R} \tag{1.25}$$

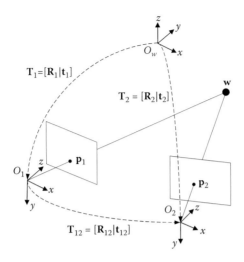

FIGURE 1.4
Stereo calibration.

1.3.2 Direct Calibration

In Figure 1.4, a world coordinate system is defined and two camera coordinate systems are also defined, which are those of the left and right cameras. The 3D transformation from the world to each camera coordinate system is represented as a matrix form $\mathbf{T}_i = \begin{bmatrix} \mathbf{R}_i \,|\, \mathbf{t}_i \end{bmatrix}$, which consists of the rotation and translation from the world to each camera, respectively. Given two transformations, it is easily known that the relation from the left to the right camera coordinate systems is

$$\mathbf{T}_{12} = \begin{bmatrix} \mathbf{R}_{12} \,\big|\, \mathbf{t}_{12} \end{bmatrix} = \mathbf{T}_2 \mathbf{T}_1^{-1} \tag{1.26}$$

Therefore, the stereo camera calibration can be regarded as a general camera calibration with respect to a fixed reference coordinate system.

A couple of camera calibration techniques that can determine external and internal parameters of a camera system are already available. In this section, a simple calibration technique called *direct method* is presented. This method calibrates the perspective projection matrix of a camera that contains the intrinsic and extrinsic camera parameters. Projection matrix is a homogeneous transform matrix that maps a 3D point in space into a 2D point in the image plane of a camera. Estimating the projection matrix is the solution of a simple and overdetermined linear system equation, and it can be solved by using the singular value decomposition (SVD) of the linear system. Calibration of a vision camera is considered an estimation of a projective transformation matrix from the world coordinate system to the camera's image coordinate system.

Suppose there is a 3D point with coordinates $\left(X_i^w, Y_i^w, Z_i^w, 1\right)$ and its projection point with coordinates $(x_c, y_c, 1)$; then a 3×4 projection matrix $\mathbf{M}$ can be written as

$$
\begin{bmatrix} u_i \\ v_i \\ w_i \end{bmatrix} = \mathbf{M} \begin{bmatrix} X_i^w \\ Y_i^w \\ Z_i^w \\ 1 \end{bmatrix}
\tag{1.27}
$$

with

$$
x_c = \frac{u_i}{w_i} = \frac{m_{11}X_i^w + m_{12}Y_i^w + m_{13}Z_i^w + m_{14}}{m_{31}X_i^w + m_{32}Y_i^w + m_{33}Z_i^w + m_{34}}
$$
$$
y_c = \frac{v_i}{w_i} = \frac{m_{21}X_i^w + m_{22}Y_i^w + m_{23}Z_i^w + m_{24}}{m_{31}X_i^w + m_{32}Y_i^w + m_{33}Z_i^w + m_{34}}
\tag{1.28}
$$

The matrix $\mathbf{M}$ is defined up to an arbitrary scale factor and has only 11 independent entries. Therefore, at least six world 3D points and their 2D projection points in the image plane are needed. If there is a calibration pattern— for example, a checkerboard pattern—$\mathbf{M}$ can be estimated through a least squares minimization technique. Assume there are N matches for a homogeneous linear system like Equation (1.28); then a linear system is given as

$$
\mathbf{Am} = 0
\tag{1.29}
$$

with

$$
\mathbf{A} = \begin{bmatrix}
X_1 & Y_1 & Z_1 & 1 & 0 & 0 & 0 & 0 & -x_1X_1 & -x_1Y_1 & -x_1Z_1 & -x_1 \\
0 & 0 & 0 & 0 & X_1 & Y_1 & Z_1 & 1 & -y_1X_1 & -y_1Y_1 & -y_1Z_1 & -y_1 \\
X_2 & Y_2 & Z_2 & 1 & 0 & 0 & 0 & 0 & -x_2X_2 & -x_2Y_2 & -x_2Z_2 & -x_2 \\
0 & 0 & 0 & 0 & X_2 & Y_2 & Z_2 & 1 & -y_2X_2 & -y_2Y_2 & -y_2Z_2 & -y_2 \\
\cdot & \cdot & \cdot & \cdot & \cdot & \cdot & \cdot & \cdot & \cdot & \cdot & \cdot & \cdot \\
\cdot & \cdot & \cdot & \cdot & \cdot & \cdot & \cdot & \cdot & \cdot & \cdot & \cdot & \cdot \\
\cdot & \cdot & \cdot & \cdot & \cdot & \cdot & \cdot & \cdot & \cdot & \cdot & \cdot & \cdot \\
X_N & Y_N & Z_N & 1 & 0 & 0 & 0 & 0 & -x_NX_N & -x_NY_N & -x_NZ_N & -x_N \\
0 & 0 & 0 & 0 & X_N & Y_N & Z_N & 1 & -y_NX_N & -y_NY_N & -y_NZ_N & -y_N
\end{bmatrix}
$$

and

$$
\mathbf{m} = \begin{bmatrix} m_{11}, m_{12}, \mathrm{L}, m_{33}, m_{34} \end{bmatrix}^T
$$

Since the rank of **A** is 11, **m** (or **M**) can be recovered from an SVD-based technique as the column of **V** corresponds to the smallest singular value of **A**, with $\mathbf{A} = \mathbf{UDV}^T$. For more information, see reference 17.

1.4 Stereo Rectification

Stereo rectification determines 2D transformations of stereo image planes such that pairs of conjugate epipolar lines become parallel to the horizontal image axes [4,5]. The rectified images can be considered as new images acquired from new parallel stereo cameras, obtained by rotating the original cameras. A great advantage of rectification is to allow convenient ways in stereo matching. After rectification, all epipolar lines become parallel to the horizontal image axis. Therefore, matching stereo correspondences can be done only on the same horizontal image axis. In this section, a simple stereo rectification method, proposed by Fusiello, Trucco, and Verri [4], is presented. This method utilizes calibration parameters to rectify stereo images and obtain new calibration parameters. This method determines the PPM of each stereo camera by virtually rotating original stereo cameras along the y-axis of each camera's coordinate system.

1.4.1 Epipolar Geometry

Let us consider a stereo vision system composed by two pinhole cameras as shown in Figure 1.5. In the figure, there are two vision cameras with optical centers $\mathbf{O}_1$ and $\mathbf{O}_2$, respectively. Let a 3D point in space be **W** and its projection to the left camera's image plane be $\mathbf{p}_1$. Then its corresponding $\mathbf{p}_2$ on the right image plane can be found on the epipolar line $\mathbf{l}_2$, which is the intersection of the right image plane and the epipolar plane of a triangle $\mathbf{WO}_1\mathbf{O}_2$. If two image planes are collinear, the epipolar line $\mathbf{l}_2$ will be collinear with the epipolar line $\mathbf{l}_1$ on the left image plane. However most stereo cameras have a toed-in angle between left and right cameras; therefore, their conjugate epipolar lines are not collinear.

A very special case is when both epipoles are at infinity. This happens when the line $\mathbf{O}_1\mathbf{O}_2$ (the baseline) is constrained in both focal planes, where the image planes are parallel to the baseline. Therefore, any stereo image pair can be transformed so that epipolar lines are parallel and horizontal in each image axis. This procedure is called *stereo rectification*.

1.4.2 Rectification of Camera Matrices

Suppose a stereo vision system is calibrated where the original PPMs $\tilde{\mathbf{M}}_{o1}$ and $\tilde{\mathbf{M}}_{o2}$ are known for the left and right cameras. Rectification estimates

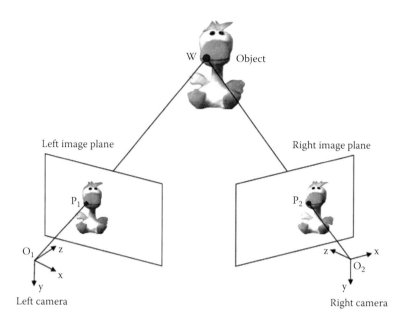

FIGURE 1.5

Epipolar geometry of a stereo vision camera. (Figure from A. Fusiello, E. Trucco, and A. Verri. A compact algorithm for rectification of stereo pairs. *Machine Vision and Applications* 12(1): 16–22, 2000.)

two new PPMs, $\tilde{\mathbf{M}}_{n1}$ and $\tilde{\mathbf{M}}_{n2}$, by rotating the old matrices around their optical centers until focal planes become coplanar.

In order to have horizontal epipolar lines, the baseline must be parallel to the new X axis of both cameras. In addition, corresponding points must have the same vertical coordinate (Y axis). Consequently, the positions of new optical centers are the same as those in the old cameras, the new matrices differ from the old ones by suitable rotations, and intrinsic parameters are the same for both cameras. Therefore, the new PPMs will differ only in their optical centers.

Let us write the new PPMs in terms of their factorization,

$$\tilde{\mathbf{M}}_{n1} = \mathbf{A}\left[\mathbf{R}\,\middle|\,-\mathbf{R}\mathbf{O}_1\right]$$
$$\tilde{\mathbf{M}}_{n2} = \mathbf{A}\left[\mathbf{R}\,\middle|\,-\mathbf{R}\mathbf{O}_2\right] \tag{1.30}$$

The intrinsic parameters matrix $\mathbf{A}$ is the same for both new PPMs and computed arbitrarily in this report as,

$$\mathbf{A} = \frac{\mathbf{A}_1 + \mathbf{A}_2}{2}$$

The rotation matrix **R** is also the same for both PPMs. It can be specified in terms of its row vectors:

$$\mathbf{R} = \begin{bmatrix} \mathbf{r}_1^T \\ \mathbf{r}_2^T \\ \mathbf{r}_3^T \end{bmatrix} \tag{1.31}$$

which are the X, Y, and Z axes of the camera coordinate system.

According to the previous descriptions, each axis is computed as follows:

1. The new X axis is parallel to the baseline: $\mathbf{r}_1 = \dfrac{(\mathbf{O}_1 - \mathbf{O}_2)}{\mathrm{PO}_1 - \mathbf{O}_2 \, \mathrm{P}}$.
2. The new Y axis is orthogonal to X and to **k**: $\mathbf{r}_2 = \mathbf{k} \wedge \mathbf{r}_1$.
3. The new Z axis is orthogonal to XY plane: $\mathbf{r}_3 = \mathbf{r}_1 \wedge \mathbf{r}_2$.

In number 2, **k** is an arbitrary unit vector and it used to be equal to the Z unit vector in the old left camera coordinate system. In other words, the new Y axis is orthogonal to both the new X and the old left Z.

In order to rectify the left and the right image, it is necessary to compute transformation mapping of the image plane $\tilde{\mathbf{M}}_{oi} = \begin{bmatrix} \mathbf{Q}_{oi} \mid \tilde{\mathbf{q}}_{oi} \end{bmatrix}$ onto the image plane $\tilde{\mathbf{M}}_{ni} = \begin{bmatrix} \mathbf{Q}_{ni} \mid \tilde{\mathbf{q}}_{ni} \end{bmatrix}$. For any 3D point **W**, we know that

$$\tilde{\mathbf{p}}_{oi} \cong \tilde{\mathbf{M}}_{oi} \tilde{\mathbf{W}}$$

$$\tilde{\mathbf{p}}_{ni} \cong \tilde{\mathbf{M}}_{ni} \tilde{\mathbf{W}}$$

According to Equation (1.25), the equations of the optical rays are the following:

$$\mathbf{W} = \mathbf{O}_i + \lambda_o \mathbf{Q}_{oi}^{-1} \tilde{\mathbf{p}}_{oi}, \quad \lambda_o \in {}^\circ$$

$$\mathbf{W} = \mathbf{O}_i + \lambda_n \mathbf{Q}_{ni}^{-1} \tilde{\mathbf{p}}_{ni}, \quad \lambda_n \in {}^\circ$$

Hence,

$$\tilde{\mathbf{p}}_{ni} = \lambda \mathbf{Q}_{ni} \mathbf{Q}_{oi}^{-1} \tilde{\mathbf{p}}_{oi}, \quad \lambda \in {}^\circ$$

$$\tilde{\mathbf{p}}_{ni} = \lambda \mathbf{T}_i \tilde{\mathbf{p}}_{oi} \tag{1.32}$$

The transformation $\mathbf{T}_i = \mathbf{Q}_{ni} \mathbf{Q}_{oi}^{-1}$ is then applied to the original stereo images to produce rectified images. Because the pixels of the rectified image correspond to noninteger positions on the original image planes, new pixel positions must be computed by using an image interpolation technique.

FIGURE 1.6
Checkerboard pattern for camera calibration.

1.5 Calibration and Rectification Examples

1.5.1 Calibration Example

To capture stereo images of an object, two cameras that have the same types of image sensors are used in common. Two cameras can be mounted in the parallel or toed-in configuration. In the parallel configuration, the optical axes are parallel to each other. In the toed-in configuration, two optical axes have some angle so that each camera can obtain the image of an object in the center of each image plane, respectively.

To calibrate the stereo camera system, a checkerboard pattern is commonly used. An example of a checkerboard is shown in Figure 1.6. The pattern has two planes that are parallel to the XY and the YZ planes of the world coordinate systems. In each pattern plane, many corners represent a set of 3D points. The top-left corner in the XY plane of the pattern is set as the origin of the world coordinate system. In addition, the coordinates of the other corners can be assigned because the width and the height of each rectangle of the pattern are known in advance. To calibrate the cameras, all 3D coordinates of the corners are assigned as 3D vectors $\mathbf{W}$ and their 2D image coordinates are $\mathbf{p}_1$ and $\mathbf{p}_2$ in the left and right images, respectively. Using the vectors, the intrinsic and extrinsic camera parameters are obtained.

(a)   (b)

FIGURE 1.7
Stereo images of a checkerboard pattern: (a) left image; (b) right image.

Examples of the stereo images of a calibration pattern are shown in Figure 1.7(a) and 1.7(b), which are the left and the right image, respectively. Each image's pixel resolution is 640 × 480. All corner points on the calibration pattern are acquired by a corner detection algorithm. The world coordinates $\mathbf{W}_i$ and the image coordinates $\mathbf{p}_i$ of the inner 96 corner points on both calibration planes are used to compute the projection matrix of each camera. The projection matrix is computed as described in an earlier section. The left and the right projection matrices, $\mathbf{M}_{o1}$ and $\mathbf{M}_{o2}$, which bring a 3D world point to the corresponding image planes, are

$$\mathbf{M}_{o1} = \begin{bmatrix} 7.172\times10^{-3} & 9.806\times10^{-5} & -5.410\times10^{-3} & 1.776\times10^{0} \\ 8.253\times10^{-4} & 8.889\times10^{-3} & 7.096\times10^{-4} & 6.427\times10^{-1} \\ 8.220\times10^{-4} & 6.439\times10^{-5} & 7.048\times10^{-4} & 1.000 \end{bmatrix} \quad (1.33)$$

$$\mathbf{M}_{o2} = \begin{bmatrix} 7.423\times10^{-3} & -4.474\times10^{-5} & -5.005\times10^{-3} & 1.638\times10^{0} \\ 7.078\times10^{-4} & 8.868\times10^{-3} & 4.957\times10^{-4} & 4.270\times10^{-1} \\ 7.750\times10^{-4} & 6.592\times10^{-5} & 7.521\times10^{-4} & 1.000 \end{bmatrix} \quad (1.34)$$

$$\left[\mathbf{R}_{o1}\,|\,\mathbf{t}_{o1}\right] = \begin{bmatrix} 6.508\times10^{-1} & 1.752\times10^{-4} & -7.591\times10^{-1} & -5.177\times10^{-2} \\ -4.517\times10^{-2} & 9.982\times10^{-1} & -3.850\times10^{-2} & -9.596\times10^{-1} \\ 7.578\times10^{-1} & 5.936\times10^{-2} & 6.497\times10^{-1} & 9.218\times10^{2} \end{bmatrix} \quad (1.35)$$

$$\left[\mathbf{R}_{o2}\,|\,\mathbf{t}_{o2}\right] = \begin{bmatrix} 6.970\times10^{-1} & -1.556\times10^{-2} & -7.168\times10^{-1} & -6.884\times10^{0} \\ -3.285\times10^{-2} & 9.980\times10^{-1} & -5.361\times10^{-2} & -9.764\times10^{1} \\ 7.163\times10^{-1} & 6.092\times10^{-2} & 6.951\times10^{-1} & 9.241\times10^{2} \end{bmatrix} \quad (1.36)$$

1.5.2 Rectification Example

Given projection matrices $\mathbf{M}_{oi}$ for the stereo cameras, new projection matrices $\mathbf{M}_{ni}$ are computed as described in Section 1.4.2:

$$\mathbf{M}_{n1} = \begin{bmatrix} -6.571 \times 10^0 & -9.838 \times 10^{-2} & 5.033 \times 10^0 & -1.581 \times 10^3 \\ -1.417 \times 10^0 & 8.030 \times 10^0 & -1.215 \times 10^0 & -2.058 \times 10^3 \\ -7.594 \times 10^{-1} & -5.936 \times 10^{-2} & -6.478 \times 10^{-1} & -9.218 \times 10^2 \end{bmatrix} \quad (1.37)$$

$$\mathbf{M}_{n2} = \begin{bmatrix} -6.571 \times 10^0 & -9.838 \times 10^{-2} & 5.033 \times 10^0 & -1.051 \times 10^3 \\ -1.417 \times 10^0 & 8.030 \times 10^0 & -1.215 \times 10^0 & -2.058 \times 10^3 \\ -7.594 \times 10^{-1} & -5.936 \times 10^{-2} & -6.478 \times 10^{-1} & -9.218 \times 10^2 \end{bmatrix} \quad (1.38)$$

$$\begin{bmatrix} \mathbf{R}_{n1} & \mathbf{t}_{n1} \end{bmatrix} = \begin{bmatrix} -6.490 \times 10^1 & 5.745 \times 10^{-4} & 7.607 \times 10^{-1} & 2.297 \times 10^0 \\ -4.478 \times 10^{-2} & 9.982 \times 10^{-1} & -3.896 \times 10^{-2} & -9.596 \times 10^1 \\ -7.594 \times 10^{-1} & -5.936 \times 10^{-2} & -6.478 \times 10^{-1} & -9.218 \times 10^2 \end{bmatrix} \quad (1.39)$$

$$\begin{bmatrix} \mathbf{R}_{n2} & \mathbf{t}_{n2} \end{bmatrix} = \begin{bmatrix} -6.490 \times 10^1 & 5.745 \times 10^{-4} & 7.607 \times 10^{-1} & 6.783 \times 10^0 \\ -4.478 \times 10^{-2} & 9.982 \times 10^{-1} & -3.896 \times 10^{-2} & -9.596 \times 10^1 \\ -7.594 \times 10^{-1} & -5.936 \times 10^{-2} & -6.478 \times 10^{-1} & -9.218 \times 10^2 \end{bmatrix} \quad (1.40)$$

Transformation matrix $\mathbf{T}_i$ to rectify the stereo pair in the image planes is then estimated as $\mathbf{T}_i = \mathbf{Q}_{ni}\mathbf{Q}_{oi}^{-1}$. For a pixel $\mathbf{p}_o$ in the original image plane and its homogeneous coordinates $\tilde{\mathbf{p}}_o$, a new pixel position $\tilde{\mathbf{p}}_n$ is estimated as

$$\tilde{\mathbf{p}}_n = \mathbf{T}_i \tilde{\mathbf{p}}_o \quad (1.41)$$

The picture coordinate $\tilde{\mathbf{p}}_n = (u', v')$ of the image point $\mathbf{p}_n$ is then obtained by multiplying the scaling and translating matrix $\mathbf{K}$ to the image coordinates:

$$\tilde{\mathbf{p}}'_n = \begin{bmatrix} k_u & 0 & 0 \\ 0 & k_v & 0 \\ 0 & 0 & 1 \end{bmatrix} \tilde{\mathbf{p}}_n \quad (1.42)$$

However when a rectified image is generated in a 2D array of a pixel plane, it is necessary to consider a translation of the principal point. Otherwise, some portions of the rectified image may be lost outside a pixel plane because of an offset between the original principal point (u_{o0}, v_{o0}) and the new principal point (u_{n0}, v_{n0}). In order to translate the rectified image back into the pixel plane, the new principal point (u_{n0}, v_{n0}) must be recomputed by adding the offset to the old principal point. The offset of the principal points can be computed by mapping the origin of the retinal plane onto the new retinal plane:

$$\tilde{\mathbf{o}}_n = \mathbf{T} \begin{bmatrix} u_o 0 \\ v_o 0 \\ 1 \end{bmatrix} \tag{1.43}$$

The new retinal coordinates are

$$\tilde{\mathbf{p}}'_n = \mathbf{K}\left(\tilde{\mathbf{p}}_n - \tilde{\mathbf{o}}_n\right) \tag{1.44}$$

The principal offset is considered only in the x direction because rectifying transformations rotate the image plane around the y axis. In this example, offsets of the principal points on the left and the right retinal planes are $(-0.020231, 0)$ and $(-0.518902, 0)$, respectively.

Figure 1.8(a) and (b) show original stereo images of an object and Figure 1.8(c) and (d) show their rectified images, which are generated by transforming the original images. In addition, the color value of new pixels is determined by using a bilinear interpolation. Rectified images are also shifted so that the new principal points are on or near the image centers.

1.6 3D Reconstruction

Using a rectified stereo image pair, a stereo matching technique can be applied to find stereo correspondences in the same horizontal lines. One of the simple stereo matching methods is normalized cross correlation, which will be described in the next section. Once the disparity of matching image points is determined, the coordinates of a 3D point associated to the matching pixels can be computed. Two simple and general reconstruction methods are presented in this section.

1.6.1 Simple Triangulation

Because stereo images are rectified already, depth computation uses a simple equation for a parallel stereo camera. When there is a disparity d'_u in x direction of the picture plane, the depth D to a 3D point from the stereo camera is

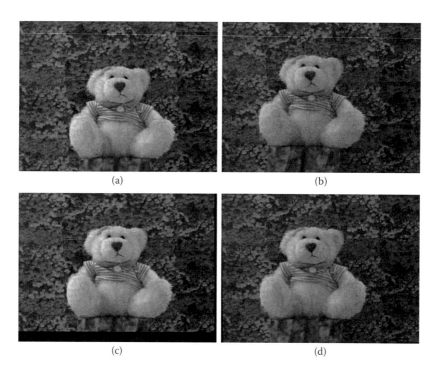

FIGURE 1.8
Rectification example. (a, b) Original left and right images; (c, d) rectified left and right images.

$$D = \frac{f \cdot B}{d_u' / k_u + \left(u_{n1}' - u_{n2}'\right)} \tag{1.45}$$

where $B = \|\mathbf{O}_1 - \mathbf{O}_2\|$ is the length of the baseline of the stereo camera. For the focal length f of the camera, the intrinsic calibration results f_u, f_v can be used.

1.6.2 Solution of a Linear Equation

The range for a pair of conjugate points is also reconstructed by using Equation (1.27). Given two conjugate points, $\tilde{\mathbf{p}}_1 = \left(u_1, v_1, 1\right)^T$ and $\tilde{\mathbf{p}}_2 = \left(u_2, v_2, 1\right)^T$, and the two projection matrices, $\tilde{\mathbf{M}}_{n1}$ and $\mathbf{M}_{n2}'$, an overconstrained linear system is written as

$$\mathbf{A}\mathbf{W} = \mathbf{y} \tag{1.46}$$

where

$$
\mathbf{A} = \begin{bmatrix} \left(\mathbf{a}_1 - u_1 \mathbf{a}_3 \right)^T \\ \left(\mathbf{a}_2 - v_1 \mathbf{a}_3 \right)^T \\ \left(\mathbf{b}_1 - u_2 \mathbf{b}_3 \right)^T \\ \left(\mathbf{b}_2 - v_2 \mathbf{b}_3 \right)^T \end{bmatrix} \quad y = \begin{bmatrix} -a_{14} + u_1 a_{34} \\ -a_{24} + v_1 a_{34} \\ -b_{14} + u_2 b_{34} \\ -b_{24} + v_2 b_{34} \end{bmatrix} \tag{1.47}
$$

Then $\mathbf{W}$ gives the position of the 3D point projected to the conjugate points. Column vectors $\mathbf{a}_i$ and $\mathbf{b}_i$ are entry vectors of the new left and the right projection matrices, respectively. Sometimes, the rectified image can be reflected along the vertical or the horizontal axis. This can be detected by checking the ordering between the two diagonal corners of the image. If a reflection occurs, the image should be reflected back to keep the original ordering.

The coordinates of $\mathbf{W}$ are represented with respect to the world system. To represent the point in one of the camera coordinate systems, it needs to convert the point. Suppose the right camera's coordinate system is a reference. Then we can transform the point by simply using the external calibration parameters $[\mathbf{R}|\mathbf{t}]$ of the right cameras. However, two transformations can be considered: One is to the old right camera's coordinates before rectification, and the other is to the new right-hand camera's coordinates after rectification. If the world 3D point is transformed to the old right-hand camera's coordinate system, then

$$
\mathbf{P}_w = \begin{bmatrix} \mathbf{R}_{o2} | \mathbf{t} \end{bmatrix} \mathbf{W} \tag{1.48}
$$

where $[\mathbf{R}_{o2}|\mathbf{t}]$ is the old external calibration parameters of the right camera. Figure 1.9 shows an example of 3D reconstruction from stereo images in Figure 1.8. In Figure 1.9(a), a reconstructed 3D model is represented by point clouds, while Figure 1.9(b) shows its mesh model.

1.7 Stereo Matching

Stereo matching is a technique to find corresponding points between stereo images captured from a pair of cameras. In general, most investigations in stereo matching are categorized as one of two methods. One is local matching and the other is global matching. The local matching method finds corresponding points by comparing image properties only. If the image properties of two image templates, one from the left image and the other from the right, are very similar to each other, the center points of the templates are considered to be matched. Therefore, in the local method, a cost function for

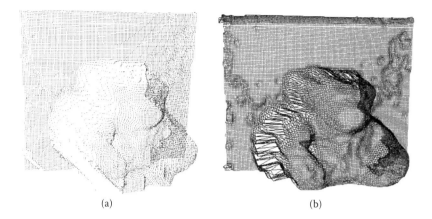

(a)　　　　　　　　　　　　　(b)

FIGURE 1.9
Experimental results of stereo matching. (a) Point clouds of the reconstructed 3D shape; (b) mesh model of the same reconstruction.

measuring the similarity of two templates is designed to use their image properties, such as intensity, color, sharpness, etc. The global matching method uses a global cost function so that it can measure the energy for a point to be matched with another point in terms of image and continuity properties. Recent stereo matching investigations have mostly focused on introducing global matching methods and some of them have yielded very successful results [14,20].

The global matching method measures the energy (or error) cost of a pair of image points in terms not only of image properties but also of disparity differences between neighboring pixels. In general, the disparity value f_p of a pixel **p** is determined in a way to minimize an energy function,

$$E(f) = \sum_{(p,q)\in N} V(f_p, f_q) + \sum_{p\in P} D_p(f_p) \qquad (1.49)$$

where N is a set of neighboring pixels and P is a set of pixels in an image. $V(f_p, f_q)$ is the cost of assigning disparity values f_p and f_q to two neighboring pixels, called the discontinuity cost. The second term, $D_p(f_p)$, is the cost of assigning disparity value f_p to pixel **p**, which is the data cost. The correspondence is therefore determined by searching for a point-pair that yields the smallest global energy.

Examples of well-known global matching methods are BP (belief propagation [3,12,21]), GC (graph cut [11,14]), DP (dynamic programming [14,18]), and SGM (semiglobal matching [6–8]). Many variants of them are also introduced to enhance matching performance. Instead of solving the np-hard problem of stereo correspondence, some global matching methods employ multiple scan lines to add cost functions in multiple directions [1,7,8]. A two-way DP technique, introduced by Kim et al. [9], is a simple example. They solve the inter-scan-line inconsistency problem using a bidirectional DP. Recently, they extended their work using several scan directions and edge orientations [16]. Hirschmüller introduced another multi-scan-line matching technique called SGM [7,8]. He generated cost volumes from multiple scan lines and aggregated them to obtain a dense disparity image. Heinrichs [6] has applied the SGM technique to a trinocular vision system for 3D modeling of real objects.

Global matching methods usually produce reasonably accurate results. However, they require more memory space and longer computation time than conventional local approaches. For example, most global matching methods generate disparity or cost space images. In SGM, (n + 1) cost volumes for (n)-scan directions are generated before they are merged to a single cost volume. The size of a cost volume is $N \times M \times D$, where $N \times M$ is the image resolution and D is the disparity range. Therefore, total $(n + 1) \times N \times M \times D$ memory space is needed to process a pair of stereo images. Computation time is therefore also increased with the increasing size of memory space.

In this section, several stereo matching methods, including local and global methods, are briefly described. In addition, their matching performance is compared to help readers understand the difference in matching methods. Using the Middlebury stereo database [20], disparity maps from different matching methods are compared.

1.7.1 Sum of Squared Difference

The SSD matching method is one of the simple stereo matching techniques. This method uses the intensity information around a pixel (x,y) in a reference image to find its matching conjugate in a target image. Since the SSD matching uses image areas around a matching pixel, it is one of the area-based stereo matching techniques. To find correspondences between stereo images, an image window (template) is defined in the reference image. Then, a cost function measures the image difference with another template in the target image.

Figure 1.10 shows a diagram of the SSD stereo matching method. In the figure, two images, $g^{(l)}$ and $g^{(r)}$, are left and right stereo images, respectively. In the left image, an image template at $g^{(l)}(x,y)$ is defined with the width and height of size m. By assuming that two images are obtained from a parallel stereo vision system, epipolar lines in both images are on the same horizontal line. Therefore, correspondence searching can be done in the horizontal epipolar lines. Now let us define a template at $g^{(r)}(x + d,y)$ in the right image, where d is a disparity value. Then, the sum of squared difference between two templates is defined as

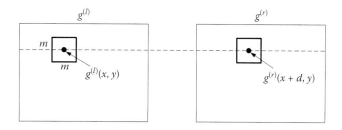

FIGURE 1.10
Stereo matching of image templates in a horizontal epipolar line.

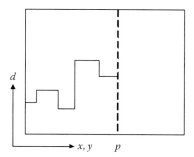

FIGURE 1.11
Scan-line optimization diagram.

$$SSD\left(g^{(l)},\ g^{(r)},\ x,y,d\right)=\sum_{k=-m/2}^{m/2}\sum_{j=-m/2}^{m/2}\left(g^{(l)}\left(x+k,\ y+j\right)-g^{(r)}\left(x+k+d,\ y+j\right)\right)^2 \quad (1.50)$$

where m is the size of the templates.

To find the matching pixel in the right image, it is necessary to compare $SSD(g^{(l)},g^{(r)},x,y,d)$ of all pixels in the horizontal line. In an ideal case, the best matching pixel in the right image yields the least SSD error.

1.7.2 Scan-Line Optimization

The scan-line optimization (SO) method is one of the global matching techniques [13]. The SO technique is very similar to the dynamic programming (DP) technique; however, it differs from the DP in that it does not consider occlusion pixels for cost aggregation. Therefore, its algorithm is simple and intuitive. The basic algorithm of the SO technique is to aggregate the energy of the previous pixel with the cost and continuity functions of the current pixel.

Figure 1.11 shows a general diagram of the SO-based stereo matching method. In the figure, the horizontal axis can be considered a 2D vector axis in which image coordinates (x,y) are two elements. For example, if the search space for correspondence in the right image is restricted to the horizontal line, the vector axis coincides with the horizontal line. In this case, only the *x* element of the vector is increased for energy aggregation. In addition, the vertical axis of the figure is the disparity axis. The range of the vertical axis is the same with the possible disparity range of pixel **p**. Suppose we need to find the disparity at pixel position **p**(*x,y*) as shown in Figure 1.11. In the SO method, the disparity of **p** is determined by searching the disparity index, which yields the least energy among the vertical axes.

Suppose the energy aggregation at pixel **p** is $E(\mathbf{p})$ and the matching cost of the pixel to disparity value *d* is $C(\mathbf{p},d)$. Then the energy aggregation of the current pixel is computed as

$$E(\mathbf{p}) = E(\mathbf{p}-1) + C(\mathbf{p},d) + \rho\left(d_{(\mathbf{p},\mathbf{p}-1)}\right) \tag{1.51}$$

where $\rho(d_{(\mathbf{p},\mathbf{p}-1)})$ is the penalty of disparity change between pixel **p** and **p** − 1. This means that disparity discontinuity increases matching energy along the scan line. Therefore, when aggregating matching energy, the discontinuity penalty should be increased if the disparity values of adjacent pixels are different. Similarly, the discontinuity penalty should be decreased if the disparity values of adjacent pixels are similar. An example of penalty values is shown in Equation (1.52). By combining data and continuity terms in global matching methods, more reliable disparity maps can be obtained:

$$\rho = \begin{cases} \rho_1 \text{ if } 1 \leq d_{(p,p-1)} < 3 \\ \rho_2 \text{ if } 3 \leq d_{(p,p-1)} \\ 0 \text{ if } d_{(p,p-1)} = 0 \end{cases} \tag{1.52}$$

1.7.3 Semiglobal Matching

In conventional global matching techniques, cost aggregation is done generally in a single matching direction, along the epipolar lines in the left or right image. However, with multiple matching directions, a more reliable disparity map can be obtained even using a simple matching algorithm. For example, Hirschmüller [7] proposes a global stereo matching technique called semiglobal matching (SGM).

Figure 1.12 shows multiple matching directions and a disparity space image in one of the directions. The left side of the figure shows matching directions through which SGM follows. Similarly to other global matching techniques, 8 or 16 matching directions are commonly used. However, the number of matching directions is not proportional to the accuracy of matching results. The right picture shows the disparity relation between **p** and **p**$_{next}$, which are in the same matching direction. In the figure, the current

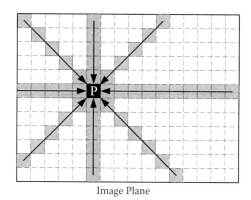

Image Plane

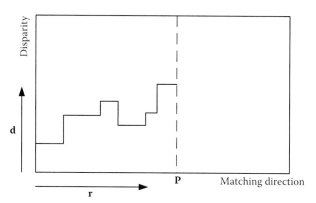

FIGURE 1.12
A general diagram of the semiglobal matching method. (Figures from H. Hirschmüller. *Proceedings of IEEE Conference on Computer Vision and Pattern Recognition*, pages 2386–2393, 2006.)

pixel's disparity d_p is used to calculate the disparity of the next pixel, d_{pnext}. The range of disparity at $\mathbf{p}_{next}$ is from $d_p - 1$ to $d_p + 1$.

In SGM, the disparity of **p** is determined so that the accumulation energy of a disparity space image D is minimized along the path depicted in the right-hand figure of Figure 1.12. Let $E(D)$ be the accumulation energy along a matching direction, and d_p and d_q be the disparities of **p** and **q**, respectively. Then the accumulation energy is defined as

$$E(D) = \sum_p C(\mathbf{p}, d_\mathbf{p}) + \sum_{q \in N_p} P_1 T \left[|d_\mathbf{p} - d_\mathbf{q}| = 1 \right] + \sum_{q \in N_p} P_2 T \left[|d_\mathbf{p} - d_\mathbf{q}| > 1 \right] \quad (1.53)$$

Here, $C(\mathbf{p}, d_\mathbf{p})$ is the cost of disparity d at pixel **p** and P_1 and P_2 are discontinuity penalties.

In Hirschmüller [7], the minimization of the accumulation energy is implemented by a cost aggregation such that

$$L_r(\mathbf{p}, d) = C(\mathbf{p}, d) + \min\left(L_r(\mathbf{p} - \mathbf{r}, d), L_r(\mathbf{p} - \mathbf{r}, d - 1) + P_1\right.$$
$$\left.L_r(\mathbf{p} - \mathbf{r}, d + 1) + P_1, \min L_r(\mathbf{p} - \mathbf{r}, i) + P_2\right) \tag{1.54}$$

where $L_r(\mathbf{p}, d)$ is the accumulated cost of $\mathbf{p}$ with disparity d along the matching direction $\mathbf{r}$. Accumulated costs of all matching directions are then accumulated again to obtain an accumulated cost image

$$S(\mathbf{p}, d) = \sum_r L_r(\mathbf{p}, d) \tag{1.55}$$

From the accumulation cost image, the disparity of $\mathbf{p}$ can be determined by searching the smallest accumulation cost at each pixel $\mathbf{p}$.

1.7.4 Belief Propagation

Belief propagation (BP) is also one of the global stereo matching methods [3]. The BP algorithm uses a MAP-MRF (maximum a posteriori-Markov random field) model as a global energy model. Let $I = \{I_L, I_R\}$ be a set of stereo images for the input of the MRF model and d^* be the disparity image to obtain from I. Then, the stereo matching problem can be considered as maximizing a posteriori probability such that

$$p(d \mid I) = \frac{p(I \mid d)p(d)}{p(I)} \tag{1.56}$$

$$d^* = \arg\max p(I \mid d) \tag{1.57}$$

In the preceding equation, a conditional probability $p(I|d)$ can be represented as

$$p(I \mid d) = e^{\left(-\Sigma_\mathbf{p} D_\mathbf{p}(d_\mathbf{p})\right)} \tag{1.58}$$

where, $D_\mathbf{p}(d_\mathbf{p})$ is the data cost of assigning disparity $d_\mathbf{p}$ at pixel $\mathbf{p}$. In addition, the *prior* term can be represented as

$$p(d) = \frac{1}{Z} e^{\left(-\Sigma V(d_p, d_q)\right)} \tag{1.59}$$

FIGURE 1.13
(See color insert.) Stereo test images from the Middlebury database (http://vision.middlebury.edu/stereo/). Top is original images and bottom is disparity images. From left: cone, teddy, and Venus.

where $V(d_p, d_q)$ is the cost of discontinuity between two adjacent pixels **p** and **q.** The MAP problem described in the preceding equation can be considered as the energy minimization problem such that

$$E(d) = \sum_{(\mathbf{p},\mathbf{q}) \in N} V\left(d_\mathbf{p}, d_\mathbf{q}\right) + \sum_{\mathbf{p} \in P} D_\mathbf{p}\left(d_\mathbf{p}\right) \qquad (1.60)$$

1.7.5 Performance Comparisons

This section presents comparisons of several stereo matching techniques. In the stereo vision community, the Middlebury stereo database is widely used to compare many matching methods [20]. The Middlebury database provides several stereo images and their ground truth disparity maps. In comparison with the ground truth disparity maps, the matching performance of any comparing method can be evaluated. In this section, three pairs of Middlebury stereo images are used: cone, teddy, and Venus. The original test images are color images; however, they are converted to gray-level images for comparison. Figure 1.13 shows the color and disparity images. The gray level of the disparity image is encoded by a scale factor of 4, from an actual disparity range from 1–63.75 to 1–255. The black pixels in the disparity images are unknown due to occlusions.

Figure 1.14 shows stereo matching results of the three test images. Six different matching methods are evaluated: CR (cooperative region), adapting

FIGURE 1.14

Comparisons of six stereo matching techniques. From top, CR, BP, SGM, GC, DP, and SSD. (Figures from the Middlebury stereo database, http://vision.middlebury.edu/stereo/.)

TABLE 1.1

Matching Errors of Different Methods

Method	Venus (nonocc, all, disc)	Teddy (nonocc, all, disc)	Cones (nonocc, all, disc)	Ref.
CR	0.11, 0.21, 1.54	5.16, 8.31, 13.0	2.79, 7.18, 8.01	19
(Rank)	(4), (3), (7)	(15), (11), (12)	(15), (4), (19)	
Adapting BP	0.10, 0.21, 1.44	4.22, 7.06, 11.8	2.48, 7.92, 7.32	10
(Rank)	(3), (4), (5)	(7), (6), (8)	(6), (11), (9)	
SGM	1.00, 1.57, 11.3	6.02, 12.2, 16.3	3.06, 9.75, 8.90	7
(Rank)	(67), (66), (76)	(21), (39), (32)	(25), (47), (37)	
GC	1.79, 3.44, 8.75	16.5, 25.0, 24.9	7.70, 18.2, 15.3	2
(Rank)	(89), (94), (72)	(105), (108), (95)	(94), (100), (89)	
DP	10.1, 11.0, 21.0	14.0, 21.6, 20.6	10.5, 19.1, 21.1	14
(Rank)	(112), (112), (101)	(99), (99), (78)	(104), (103), (101)	
SSD + MF	3.74, 5.16, 11.9	16.5, 24.8, 32.9	10.6, 19.8, 26.3	
(Rank)	(100), (101), (79)	(106), (107), (109)	(105), (105), (110)	

Note: Error percentages are calculated over three different areas: nonocclusion (nonocc), discontinuous (disc), and all (all) image areas.

BP (belief propagation), SGM (semiglobal matching), GC (graph cut), DP (dynamic programming), and SSD (sum of squared difference). Disparity results in the top three rows are very close to the ground truth in Figure 1.13. The three methods are all based on global cost minimization and have been introduced recently. Conventional global matching methods in the fourth and fifth rows show reasonable results. The last row is the result of an SSD-based local matching technique. In this result, disparity images are median filtered to remove matching errors.

Table 1.1 shows stereo matching errors evaluated with respect to the ground truth of the Middlebury stereo database. In addition, the performance ranking of each matching method is shown in the table. On the Middlebury stereo vision homepage [20], hundreds of stereo matching methods are compared and their rankings are shown. As mentioned in the previous paragraph, three recent global matching techniques yield very small errors compared with two conventional global and one local matching technique.

1.8 Conclusions

This chapter presents fundamental theories and inherent problems of stereo vision. The main goal of stereo vision is to acquire 3D depth information of a scene using two images obtained from slightly different views. Depth information of a pixel in an image is derived directly from its disparity, which is the coordinate difference with the corresponding pixel in the other image. The epipolar geometry of a stereo vision system and a rectification technique are also presented to explain the epipolar constraint. Several well-known stereo matching techniques that solve the inherent stereo problem in different manners are briefly introduced. Recent stereo matching techniques mostly use global cost functions, which consist of data and continuity terms. It is known that global matching techniques provide better performance than local techniques.

In addition to many existing application areas, stereo vision will bring many new application areas in the future. With the increasing computation power of computers and mobile devices, stereo vision can be applied to autonomous vehicle driving, robot navigation, real-time 3D modeling, etc. Time complexity in stereo matching will not be a problem in the future as it becomes common to use special processing units such as graphics processing units. In the near future, stereo vision will be one of the most used 3D scanning techniques and will be adopted in many commercial products.

References

1. S. Baek, S. Park, S. Jung, S. Kim, and J. Kim. Multi-directional greedy stereo matching. *Proceedings of the International Technical Conference on Circuits/Systems, Computers and Communications,* (1): 753–756, 2008.
2. Y. Boykov, O. Veksler, and R. Zabih. Fast approximate energy minimization via graph cuts. *IEEE Transactions on Pattern Analysis and Machine Intelligence,* 23 (11): 1222–1239, 2001.
3. P. Felzenszwalb and D. Huttenlocher. Efficient belief propagation for early vision. *Proceedings of the IEEE Conference on Computer Vision and Pattern Recognition,* (1): 261–268, 2004.
4. A. Fusiello, E. Trucco, and A. Verri. A compact algorithm for rectification of stereo pairs. *Machine Vision and Applications,* 12 (1): 16–22, 2000.
5. J. Gluckman and S Nayar. Rectifying transformations that minimize resampling effects. *Proceedings of the International Conference on Computer Vision and Pattern Recognition,* (1): 111–117, 2001.
6. M. Heinrichs, V. Rodehorsta, and O. Hellwich. Efficient semi-global matching for trinocular stereo. *Proceedings of Photogrammetric Image Analysis (PIA07),* 185–190, 2007.
7. H. Hirschmüller. Stereo vision in structured environments by consistent semi-global matching. *Proceedings of IEEE Conference on Computer Vision and Pattern Recognition,* 2386–2393, 2006.
8. H. Hirschmüller. Stereo processing by semi-global matching and mutual information. *IEEE Transactions on Pattern Analysis and Machine Intelligence,* 30 (2): 328–341, 2007.
9. C. Kim, K. M. Lee, B. T. Choi, and S. U. Lee. A dense stereo matching using two-pass dynamic programming with generalized ground control points. *Proceedings of IEEE Conference on Computer Vision and Pattern Recognition,* (2): 1075–1082, 2005.

10. A. Klaus, M. Sormann, and K. Karner. Segment-based stereo matching using belief propagation and a self-adapting dissimilarity measure. *Proceedings of IEEE International Conference on Pattern Recognition (ICPR2006)*, (3): 15–18, 2006.

11. V. Kolmogorov and R. Zabih. Computing visual correspondence with occlusions using graph cuts. *Proceedings of the 8th IEEE International Conference on Computer Vision*, (2): 508–515, 2001.

12. S. Larsen, P. Mordohai, M. Pollefeys, and H. Fuchs. Temporally consistent reconstruction from multiple video streams using enhanced belief propagation. *Proceedings of the 11th IEEE International Conference on Computer Vision*, 1–8, 2007.

13. S. Mattoccia, F. Tombari, and L. Di Stefano. Stereo vision enabling precise border localization within a scan line optimization framework. *Proceedings of the 8th Asian Conference on Computer Vision (ACCV2007)*, (2): 517–527, 2007.

14. D. Scharstein and R. Szeliski. A taxonomy and evaluation of dense two-frame stereo correspondence algorithms. *International Journal of Computer Vision*, 47 (1–3): 7–42, 2002.

15. S. D. Sharghi and F. A. Kamangar. Geometric feature-based matching in stereo images. *Proceedings of Information, Decision and Control Conference (IDC99)*, 65–70, 1999.

16. M. Sung, S. Lee, and N. Cho. Stereo matching using multi-directional dynamic programming and edge orientations. *Proceedings of IEEE International Conference on Image Processing*, (1): 233–236, 2007.

17. E. Trucco and A. Verri. *Introductory techniques for 3D computer vision.* Prentice Hall, Englewood Cliffs, NJ, 1998.

18. O. Veksler. Stereo correspondence by dynamic programming on a tree. *Proceedings of IEEE Conference on Computer Vision and Pattern Recognition*, (2): 384–390, 2005.

19. L. Wang, M. Liao, M. Gong, R. Yang, and D. Nistér. High-quality real-time stereo using adaptive cost aggregation and dynamic programming. *Proceedings of the Third International Symposium on 3D Data Processing, Visualization, and Transmission (3DPVT'06)*, 798–805, 2006.

20. Middlebury Stereo webpage. http://vision.middlebury.edu/stereo/

21. Q. Yang, L. Wang, R. Yang, H. Stewenius, and D. Nister. Stereo matching with color-weighted correlation, hierarchical belief propagation and occlusion handling. *Proceedings of the IEEE Conference on Computer Vision and Pattern Recognition*, (2): 17–22, 2006.

2

3D Shapes from Speckle

Yuan Hao Huang, Yang Shang, Yusheng Liu, and Hujun Bao

CONTENTS

One of the major difficulties of reconstructing three-dimensional (3D) shapes from stereo vision is the correspondence problem. Speckle can help in this aspect by providing unique patterns for area-based matching. After applying a high-contrast random speckle pattern on the object surface, disparities map with accuracy up to 0.02 pixels and ultrahigh spatial resolution can be established. The disparity information can then be used for accurate 3D reconstruction by triangulation. This chapter will introduce various algorithms of digital image correlation (DIC) for speckle pattern matching and various algorithms for 3D shape reconstruction from the disparity map. Examples of 3D shape measurement in various fields will also be given.

2.1 Introduction

Conventional stereo vision algorithms (refer to Chapter 1) use image features such as corner, edge, etc., for establishing correspondence in stereo image pairs and reconstructing a 3D scene by triangulation. These methods are efficient and can be implemented in real time for some machine vision applications. In 3D reverse engineering, experimental mechanics, and some other fields where researchers are more concerned about measurement accuracy and spatial resolution, these methods may not be able to satisfy these stringent requirements due to the limited number of feature points available and the matching errors of feature points. Thus, a more accurate 3D reconstruction technique is in demand. As a result, metrology based on speckle has been given intensive attention and is found to provide much better spatial resolution with improved accuracy.

The advantage of using speckle lies in the fact that it provides abundant high-frequency features for establishing accurate correspondence between a stereo image pair. After creating a high-contrast, random speckle pattern on an object surface, each point on the object will be surrounded by a unique intensity pattern that acts as the fingerprint of that point. Thus, accurate and high spatial resolution correspondence can then be established by correlating the intensity patterns. Normally, subpixel matching accuracy of 0.02–0.05 pixel with ultrahigh spatial resolution of 1 pixel can be achieved by proper algorithm design.

The process of establishing accurate correspondence between images with similar speckle patterns is normally called digital image correlation (DIC). Sometimes it is also called electronic speckle photography (ESP), digital speckle correlation (DSC), or texture correlation. There are multiple difficulties in using DIC for accurate 3D reconstruction. Firstly, a proper criterion should be defined to evaluate the level of similarity between speckle patterns. Secondly, the captured stereo speckle patterns are not exactly identical since they are captured from different viewing angles; thus, proper shape adjustment should be enabled to achieve an accurate matching. Thirdly, to achieve subpixel accuracy, proper interpolation schemes should be implemented to construct the intensity values at noninteger pixel positions. Finally, direct searching for the best matching position will require a large amount of computation, especially when accuracy at the order of 0.02 pixel is required; thus, advanced optimization schemes are needed to speed up the algorithm.

In this chapter, the technique of digital image correlation for accurate and efficient speckle matching will be introduced in detail in Section 2.2. Section 2.3 will introduce several triangulation methods for reconstructing a 3D shape from the disparity information obtained in Section 2.2. Section 2.4 gives a few examples for 3D shape measurement and Section 2.5 summarizes the chapter with references for further reading.

2.2 Principles of Digital Image Correlation

It is well known in signal processing that the cross correlation of two wave-forms provides a quantitative measure of their similarity. This basic concept has been explored to compare digital images (two-dimensional [2D] wave-forms) taken from different viewing directions or at different deformed stages to establish point-to-point correspondence and has been named digital image correlation [1–4]. After about 30 years of development, digital image correlation has been well developed and widely applied for in-plane displacement and strain measurement, material characterization, experimental stress analysis, and 3D shape reconstruction from macro to micro scales. Over 1,000 papers on this topic have been published, and the number is increasing every year.

2.2.1 Speckle Generation and Imaging

Before applying digital image correlation algorithm for 3D shape measurement, it is necessary to produce a high-contrast, random speckle pattern on the object surface. For a macro-object, this can be done by first spraying a flat-white paint followed by a flat-black paint on the object, or by using watercolor, ink, or other markers. A more convenient way without manual speckle preparation is to project a computer-generated speckle pattern via a digital projector or to direct a laser interference speckle pattern via a diffuser onto the object surface. However, due to brightness deficiency and various electronic noises, a projecting system normally renders a speckle pattern with lower contrast and larger noise than painted speckle and results in lower measurement accuracy. A random laser interference speckle pattern also suffers from low contrast and an additional drawback of requiring a relatively small aperture for capturing larger laser speckle, which will reduce the light intensity and signal-to-noise ratio. Thus, manually painted speckle patterns are preferred for most applications.

For a micro-object, fine toner or other micro- and nanomarking techniques can be applied on the object surface to form a certain pattern. Alternatively, if the natural surface provides apparent intensity variations under high magnification (e.g., some atomic force microscopes, scanning electron microscopes, or optical microscope images), it can be used directly for correlation at the cost of slightly lower measurement accuracy due to the lack of high-frequency speckle components. Normally, resolution of 0.1–0.2 pixel is still achievable with a magnified natural surface. For more details of various practical speckle generation methods, please refer to references [5–10].

After generating a high-contrast speckle pattern, the next step is to image the object with the appropriate optical system. For in-plane displacement and strain measurement of planar objects, normally only one camera is required; it is placed normal to the object surface. Speckle patterns of the planar object before and after a deformation are captured and processed by DIC algorithm to deliver the displacement and strain information. Sometimes a telecentric lens is used if out-of-plane deformation cannot be ignored. For 3D shape or deformation measurement of a nonplanar object, a binocular system is used to capture speckle images of the object simultaneously from two different viewing directions. More details of binocular imaging systems are given in Section 2.3.

2.2.2 Speckle Cell Definition and Matching

After the stereo speckle pairs are captured, the next step is to define a series of points for determination of disparities and a series of speckle cells centered at these points for area-based intensity matching between the left and right stereo images. The size of the speckle cell should be large enough to contain unique speckle features so that correct correspondence can be established. However, it cannot be too large as the disparity information of points inside the cell will be averaged and larger cells will cause more apparent system error. Optimal speckle cells should contain 10–20 speckles.

Figure 2.1(a) shows a pair of typical speckle images for correlation. The left image indicates a point to determine the disparity information and a surrounding speckle cell with proper size. To find out the best matched cell in the right image, the intensity profile of the left speckle cell is shifted within the right-hand image to correlate with corresponding speckle cells and a correlation coefficient (to be defined in Section 2.2.4) is computed and plotted against the disparity values (u,v) as shown in Figure 2.1(b). It can be clearly

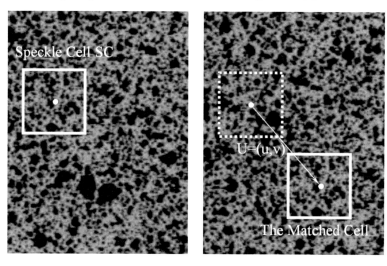

(a)

FIGURE 2.1
(a) Illustration of speckle-cell-based matching; (b) distribution of correlation coefficient by varying disparity (u, v); (c) schematic diagram of speckle cell deformation. (Reproduced from Huang, Y. H. et al. *Optics and Laser Technology* 41 (4): 408–414, 2009. With permission.)

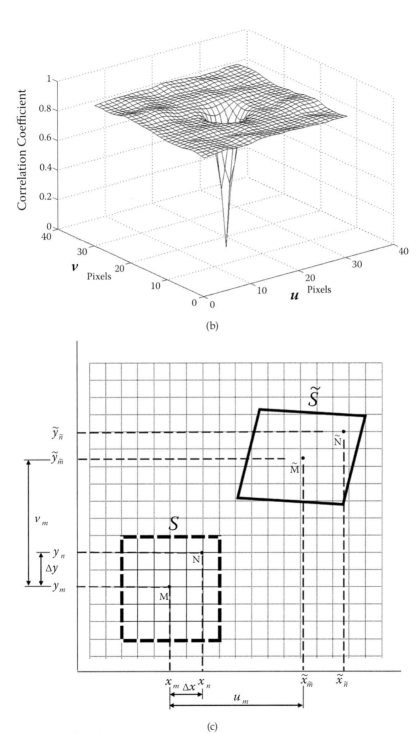

(b)

(c)

FIGURE 2.1 *(Continued)*

observed that the coefficient map presents a sharp minimum that indicates the best matching status. Thus, the corresponding disparity values can be determined from the minimum position for 3D reconstruction

2.2.3 Shape Transformation Function

In a practical imaging system, as the left and right speckle images are captured from different viewing angles or positions, the speckle patterns will undergo certain relative scaling, rotation, and higher order distortion. This will greatly flatten the correlation coefficient peak in Figure 2.1(b) or even cause decorrelation (no apparent peak is found) if pixel-by-pixel comparison is carried out. To avoid such decorrelation problems and improve measurement accuracy, a proper shape transformation function should be considered in the theoretical modeling.

For a point N in a small reference speckle cell S centered at point M (as shown in Figure 2.1c), it would be reasonable to formulate the coordinates $(\tilde{x}_{\tilde{n}}, \tilde{y}_{\tilde{n}})$ of the corresponding point $\tilde{N}$ in the objective speckle cell $\tilde{S}$ using second-order Taylor expansion as

$$\tilde{x}_{\tilde{n}} = x_m + P_1 + P_3 \Delta x + P_5 \Delta y + P_7 \Delta x^2 + P_9 \Delta y^2 + P_{11} \Delta x \Delta y \tag{2.1}$$

$$\tilde{y}_{\tilde{n}} = y_m + P_2 + P_4 \Delta x + P_6 \Delta y + P_8 \Delta x^2 + P_{10} \Delta y^2 + P_{12} \Delta x \Delta y \tag{2.2}$$

where
(x_m, y_m) are the coordinates of central point M in, the reference speckle cell
Δx and Δy are the distance of point N from the central point M
$P_1 \cdots P_{12}$ are parameters defining the shape transformation

In particular, P_1 and P_2 represent the central point disparity u_m and v_m (as indicated in Figure 2.1c) and are the desired information for 3D reconstruction.

After establishing such point-to-point correspondence between the reference and objective speckle cells, the control parameters $P_1 \cdots P_{12}$ can be tuned to achieve a sharp minimum correlation coefficient. However, this direct approach involves tremendous computation as the correlation coefficient map is now established on a 12-dimensional base. Thus, a more efficient method should be implemented to speed up the calculation; this will be shown in Section 2.2.5.

It should be noted that in some ideal applications where the shape change between reference and objective image is quite small, then Equations (2.1) and (2.2) may be simplified to first-order Taylor expansion involving only parameters $P_1 \cdots P_6$ to reduce computation. However, in a normal binocular system, if the angle between the two cameras is larger than 30°, the second-order effect will be apparent, especially for term $\Delta x \Delta y$ in Equations (2.1) and (2.2). Thus, the second-order terms should be included for optimal accuracy.

2.2.4 Correlation Coefficient

As shown in previous sections, a correlation coefficient should be defined to indicate the similarity of reference and objective speckle cells. In the literature, different kinds of coefficients have been defined with different performance and computation costs [7,12]. Basically, there are two types of correlation coefficient: one based on the cross-correlation function and the other on the sum of squared differences. It has been proved that these two types of coefficients are linearly related and exchangeable [13]. For the case of stereo matching, if both cameras and the A/D converters are the same, it can be assumed that the intensity values of the left- and right-hand speckle patterns are roughly the same. Thus, a simple correlation coefficient can be defined based on the direct sum of the squared difference formula as

$$C(\vec{P}) = \sum_{N \in S} [I(x_n, y_n) - \tilde{I}(\tilde{x}_{\tilde{n}}, \tilde{y}_{\tilde{n}})]^2 \qquad (2.3)$$

where $I(x_n, y_n)$ is the intensity of point N in the reference speckle cell S, and $\tilde{I}(\tilde{x}_{\tilde{n}}, \tilde{y}_{\tilde{n}})$ is the intensity of the corresponding point in the objective speckle cell $\tilde{S}$ (refer to Figure 2.1c). The coordinates of $(\tilde{x}_{\tilde{n}}, \tilde{y}_{\tilde{n}})$ are determined by Equations (2.1) and (2.2) and $\vec{P}$ is a vector containing the control parameters $P_1 \cdots P_{12}$.

This simple coefficient offers very good matching accuracy with reduced computation for stereo images. In a case when the left and right images have apparent intensity scaling and offset with the relationship of $\tilde{I} = \gamma I + \beta$, the zero-mean normalized form of Equation (2.3) would be preferred at the cost of larger computation [13], which can be expressed as

$$C(\vec{P}) = \sum_{N \in S} \left[\frac{I(x_n, y_n) - E(I)}{\sigma(I)} - \frac{\tilde{I}(\tilde{x}_{\tilde{n}}, \tilde{y}_{\tilde{n}}) - E(\tilde{I})}{\sigma(\tilde{I})} \right]^2 \qquad (2.4)$$

where $E(I)$, $\sigma(I)$ and $E(\tilde{I})$, $\sigma(\tilde{I})$ are the mean intensity and standard deviation values for the reference and objective speckle cells, respectively.

For coefficients in the form of the sum of squared difference, it is obvious that a smaller coefficient indicates a better matching between the reference and objective speckle cells. During DIC computation, quite frequently a criterion is required to judge whether or not the parameters $P_1 \cdots P_{12}$ are close enough to the optimal values. Obviously, the correlation coefficient can serve as such a criterion. As shown in Figure 2.1(b), where a normalized coefficient is plotted, it can be seen that the global minimum is quite close to zero, while the other coefficients outside the minimum cone are quite close to one. Thus, it would be reasonable to predefine a largest allowed coefficient C_{max} as half of the average coefficients over a large area. For parameters $P_1 \cdots P_{12}$, which satisfy the equation $C(\vec{P}) < C_{\max}$, it can be presumed that the optimal parameters are close to the current values and can be determined by a nonlinear iteration process (to be introduced in Section 2.2.5).

2.2.5 Nonlinear Optimization

Direct minimization of Equation (2.3) or (2.4) by varying the 12 parameters is $P_1 \cdots P_{12}$ computation intensive. Thus, a more efficient approach has been developed [14]. To determine the minimum position of the correlation coefficient, Fermat's theorem can be applied. This theorem states that every local extremum of a differentiable function is a stationary point (the first derivative in that point is zero), so the following equation should be satisfied at the minimum position:

$$\nabla C(\vec{P}) = 0 \qquad (2.5)$$

Equation (2.5) can be solved using the Newton–Raphson (N–R) iteration method as follows:

$$\vec{P}^{k+1} = \vec{P}^k - \frac{\nabla C(\vec{P}^k)}{\nabla\nabla C(\vec{P}^k)} \qquad (2.6)$$

where the expression of $\nabla C(\vec{P})$ and $\nabla\nabla C(\vec{P})$ can be derived from Equation (2.3) as

$$\nabla C(\vec{P}) = \left(\frac{\partial C}{\partial P_i}\right)_{i=1\cdots12} = \left\{\sum_{N \in S} -2[I(x_n, y_n) - \tilde{I}(\tilde{x}_{\tilde{n}}, \tilde{y}_{\tilde{n}})]\frac{\partial \tilde{I}}{\partial P_i}\right\}_{i=1\cdots12} \qquad (2.7)$$

$$\nabla\nabla C(\vec{P}) = \left(\frac{\partial^2 C}{\partial P_i \partial P_j}\right)_{\substack{i=1\cdots12 \\ j=1\cdots12}} = \left\{\sum_{N \in S} -2[I(x_n, y_n) - \tilde{I}(\tilde{x}_{\tilde{n}}, \tilde{y}_{\tilde{n}})]\frac{\partial^2 \tilde{I}}{\partial P_i \partial P_j}\right\}_{\substack{i=1\cdots12 \\ j=1\cdots12}}$$

$$+ \left\{\sum_{N \in S} 2\frac{\partial \tilde{I}}{\partial P_i} \cdot \frac{\partial \tilde{I}}{\partial P_j}\right\}_{\substack{i=1\cdots12 \\ j=1\cdots12}} \qquad (2.8)$$

In Equation (2.8), when $\vec{P}$ is close to the correct matching position, $I(x_n, y_n) \approx \tilde{I}(\tilde{x}_{\tilde{n}}, \tilde{y}_{\tilde{n}})$, the first term can be reasonably ignored, which leads to an approximated Hessian matrix of

$$\nabla\nabla C(\vec{P}) = \left(\frac{\partial^2 C}{\partial P_i \partial P_j}\right)_{\substack{i=1\cdots12 \\ j=1\cdots12}} \approx \left\{\sum_{N\in S} 2\frac{\partial \tilde{I}}{\partial P_i} \cdot \frac{\partial \tilde{I}}{\partial P_j}\right\}_{\substack{i=1\cdots12 \\ j=1\cdots12}} \tag{2.9}$$

This approximation has been proved to save a large amount of computation and yet deliver a similar result. The expression of

$$\left(\frac{\partial \tilde{I}}{\partial P_i}\right)_{i=1\cdots12}$$

could be further expressed as

$$\left(\frac{\partial \tilde{I}}{\partial P_i}\right)_{i=1\cdots12} = \left(\frac{\partial \tilde{I}}{\partial \tilde{x}_{\tilde{n}}} \cdot \frac{\partial \tilde{x}_{\tilde{n}}}{\partial P_i} + \frac{\partial \tilde{I}}{\partial \tilde{y}_{\tilde{n}}} \cdot \frac{\partial \tilde{y}_{\tilde{n}}}{\partial P_i}\right)_{i=1\cdots12} \tag{2.10}$$

where

$$\frac{\partial \tilde{x}_{\tilde{n}}}{\partial P_i} \quad \text{and} \quad \frac{\partial \tilde{y}_{\tilde{n}}}{\partial P_i}$$

are readily determined from Equations (2.1) and (2.2) and

$$\frac{\partial \tilde{I}}{\partial \tilde{x}_{\tilde{n}}} \quad \text{and} \quad \frac{\partial \tilde{I}}{\partial \tilde{y}_{\tilde{n}}}$$

are the intensity derivatives in the objective speckle cell, which can be obtained by an interpolation scheme to be introduced in Section 2.2.6.

It should be noted that Equation (2.6) requires an initial value $\vec{P}^0$ to start the iteration, and the given $\vec{P}^0$ should be close to the correct value; otherwise, the iteration will converge to a local extremum rather than the global minimum. In most cases, the shape transformation between the left and right images is not severe, so only the initial values P_1 and P_2 are provided as (u_m^0, v_m^0), while P_3, P_6 are set to one and the other eight parameters are set to zero (i.e., pure translation). Looking back to the example in Figure 2.1(b), if the initial (u_m^0, v_m^0) are given within the cone of global minimum (the base diameter of the cone is about 12–14 pixels), then the N–R algorithm will quickly converge to the correct disparity values. However, if the initial (u_m^0, v_m^0) are given outside the cone, it will converge to a local minimum and lead to wrong results. Thus, it is essential to provide a proper initial guess for the iteration algorithm.

One way to provide a good initial guess automatically is to evaluate the correlation coefficients of a coarse grid with steps smaller than the cone radius and subsequently a finer grid with a step of 1 pixel around the smallest coefficient obtained from the coarse grid [11]. The location of the minimal coefficient obtained this way is within 1 pixel of the global minimum and can be fed to the algorithm as an initial guess. In the case when the stereo system has been properly calibrated, a search for an initial guess can be conducted along the epipolar line and the computation can be greatly reduced.

When a binocular system with a large intersection angle is used, the second-order effect is apparent and the procedure proposed before that searches for an initial guess by varying P_1 and P_2 only may not be able to give a correct guess (i.e., the criterion $C < C_{max}$ is not satisfied). In such cases, the positions of five local minimum coefficients can be picked up and P_{11} and P_{12} are included in the searching processes for these five points. Hopefully, this will give an initial guess including P_1, P_2, P_{11}, P_{12}, which satisfies $C < C_{max}$.

If the inclusion of P_{11} and P_{12} still fails to give a correct initial guess, manual correspondence must be used for this purpose. By selecting 6–10 feature points around the reference speckle cell and manually identifying their corresponding points in the objective image, the 12 parameters $P_1 \cdots P_{12}$ can then be calculated and provided to the DIC algorithm.

Once a correct speckle cell matching is achieved, the determined parameters $P_1 \cdots P_{12}$ can then serve as the initial guess for adjacent speckle cells. Thus, for a region with smooth disparity change, only one proper initial guess is required. If a region contains some steep disparity changes, the algorithm may result in tremendous error at some points, leading to abrupt increase of the correlation coefficient. A new searching process should be conducted to assign a correct initial guess to those points.

2.2.6 Interpolation Schemes

During the process of stereo matching, the points in the object speckle cell may not reside in integral pixel positions, so it is necessary to interpolate for Gray value and its derivatives (refer to Equation [2.10]) at nonintegral pixel positions. Normally, a higher-order interpolation scheme will give better accuracy and faster convergence, but require more computation. In the literature, bicubic and biquintic spline interpolation have been widely used to render good accuracy (at the order of 0.01 pixel); lower-order schemes such as bilinear interpolation are only used when the accuracy requirement is low.

As the interpolation process accounts for a very large amount of computation for digital image correlation, special attention should be paid to the design of the interpolation scheme. Here a specially designed biquadratic spline interpolation scheme proposed by Spath [15] is recommended where the knots are shifted a half pixel from the nodes, resulting in similar accuracy with normal bicubic spline interpolation while requiring only half of the computation.

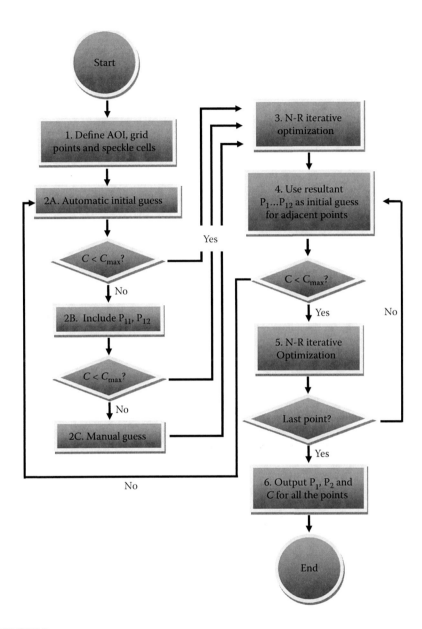

FIGURE 2.2
Flow chart for digital image correlation algorithm design.

2.2.7 Flow Chart Illustration of DIC Algorithm

Figure 2.2 summarizes the whole process of the digital image correlation algorithm described in previous sections. Readers can refer to Sections 2.2.2–2.2.6 for details of each step.

2.2.8 Other Versions of DIC Algorithm

In the literature, researchers have proposed some other versions of DIC algorithms. One of the most intuitive algorithms is the coarse–fine searching algorithm [16], where the integral-pixel correlation coefficients are first calculated by varying u_m, v_m and then subpixel coefficients are further computed around the minimum integral-pixel coefficient by interpolation. Simple as it is, the direct coarse–fine searching algorithm requires a large amount of computation due to the intensive Gray value interpolation involved. Thus, peak-fitting algorithms have been proposed [3,17] to reduce the computation where curve fitting or interpolation is used to establish a continuous correlation coefficient map based on integral-pixel coefficients and the minimum coefficient position is then determined with subpixel accuracy. These algorithms, based on coarse–fine searching and peak fitting, are intuitive and simple. However, because they ignore the shape change between the reference and objective images, they would be most suitable for disparity computation with stereo images captured from a canonical imaging system (parallel stereo cameras). For applications with a nonparallel setup, shape transformation should be considered for optimized accuracy.

Most of the DIC algorithms employ the concept of speckle-cell-based matching. Yet, there is another group of algorithms that are not speckle cell based; they establish whole-field mapping between the reference and objective speckle images using a continuous function [18,19]. Theoretically, such algorithms may eliminate mismatching for individual points due to the continuous constraint and improve the matching accuracy as more information is involved. However, these methods seem not to provide better accuracy according to the reported results, probably because the prescribed continuous function cannot genuinely represent the true displacement or disparity function. Thus, the speckle-cell-based algorithm described earlier dominates in most applications.

2.3 3D Shape Measurement from Stereo DIC

Section 2.2 introduced the DIC method for establishing accurate correspondence between stereo image pairs. This section will introduce the process of 3D reconstruction, which consists of modeling and calibrating the stereo system and subsequently calculating the 3D coordinates of a point from its correspondence information.

2.3.1 Stereo Camera Modeling

A simplified real camera comprises a pinhole and an imaging screen behind it. Rays emitted from the 3D scene will have to travel through the pinhole before

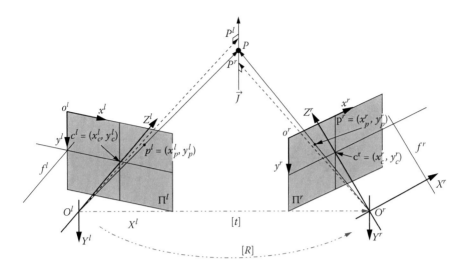

FIGURE 2.3
Pinhole model of a stereo system.

they reach the imaging screen. This will establish a unique projection from a 3D point in the scene to a 2D point on the screen. On the other hand, a 2D point on the screen cannot be uniquely identified in the 3D scene, but rather forms a line that connects the 2D point and the focal point. Thus, stereo cameras are required to uniquely identify a point in the 3D scene from its projections into the left and right imaging screens. The purpose of stereo camera modeling is to establish mathematically the projection of a point in a 3D scene onto the imaging screens and to establish the relation between the left and right cameras.

As shown in Figure 2.3, the imaging screens are placed in front of the pinholes at the symmetric positions for simplicity. Two Cartesian coordinate systems, which are denoted left and right camera coordinates, are established with their origins O^l and O^r residing at the focal points of the left and right cameras (From now on, we use superscripts l and r to represent elements in the left and right camera coordinate systems, respectively). The Z-axes are aligned along the optical axes, and the X-, Y-axes are chosen paralleled to the edges of the corresponding imaging screen Π (refer to Figure 2.3). The imaging screen Π is sampled with periods of h_x and h_y along the x- and y-directions and forms a grid of sensing elements called "pixel," which is the unit for the image coordinate system xoy.

The optical axis intersects with the imaging screen at principal point c with image coordinate (x_c, y_c) in units of pixels. The distance between the focal point O and the principal point c is known as the focal length f. The parameters f, x_c, y_c, h_x, h_y are intrinsic with the camera and called intrinsic parameters (the geometric distortion is ignored temporally for simplicity). These parameters fully determine the projection of a point from the 3D camera coordinate system onto the 2D image coordinate system.

Take the left camera coordinate system $O^l X^l Y^l Z^l$ as an example. A point P in the 3D scene is represented by its coordinates (X_P^l, Y_P^l, Z_P^l). The projection of P onto the imaging screen results in an image point p^l with image coordinates (x_p^l, y_p^l) expressed in the unit of pixels. Using simple triangulation (refer to Figure 2.3), we have

$$\frac{(x_p^l - x_c^l)h_x^l}{X_P^l} = \frac{(y_p^l - y_c^l)h_y^l}{Y_P^l} = \frac{f^l}{Z_P^l} \qquad (2.11)$$

The image coordinates of point p^l can then be expressed in terms of the camera intrinsic parameters and the camera coordinate of point P as

$$x_p^l = \frac{f^l X_P^l}{Z_P^l h_x^l} + x_c^l \qquad (2.12)$$

$$y_p^l = \frac{f^l Y_P^l}{Z_P^l h_y^l} + y_c^l \qquad (2.13)$$

Similarly, the same point P in the 3D scene can also be represented in the right-camera coordinate system as (X_P^r, Y_P^r, Z_P^r), which results in similar projection onto the right imaging screen as

$$x_p^r = \frac{f^r X_P^r}{Z_P^r h_x^r} + x_c^r \qquad (2.14)$$

$$y_p^r = \frac{f^r Y_P^r}{Z_P^r h_y^r} + y_c^r \qquad (2.15)$$

Equations (2.12)–(2.15) have fully established the projection of a point P in the 3D scene onto the left and right imaging screens. However, the same point P has been simultaneously represented by (X_P^l, Y_P^l, Z_P^l) in the left camera coordinate system and (X_P^r, Y_P^r, Z_P^r) in the right camera coordinate system. Thus, the relationship between the left and right camera coordinates should be established to relate (X_P^l, Y_P^l, Z_P^l) and (X_P^r, Y_P^r, Z_P^r). This has been done in numerous textbooks [18,19].

Let $[t]$ specify the column vector representing the translation from origin O^l to O^r and $[R]$ specify the orthogonal rotation matrix from left camera coordinate system $O^l X^l Y^l Z^l$ to right camera coordinate system $O^r X^r Y^r Z^r$ (refer to Figure 2.3). Then the coordinates of point P in these two systems, $P^l = (X_P^l, Y_P^l, Z_P^l)$ and $P^r = (X_P^r, Y_P^r, Z_P^r)$, are related by the following formula:

$$P^r = [R](P^l - [t]) \tag{2.16}$$

Matrix $[R]$ and column vector $[t]$ characterize the relation between the left and right coordinate systems, which is independent of the individual camera projection model. Thus, they are normally called extrinsic parameters. The intrinsic parameters $f^l, x_c^l, y_c^l, h_x^l, h_y^l$ and $f^r, x_c^r, y_c^r, h_x^r, h_y^r$, as well as the extrinsic parameters $[R]$ and $[t]$, fully characterize the stereo pinhole system.

2.3.2 System Calibration

The calibration process intends to determine the intrinsic camera parameters and the extrinsic parameters of a stereo system. These parameters are a prerequisite for full 3D reconstruction based on the disparity data extracted from the stereo image pairs. The accuracy of the calibration parameters has been found to place an essential and nonlinear effect on the reconstruction accuracy [20].

There are generally two categories of camera calibration techniques: One is based on a known reference object and the other is self-calibration. For the reference-object-based calibration, normally a 3D object with known geometry or a 2D object undergoing known translations is used to establish a series of points with precisely known coordinates in a world coordinate system [21,22]. A series of linear equations are then constructed to determine the intrinsic and extrinsic parameters. The self-calibration technique [23,24], on the other hand, does not require any known calibration object, but rather moves the camera on a static scene and captures images. The rigidity of the scene poses some constraints on the camera's intrinsic parameters, and correspondence information between three or more images is sufficient to determine the calibration parameters up to a scaling factor. Other techniques are also available using two- or one-dimensional known objects [25,26] for flexible calibration. Here we introduce the standard calibration method based on 3D objects with known geometry. Readers can refer to Gruen and Huang [27] for a more complete study in the calibration methods and Sun and Cooperstock [28] for an empirical comparison of three widely used calibration techniques.

For calibrating the stereo system shown in Figure 2.3 using a 3D known object, an auxiliary world coordinate system is introduced and the coordinates of N known 3D points are defined based on the world coordinate system as $P_i^w = (X_{pi}^w, Y_{pi}^w, Z_{pi}^w)$, $i = 1 \cdots N$. The rotation matrixes and translation column vectors from the world coordinate system to the left and right camera coordinate systems are defined as $[R^l], [t^l]$ and $[R^r], [t^r]$, respectively, which help to transform point P into the left and right camera coordinates as

$$P^l = [R^l](P^w - [t^l]) \tag{2.17}$$

$$P^r = [R^r](P^w - [t^r])$$ (2.18)

By projecting the N points from the left camera coordinate system onto the left image coordinate system according to Equations (2.12) and (2.13), we have $2N$ equations with 14 unknowns (five intrinsic parameters, six independent components in rotation matrix $[R^l]$, and three independent components in the translation vector $[t^l]$). Thus, if the reference object contains more than seven known points, the 14 calibration parameters $f^l, x_c^l, y_c^l, h_x^l, h_y^l$ and $[R^l], [t^l]$ for the left camera system can be determined from the overdetermined linear equation group by singular value decomposition [29]. Similarly, the 14 calibration parameters $f^r, x_c^r, y_c^r, h_x^r, h_y^r$ and $[R^r], [t^r]$ from the right camera system can also be determined from the same series of known points using Equations (2.18), (2.14), and (2.15). The rotation matrix $[R]$ and translation vector $[t]$ can be subsequently obtained from Equations (2.16)–(2.18) as

$$[R] = [R^r][R^l]^T$$ (2.19)

$$[t] = [R^l]([t^r] - [t^l])$$ (2.20)

So far the stereo camera system is fully calibrated and ready for measuring a 3D scene from the stereo 2D projections.

2.3.3 Conventional 3D Shape Reconstruction

Three-dimensional reconstruction using the correspondence information of the left and right images and the calibration parameters of the stereo system is straightforward. Grouping Equations (2.12)–(2.15) and the vector in Equation (2.16), we have seven linear equations in total. Within all the parameters in these seven equations, since the correspondence positions $p^l = (x_p^l, y_p^l)$, $p^r = (x_p^r, y_p^r)$, the intrinsic parameters $f^l, x_c^l, y_c^l, h_x^l, h_y^l, f^r, x_c^r, y_c^r, h_x^r, h_y^r$, and the extrinsic parameters $[R], [t]$ are all known, we have only six unknowns: (X_P^l, Y_P^l, Z_P^l) and (X_P^r, Y_P^r, Z_P^r). Thus, the six unknowns can be determined from the overdetermined equation group by least-squares optimization.

A more intuitive geometric solution is also given in Cyganek and Siebert [18] and Trucco and Verri [19]. Referring to Figure 2.3, based on the calibration parameters and the correspondence positions p^l and p^r, the equations of lines $O^l p^l$ and $O^r p^r$ can be written (in the left camera coordinate system, for example). These two lines may not intersect with each other due to various modeling and measurement errors associated with a practical system. Let the two reconstructed lines $O^l p^l$ and $O^r p^r$ be represented by the two dashed lines (as shown in Figure 2.3) adjacent to their perfect positions $O^l P$ and $O^r P$. It is intuitive and straightforward that there exists a unique line perpendicular to both dashed lines simultaneously and intersecting the two dashed lines at points P^l and P^r. The middle point of segment P^lP^r is then chosen as the optimal 3D estimation P_E, which forms minimal distances to both reconstructed dashed lines.

The coordinates of points P^l and P^r can be determined by the following equation:

$$k^l[O^lp^l]^T + k\vec{J} = [t] + k^r[R]^T[O^rp^r]^T \qquad (2.21)$$

where
$[O^lp^l]^T$ is the column vector connecting O^l and p^l expressed in the left camera coordinate system
$[O^rp^r]^T$ is the column vector connecting O^r and p^r expressed in the right camera coordinate system
$[t] + k^r[R]^T[O^rp^r]^T$ is the column vector connecting O^r and P^r expressed in the left camera coordinate
$\vec{J}$ is a column vector perpendicular to both O^lP^l and O^rP^r, as shown in Figure 2.3, and can be expressed in the left camera coordinate system as

$$\vec{J} = [O^lp^l]^T \times [R]^T[O^rp^r]^T \qquad (2.22)$$

and the scaling factors k^l, k^r, k can be determined by solving vector Equation (2.21).

Finally, the coordinate of P^l expressed in the left camera coordinate system is $k^l[O^lp^l]^T$, the coordinate of P^r expressed in the left camera coordinate system is $[t] + k^r[R]^T[O^rp^r]^T$, and the reconstructed 3D point P_E is

$$P_E = (k^l[O^lp^l]^T + [t] + k^r[R]^T[O^rp^r]^T)/2 \qquad (2.23)$$

Once P_E is determined in the left camera coordinate system, it can be transformed to any other coordinate system by a proper rotation matrix and translation vector. An algorithm for this approach can be found in Ahuja [30].

2.3.4 Improved 3D Shape Reconstruction Using Back-Projection

The reconstruction methods introduced in Section 2.3.3 tend to obtain the correspondence information beforehand by the stereo matching method introduced in Section 2.2 and then conduct the 3D reconstruction accordingly. As the stereo images captured by the left and right cameras are related by the intrinsic and the extrinsic parameters of the stereo system, the incorporation of such information during the speckle matching process will help to improve the matching accuracy as well as reduce computation time.

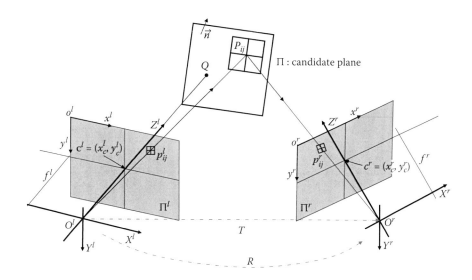

FIGURE 2.4
Process of back-projection of pixels within a speckle cell.

The incorporation of the calibration parameters will help to design a specific shape function from the constraint posed by the epipolar geometry and perspective transformation of the stereo system. This can be fulfilled by a back-propagation model [7,31].

As shown in Figure 2.4, after capturing the left and right speckle image, the DIC process is not directly applied to determine the disparities. On the contrary, a small speckle cell from the left speckle image is back-projected into the 3D scene to hit the object surface. As the cell has a very small area, the object surface at this area can be approximated by a plane Π. The intersection between the candidate plane and each ray emitted from the left speckle cell will then be determined and projected onto the right imaging screen to form a corresponding objective speckle cell. The correlation coefficient between the reference speckle cell in the left image and the back-projected objective speckle cell in the right image is then minimized by varying the parameters for the candidate plane Π.

The candidate plane Π can be defined in the left camera coordinate system using a unit vector $\vec{n} = (n_1, n_2, n_3)$, denoting its surface normal, and a point $Q = (0, 0, Z_Q)$, denoting the intersection position of the left optical axis with the candidate plane Π. For each point P^l within the candidate plane Π, its coordinates (X_P^l, Y_P^l, Z_P^l) should satisfy the following equation:

$$[P^l Q^l] \cdot \vec{n} = 0 \qquad (2.24)$$

where $[P^l Q^l]$ is the vector connecting P^l and Q^l.

Expanding Equation (2.24) explicitly yields

$$n_1 X_P^l + n_2 Y_P^l + n_3 (Z_P^l - Z_Q^l) = 0 \tag{2.25}$$

By combining Equations (2.12), (2.13), and (2.25), the coordinates of the back-projected point $P^l = (X_P^l, Y_P^l, Z_P^l)$ can be expressed explicitly in the left camera system as

$$X_P^l = \frac{n_3 Z_Q^l (x_p^l - x_c^l) h_x^l}{n_1 (x_p^l - x_c^l) h_x^l + n_2 (y_p^l - y_c^l) h_y^l + n_3 f^l} \tag{2.26}$$

$$Y_P^l = \frac{n_3 Z_Q^l (y_p^l - y_c^l) h_y^l}{n_1 (x_p^l - x_c^l) h_x^l + n_2 (y_p^l - y_c^l) h_y^l + n_3 f^l} \tag{2.27}$$

$$Z_P^l = \frac{n_3 Z_Q^l f^l}{n_1 (x_p^l - x_c^l) h_x^l + n_2 (y_p^l - y_c^l) h_y^l + n_3 f^l} \tag{2.28}$$

These coordinates are then transformed into the right camera coordinate system using Equation (2.16) to obtain $P^r = (X_P^r, Y_P^r, Z_P^r)$; the coordinates of the back-projected point $p^r = (x_p^r, y_p^r)$ on the right imaging screen are subsequently obtained using Equations (2.14) and (2.15).

After all pixels p_{ij}^l within the left speckle cell are back-projected onto the right imaging screen to obtain the corresponding position p_{ij}^r, a sum of the square difference correlation coefficient $C(\vec{P})$ can then be established using Equation (2.3) and minimized using the N–R iteration method discussed in Section 2.2.5. The controlling parameters $\vec{P}$ now contain only three independent parameters, $n_1, n_2,$ and Z_Q^l (n_3 can be determined from n_1, n_2); thus, less computation is required by the nonlinear optimization process. In addition, the back-projection model has incorporated the epipolar constraint and an improved shape function. This approach can potentially deliver 3D reconstruction results with higher accuracy.

It should be noted that the procedure of back-projection is unique to the area-based matching process. Since the feature points are sparse, this approach cannot be implemented for feature-based stereo matching.

2.3.5 A Simplified Case: Canonical Stereoscopic System

When the two cameras in the stereo setup have only a relative translation t_x in X direction without any rotation [32], or a single camera is given a pure translation t_x in X direction [33] to capture two different images, a simplified system called canonical setup results. In this simplified case, if identical cameras are used, we have $x_c^l = x_c^r = x_c$ (same principal point position), $h_x^l = h_x^r = h_x$ (same pixel dimension), $f^l = f^r = f$ (same focal length), $Z_P^l = Z_P^r = Z_P$ (same

Z coordinate), and $X_P^l - X_P^r = t_x$ (X coordinate difference is equal to baseline length t_x). Subtracting Equation (2.12) from Equation (2.14) and taking into account the previously mentioned identity, we have

$$Z_p = \frac{t_x f}{(x_p^l - x_p^r)h_x} = \frac{\text{Constant}}{x_p^l - x_p^r} \qquad (2.29)$$

This means that the Z coordinate (which is the shape) of a point P is in inverse proportion to the X disparity of the image points, implying a much simpler calibration and measurement process. This simple relationship can also be deduced from a geometric point of view [18].

2.4 Application Examples

Thanks to the high-contrast speckle pattern and accurate speckle matching algorithm, 3D DIC has enabled stereo vision systems for reconstruction of 3D shapes with high accuracy. During the last two decades, 3D DIC has found numerous applications in various fields.

Pan and coauthors [34] used 3D DIC for measuring the surface profile of a carbon fiber composite satellite antenna with a diameter of 730 mm. The reconstructed profile in Figure 2.5(a) shows good agreement with measurements taken from a commercial 3D coordinate measuring machine (CMM) and the maximum discrepancy is less than 2% of the maximum height. Orteu et al. [35] developed a four-camera system with the DIC algorithm and applied the system for measurement of a 3D profile (refer to Figure 2.5b) and deformation of sheet metal during the process of single point incremental forming. In comparison with the commercial HandyScan laser scanning system, accuracy of better than 0.05 mm has been reported.

A very interesting application of 3D DIC had also been reported by Morgan, Liu, and Yan [36], where a canonical setup with a narrow baseline was employed. A narrow baseline system is compact and suitable for measuring large-scale objects; it also helps to avoid occlusion in a steep region. However, measurement sensitivity has been inevitably reduced by the employment of a narrow baseline. Thus, high-accuracy subpixel methods should be enabled for disparity computation. In their experiment measuring the surface profile of a model landscape (as shown in Figure 2.6), accuracy as high as 0.015 pixel was obtained and the root mean square error of the reconstructed 3D shape was less than 1% of the maximum height. Real 3D terrain had also been obtained by the authors using satellite stereo images;

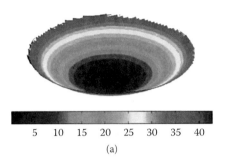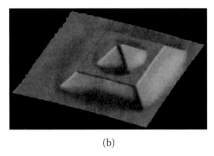

5 10 15 20 25 30 35 40	
(a)	(b)

FIGURE 2.5
Three-dimensional profile measurement of (a) a satellite antenna (reproduced from Pan, B. et al. *Strain* 45:194–200, 2009; with permission); (b) a sheet metal part produced by single point incremental forming (reproduced from Orteu, J. J. et al. *Experimental Mechanics* 51:625–639, 2011; with permission).

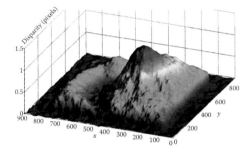

FIGURE 2.6
Surface profile measurement of (a) a model landscaped; (b) its photography for comparison. (Reproduced from Morgan, G. L. K. et al. *IEEE Transactions on Geoscience and Remote Sensing* 48 (9): 3424–3433, 2010. With permission.)

the results agreed well with publically available measurement results and showed better spatial resolution.

Due to its inherent nature of high accuracy, the 3D DIC method has been extensively used in the fields of experimental mechanics for 3D profile, 3D deformation measurement, and material characterization [37–39]. It has also been extended for microscale measurement using stereo images for light microscopes [40] or scanning electron microscopes [41].

2.5 Summary

The techniques for reconstructing an accurate 3D profile from stereo speckle image pairs have been introduced and the algorithms described in fine detail to facilitate the work of researchers interested in applying 3D DIC for measurement. Due to length limits, we have focused on the most essential part of this topic while ignoring some other contents. Readers can refer to other literature for error analysis and elimination [42] and 3D shape measurement from speckle projection [43]. A more complete description of 3D DIC can also be found in Sutton, Orteu, and Schreier [44].

References

1. M. A. Sutton, W. J. Wolters, W. H. Peters, W. F. Ransons, and S. R. McNeill. Determination of displacements using an improved digital correlation method. *Image Vision Computing* 1:133, 1983.
2. M. A. Sutton, H. A. Bruck, and S. R. McNeill, Determination of deformations using digital correlation with the Newton-Raphson method for partial differential corrections. *Experimental Mechanics* 29:261, 1989.
3. M. Sjodahl and L. R. Benchert. Electronic speckle photography: Analysis of an algorithm giving the displacement with subpixel accuracy. *Applied Optics* 32:2278–2284, 1993.
4. H. Lu and P. D. Cary. Deformation measurements by digital image correlation: Implementation of a second-order displacement gradient. *Experimental Mechanics* 40:393–400, 2000.
5. D. J. Chen, F. P. Chiang, Y. S. Tan, and H. S. Don. Digital speckle-displacement measurement using a complex spectrum method. *Applied Optics* 32:1839–1849, 1993.
6. M. Sjodahl. Electronic speckle photography—Increased accuracy by nonintegral pixel shifting. *Applied Optics* 33:6667–6673, 1994.
7. M. A. Sutton, S. R. McNeill, J. D. Helm, and Y. J. Chao. 2000, Advances in two-dimensional and three-dimensional computer vision. In *Photomechanics*, ed. P K Rastogi, 323–372. Berlin: Springer, 2000.
8. D. Garcia, J. J. Orteu, and L. Penazzi. A combined temporal tracking and stereo-correlation technique for accurate measurement of 3D displacements: Application to sheet metal forming. *Journal of Materials Processing Technology* 125–126:736–742, 2002.
9. T. A. Berfield, J. K. Patel., and R. G. Shimmin. Micro- and nanoscale deformation measurement of surface and internal planes via digital image correlation. *Experimental Mechanics* 47 (1):51–62, 2007.
10. M. Dekiff, P. Berssenbrügge, B. Kemper, C. Denz, and D. Dirksen. Three-dimensional data acquisition by digital correlation of projected speckle patterns. *Applied Physics* B 99:449–456, 2010.
11. Y. H. Huang, L. Liu, T. W. Yeung, and Y. Y. Hung. Real time monitoring of clamping force of a bolted joint by use of automatic digital image correlation. *Optics and Laser Technology* 41 (4):408–414, 2009.
12. B. Pan, K. Qian, H. Xie, and A. Asundi. Two-dimensional digital image correlation for in-plane displacement and strain measurement: A review. *Measurement Science Technology* 20:062001, 2009.

13. B. Pan, H. M. Xie, Z. Q. Guo, and T. Hua. Full-field strain measurement using a two-dimensional Savitzky–Golay digital differentiator in digital image correlation. *Optical Engineering* 46:033601, 2007.
14. G. Vendroux and W. G. Knauss, Submicron deformation field measurements: Part 2. Improved digital image correlation. *Experimental Mechanics* 38 (2):86–92, 1998.
15. H. Spath. *Two-dimensional spline interpolation algorithms.* Wellesley, MA: A. K. Peters, pp. 49–67, 1995.
16. W. H. Peters and W. F. Ranson. Digital imaging techniques in experimental stress analysis. *Optical Engineering* 21:427–431, 1981.
17. B. C. Wattrisse, A. Muracciole, and J. M. Nemoz-Gaillard. Analysis of strain localization during tensile tests by digital image correlation. *Experimental Mechanics* 41:29–39, 2001.
18. B. Cyganek and J. P. Siebert. *An introduction to 3D computer vision techniques and algorithms.* New York: Wiley, 2009.
19. E. Trucco and A. Verri. *Introductory techniques for 3-D computer vision.* Englewood Cliffs, NJ: Prentice Hall, 1998.
20. W. E. L. Grimson. Why stereo vision is not always about 3D reconstruction. Memo 1435, MIT Artificial Intelligence Laboratory, 1993.
21. O. Faugeras. *Three-dimensional computer vision: A geometric viewpoint.* Cambridge, MA: MIT Press, 1993.
22. R. Y. Tsai. A versatile camera calibration technique for high-accuracy 3D machine vision metrology using off-the-shelf TV cameras and lenses. *IEEE Journal Robotics and Automation* 3 (4):323–344, 1987.
23. Q. T. Luong and O. D. Faugeras, Self-calibration of a stereo rig from unknown camera motions and point correspondences. INRIA Technical Report 2014, 1993.
24. M. I. A. Lourakis and R. Deriche, Camera self-calibration using the Kruppa equations and the SVD of the fundamental matrix: The case of varying intrinsic parameters. INRIA Technical Report 2121, 2000.
25. Z. Y. Zhang. A flexible new technique for camera calibration. *IEEE Transactions on Pattern Analysis Machine Intelligence* 22:1330–1334, 2000.
26. Z. Y. Zhang. Camera calibration with one-dimensional objects. *IEEE Transactions on Pattern Analysis and Machine Intelligence* 26 (7):892–899, 2004.
27. A. Gruen and T. S. Huang, eds. *Calibration and orientation of cameras in computer vision,* Springer Series in Information Sciences. New York: Springer, 2001.
28. W. Sun and J. R. Cooperstock. An empirical evaluation of factors influencing camera calibration accuracy using three publicly available techniques. *Machine Vision and Applications* 17 (1):51–67, 2006.
29. W. H. Press, S. A. Teukolsky, W. T. Vetterling, and B. P. Flannery. *Numerical recipes in C: The art of scientific computing,* 3rd ed. Cambridge, England: Cambridge University Press, 2007.
30. N. Ahuja. *Motion and structure from image sequence.* New York: Springer–Verlag, 1993.
31. J. D. Helm, S. R. McNeil, and M. A. Sutton. Improved three-dimensional image correlation for surface displacement measurement. *Optical Engineering* 35:1911–1920, 1996.
32. P. Synnergren. Measurement of three-dimensional displacement fields and shape using electronic speckle photography. *Optical Engineering* 36 (8):2302–2310, 1997.

33. Y. H. Huang, C. Quan, C. J. Tay, and L. J. Chen. Shape measurement by the use of digital image correlation. *Optical Engineering* 44 (8):087011, 2005.
34. B. Pan, H. M. Xie, L. H. Yang, and Z. Y. Wang. Accurate measurement of satellite antenna surface using three-dimensional digital image correlation technique. *Strain* 45:194–200, 2009.
35. J. J. Orteu, F. Bugarin, J. Harvent, L. Robert, and V. Velay. Multiple-camera instrumentation of a single point incremental forming process pilot for shape and 3D displacement measurements: Methodology and results. *Experimental Mechanics* 51:625–639, 2011.
36. G. L. K. Morgan, J. G. Liu, and H. S. Yan. Precise subpixel disparity measurement from very narrow baseline stereo. *IEEE Transactions on Geoscience and Remote Sensing* 48 (9):3424–3433, 2010.
37. J. D. Helm, M. A. Sutton, and S. R. McNeill. Deformations in wide, center-notched, thin panels, part I: Three-dimensional shape and deformation measurements by computer vision. *Optical Engineering* 42 (5):1293–1305, 2003.
38. J. J. Orteu. 3-D computer vision in experimental mechanics. *Optics and Lasers in Engineering* 47:282–291, 2009.
39. S. B. Park, C. Shah, J. B. Kwak, C. Jang, S. Chung, and J. M. Pitarresi. Measurement of transient dynamic response of circuit boards of a handheld device during drop using 3D digital image correlation. *Journal of Electronic Packaging—Transactions of the ASME* 130–044502-1-3, 2008.
40. H. W. Schreier, D. Garcia, and M. A. Sutton. Advances in light microscope stereo vision. *Experimental Mechanics* 44 (3):278–288, 2004.
41. N. Cornille. Accurate 3D shape and displacement measurement using a scanning electron microscope. PhD thesis. INSA, France, and University of South Carolina, Columbia, June 2005.
42. M. Bornert, F. Brémand, P. Doumalin, J. C. Dupré, M. Fazzini, M. Grédiac, F. Hild, et al., Assessment of digital image correlation measurement errors: Methodology and results. *Experimental Mechanics* 49:353–370, 2009.
43. H. J. Dai and X. Y. Su. Shape measurement by digital speckle temporal sequence correlation with digital light projector. *Optical Engineering* 40 (5):793–800, 2001.
44. M. A. Sutton, J. J. Orteu, and H. W. Schreier. *Image correlation for shape, motion and deformation measurements: Basic concepts, theory and applications.* New York: Springer.

3

Spacetime Stereo

Li Zhang, Noah Snavely, Brian Curless, and Steven M. Seitz

CONTENTS

3.1 Introduction

Very few shape-capture techniques work effectively for rapidly moving scenes. Among the few exceptions are depth from defocus [9] and stereo [5]. Structured-light stereo methods have shown particularly promising results for capturing depth maps of moving faces [6,11]. Using projected light patterns to provide dense surface texture, these techniques compute pixel correspondences and then derive depth maps by triangulation. Products based on these triangulation techniques are commercially available.[*]

Traditional one-shot triangulation methods [3,12] treat each time instant in isolation and compute spatial correspondences between pixels in a single pair of images for a static moment in time. While they enable reconstructing moving scenes, they typically provide limited shape detail and resolution. Better results may be obtained by considering how each pixel varies over time and using this variation as a cue for correspondence, an approach we call *spacetime stereo.*

---

[*] For example, see www.3q.com, www.eyetronics.com, and www.xbox.com/en-US/kinect.

3.2 Spacetime Stereo Matching Metrics

In this section, we formulate the spacetime stereo problem and define the metrics that are used to compute correspondences. Consider a Lambertian scene observed by two synchronized and precalibrated video cameras.[*] Spacetime stereo takes as input the two rectified image streams $I_l(x,y,t)$ and $I_r(x,y,t)$ from these cameras. To recover the time-varying three-dimensional (3D) structure of the scene, we wish to estimate the disparity function $d(x,y,t)$ for each pixel (x,y) in the left image at each time t. Many existing stereo algorithms solve for $d(x,y,t)$ at some position and moment (x_0,y_0,t_0) by minimizing the following error function:

$$E(d_0) = \sum_{(x,y)\in W_0} e\left(I_l(x,y,t_0), I_r(x-d_0,y,t_0)\right) \tag{3.1}$$

where d_0 is a shorthand notation for $d(x_0,y_0,t_0)$, W_0 is a spatial neighborhood window around (x_0,y_0), and $e(p,q)$ is a similarity measure between pixels from two cameras. Depending on the specific algorithm, the size of W_0 can vary from being a single pixel to, say, a 10×10 neighborhood, or it can be adaptively estimated for each pixel [8]. A common choice for $e(p,q)$ is simply

$$e(p,q) = (p-q)^2 \tag{3.2}$$

In this case, Equation (3.1) becomes the standard sum of squared difference (SSD). Better results can be obtained in practice by defining $e(p,q)$ to compensate for radiometric differences between the cameras:

$$e(p,q) = (s\cdot p + o - q)^2 \tag{3.3}$$

where s and o are window-dependent scale and offset constants to be estimated. Other forms of $e(p,q)$ are summarized in Baker and Matthews [1].

Spacetime stereo seeks to incorporate *temporal appearance variation* to improve stereo matching and generate more accurate depth maps. In the next two subsections, we will consider how multiple frames can help to recover static and nearly static shapes and then extend the idea for moving scenes.

---

[*] Here we assume the two cameras are offset horizontally. However, our formulation can be used for any stereo pair orientation.

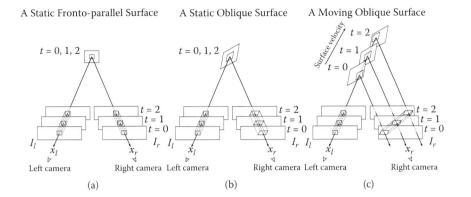

FIGURE 3.1

Illustration of spacetime stereo. Two stereo image streams are captured from stationary cameras. The images are shown spatially offset at three different times, for illustration purposes. For a static surface (a, b), the spacetime windows are "straight," aligned along the line of sight. For an oblique surface (b), the spacetime window is horizontally stretched and vertically sheared. For a moving surface (c), the spacetime window is also temporally sheared (i.e., "slanted"). The best affine warp of each spacetime window along epipolar lines is computed for stereo correspondence.

3.2.1 Static Scenes

Scenes that are geometrically static may still give rise to images that change over time. For example, the motion of the sun causes shading variations over the course of a day. In a laboratory setting, projected light patterns can create similar but more controlled changes in appearance.

Suppose that the geometry of the scene is static for a period of time $T_0 = [t_0 - \Delta t, t_0 + \Delta t]$. As illustrated in Figure 3.1(a), we can extend the spatial window to a spatiotemporal window and solve for d_0 by minimizing the following sum of SSD (SSSD) cost function:

$$E(d_0) = \sum_{t \in T_0} \sum_{(x,y) \in W_0} e\left(I_l(x, y, t), I_r(x - d_0, y, t)\right) \tag{3.4}$$

This error function reduces matching ambiguity in any single frame by simultaneously matching intensities in multiple frames. Another advantage of the spacetime window is that the spatial window can be shrunk and the temporal window can be enlarged to increase matching accuracy. This principle was originally formulated as spacetime analysis in Curless and Levoy [4] and Kanade, Gruss, and Carley [7] for laser scanning and was applied by several researchers [2,13] for structured-light scanning. However, they only consider either a single-frame spatial window or a single-pixel temporal window and their specialized formulation is only effective for single stripe illumination. Here we are casting this principle in a general spacetime stereo framework that is valid for any illumination variation.

We should point out that Equations (3.1) and (3.4) treat disparity as being constant within the window W_0, which assumes the corresponding surface is frontoparallel. For a static but oblique surface, as shown in Figure 3.1(b), a more accurate (first order) local approximation of the disparity function is

$$d(x,y,t) \approx \hat{d}_0(x,y,t) \overset{\text{def}}{=} d_0 + d_{x_0} \cdot (x - x_0) + d_{y_0} \cdot (y - y_0) \tag{3.5}$$

where d_{x_0} and d_{y_0} are the partial derivatives of the disparity function with respect to spatial coordinates x and y at (x_0, y_0, t_0). This local spatial linearization results in the following SSSD cost function being minimized:

$$E(d_0, d_{x_0}, d_{y_0}) = \sum_{t \in T_0} \sum_{(x,y) \in W_0} e\big(I_l(x,y,t), I_r(x - a_0, y, t)\big) \tag{3.6}$$

where $\hat{d}_0$ is a shorthand notation for $\hat{d}_0(x, y, t)$, which is defined in Equation (3.5) in terms of (d_0, d_{x_0}, d_{y_0}) and is estimated for each pixel. Nonzero d_{x_0} and d_{y_0} will cause a horizontal stretch or shrinkage and vertical shear of the spacetime window, respectively, as illustrated in Figure 3.1(b).

Figure 3.2 shows the experimental results obtained by imaging a small sculpture (a plaster bust of Albert Einstein) using structured light. Specifically, stripe patterns based on a modified Gray code [14] are projected onto the bust using a digital projector and 10 stereo image pairs are captured. The 10 image pairs are matched with a spacetime window of 5 × 5 × 10 (5 × 5 pixels per frame by 10 frames). The shaded rendering reveals details visually comparable to those obtained with a laser range scanner.

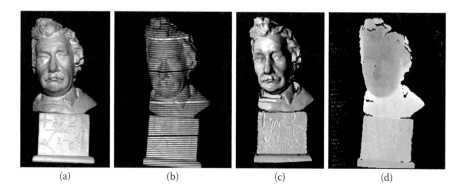

(a)   (b)   (c)   (d)

FIGURE 3.2
Spacetime stereo reconstruction with structured light. (a) Einstein bust under natural lighting. (b) One image taken from the set of 10 stereo image pairs captured when the bust is illuminated with modified Gray code structured-light patterns [14]. (c) Shaded rendering of geometric reconstruction. (d) Reconstructed disparity map where pixel intensities encode surface depth.

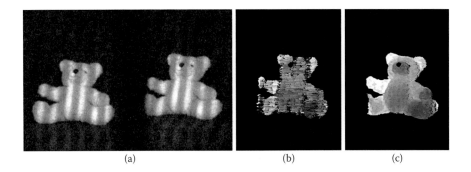

FIGURE 3.3
Spacetime stereo reconstruction with loosely structured light using transparency and desk lamp. (a) One out of the 125 stereo pair images. (b) Disparity map for a traditional stereo reconstruction using a single stereo pair. (c) Disparity map for spacetime stereo using all the 125 stereo pairs.

Figure 3.3 shows the experimental results obtained by a simpler imaging system with much looser structured light. Specifically, for illumination, an ordinary desk lamp is shined through a transparency printed with a black and white square wave pattern onto the subject (a teddy bear) and the pattern is moved by hand in a free-form fashion. During the process, 125 stereo pairs are captured. Both single-frame stereo for one of the stereo pairs using a 5×1 window and spacetime stereo over all frames using a $5 \times 1 \times 125$ window are tried. Figure 3.3 shows marked improvement of spacetime stereo over regular single-frame stereo.

3.2.2 Quasi-Static Scenes

The simple SSSD method proposed in the previous section can also be applied to an interesting class of time-varying scenes. Although some natural scenes, like water flow in Figure 3.4, have spatially varying texture and motion, they have an overall shape that is roughly constant over time. Although these natural scenes move stochastically, people tend to fuse the image stream into an average shape over time. We refer to this class of natural scenes as *quasi-static*. By applying the SSSD method from the previous section, we can compute a temporally averaged disparity map that corresponds roughly to the "mean shape" of the scene. In graphics applications where a coarse geometry is sufficient, one could, for instance, use the mean shape as static geometry with time-varying color texture mapped over the surface.

Figure 3.4 shows experimental results of applying spacetime stereo to a quasi-static scene: a small but fast moving waterfall. Specifically, 45 stereo image pairs are captured and both traditional stereo for one of the image pairs and spacetime stereo for all 45 image pairs are tried. As Figure 3.4 shows, spacetime stereo produces much more consistent results than traditional stereo does.

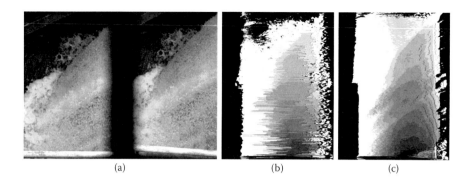

(a)   (b)   (c)

FIGURE 3.4
Spacetime stereo reconstruction of quasi-static rushing water. (a) One out of the 45 stereo pair images. (b) Disparity map for a traditional stereo reconstruction on one image pair. (c) Disparity map for spacetime stereo on all 45 stereo image pairs.

3.2.3 Moving Scenes

Now we consider the case where the object is moving in the time interval T_0 = $[t_0 - \Delta t, t_0 + \Delta t]$, as illustrated in Figure 3.1(c). Because of the object motion, the window in the left video is deformed in the right sequence. The temporal trajectory of window deformation in the right video is determined by the object motion and could be arbitrarily complex. However, if the camera has a high enough frame rate relative to the object motion and there are no changes in visibility, we can locally linearize the temporal disparity variation in much the same way we linearized spatial disparity in Equation (3.5). Specifically, we take a first-order approximation of the disparity variation with respect to both spatial coordinates x and y and temporal coordinate t as

$$d(x,y,t) \approx \tilde{d}_0(x,y,t) \overset{\text{def}}{=} d_0 + d_{x_0} \cdot (x - x_0) + d_{y_0} \cdot (y - y_0) + d_{t_0} \cdot (t - t_0) \quad (3.7)$$

where d_{t_0} is the partial derivative of the disparity function with respect to time at (x_0, y_0, t_0). This local spatial–temporal linearization results in the following SSSD cost function to be minimized:

$$E(d_0, d_{x_0}, d_{y_0}, d_{t_0}) = \sum_{t \in T_0} \sum_{(x,y) \in W_0} e\big(I_l(x,y,t), I_r(x - d_0', y, t)\big) \quad (3.8)$$

where $\tilde{d}_0$ is a shorthand notation for $\tilde{d}_0(x,y,t)$, which is defined in Equation (3.7) in terms of $(d_0, d_{x_0}, d_{y_0}, d_{t_0})$ and is estimated for each pixel at each time. Note that Equation (3.7) assumes a linear model of disparity within the spacetime window (i.e., $(d_0, d_{x_0}, d_{y_0}, d_{t_0})$ is constant within $W_0 \times T_0$).

We use the term *straight window* to refer to a spacetime window whose position and shape are fixed over time, such as the windows shown in

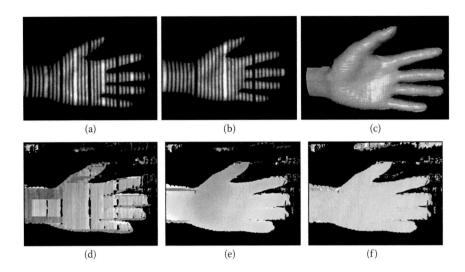

FIGURE 3.5
Spacetime stereo reconstruction of moving hand with structured light. (a, b) Two images taken from one of the cameras. The hand is moving away from the stereo rig, which is why it is getting smaller. (c) Shaded rendering of the reconstructed model using slanted window spacetime stereo. (d) Disparity map with straight spacetime windows. (e) Disparity map with slanted spacetime windows. (f) Temporal derivative of disparity. Since the hand is translating at roughly a constant speed, the disparity velocity is fairly constant over the hand.

Figure 3.1(a, b). If the position of the window varies over time, we say that the spacetime window is *slanted,* such as the one in the right camera in Figure 3.1(c).

Figure 3.5 and Figure 3.6 show two experimental results for moving scenes. In Figure 3.5, the modified Gray code patterns are projected onto a human hand that is moving fairly steadily away from the stereo rig. In this case, the disparity is changing over time, and the straight spacetime window approach fails to reconstruct a reasonable surface. By estimating the temporal derivative of the disparity using slanted windows, we obtain a much better reconstruction, as shown in Figure 3.5.

In Figure 3.6, spacetime stereo and one-shot stereo are compared for reconstructing moving scenes. Specifically, a short sequence of stripe patterns is generated randomly and projected repeatedly to a deforming human face. To demonstrate the effect of extruding the matching window over time, a $9 \times 5 \times 5$ spacetime window and a 15×15 spatial window, which have the same number of pixels, are chosen to compare spacetime stereo and frame-by-frame stereo on reconstructing facial expressions. Notice that the spacetime reconstruction generates significantly more accurate results, as shown in Figure 3.6. The video format of this comparison is available at http://grail.cs.washington.edu/projects/ststereo/, which also shows that spacetime stereo results in temporally much more stable reconstruction than frame-by-frame stereo.

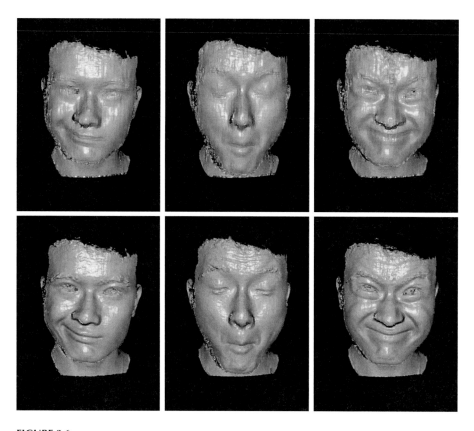

FIGURE 3.6
Comparison of spacetime and frame-by-frame stereo for a moving face reconstruction. Top row: three face expressions reconstructed using a 15×15 spatial window. Bottom row: the same face expressions reconstructed using a $9 \times 5 \times 5$ spacetime window. Notice that although both the spatial and spacetime windows have the same number of pixels, spacetime stereo results in a more detailed reconstruction.

3.3 Global Spacetime Stereo Matching

One major artifact of spacetime stereo matching is that it produces quite noticeable ridging artifacts, evident in Figure 3.8(e). It has been suggested [16] that these artifacts are due primarily to the fact that Equation (3.8) is minimized for each pixel independently, without taking into account constraints between neighboring pixels. Specifically, computing a disparity map with M pixels introduces $4M$ unknowns: M disparities and $3M$ disparity gradients. While this formulation results in a system that is convenient computationally, it is clearly overparameterized, since the $3M$ disparity gradients are a function of only M disparities. Indeed, the estimated disparity gradients may not agree with the estimated disparities.

For example, $d_x(x,y,t)$ may be quite different from its central difference approximation, $1/2\big(d(x+1,y,t)-d(x-1,y,t)\big)$, because $d_x(x,y,t)$, $d(x+1,y,t)$, and $d(x-1,y,t)$ are independently estimated for each pixel. This inconsistency between disparities and disparity gradients results in inaccurate depth maps, as shown in Figure 3.8(e, i).

In this section, spacetime stereo is reformulated as a global optimization problem to overcome this inconsistency deficiency. Specifically, the global spacetime stereo method computes the disparity function while taking into account gradient constraints between pixels that are adjacent in space and time. Given image sequences $I_l(x,y,t)$ and $I_r(x,y,t)$, the desired disparity function $d(x,y,t)$ minimizes

$$\Gamma\Big(\big\{d(x,y,t)\big\}\Big)\overset{\Delta}{=}\sum_{x,y,t}E\big(d,d_x,d_y,d_t\big)\tag{3.9}$$

This is subject to the following constraints[*]:

$$d_x(x,y,t)=\frac{1}{2}\big(d(x+1,y,t)-d(x-1,y,t)\big)$$

$$d_y(x,y,t)=\frac{1}{2}\big(d(x,y+1,t)-d(x,y-1,t)\big)\tag{3.10}$$

$$d_t(x,y,t)=\frac{1}{2}\big(d(x,y,t+1)-d(x,y,t-1)\big)$$

Equation (3.9) defines a nonlinear least squares problem with linear constraints. This problem can be solved using the Gauss–Newton method [10].

Figure 3.8 shows the improvement using global spacetime stereo by comparing it to the spacetime stereo matching method presented in the previous section, as well as standard frame-by-frame stereo. The experiment's data are captured by a camera rig of six synchronized video streams (four monochrome and two color) running at 60 Hz, shown in Figure 3.7. Three of the cameras capture the left side of the face, and the other three capture the right side. To facilitate depth computation, two video projectors are used that project grayscale random stripe patterns onto the face. The projectors send a solid black pattern every three frames, and surface color texture is captured at these frames. A video of a pair of depth map sequences reconstructed from both left and right sides of a moving face using global spacetime stereo is available at http://grail.cs.washington.edu/projects/stfaces/.

---

[*] At spacetime volume boundaries, we use forward or backward differences instead of central differences.

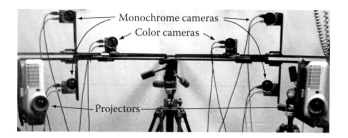

FIGURE 3.7

A camera rig consists of six video cameras and two data projectors. The two monochrome cameras on the left constitute one stereo pair, and the two on the right constitute a second stereo pair. The projectors provide stripe pattern textures for high-quality shape estimation. The color cameras record video streams used for optical flow and surface texture.

3.4 From Depth Map Videos to Dynamic Mesh Models

Using the camera rig in Figure 3.7, spacetime stereo methods capture both shape and texture for a dynamic surface. The captured shapes are represented as depth maps. These depth maps have missing data due to occlusions. Also, they lack correspondence information that specifies how a point on the 3D surface moves over time. For face modeling applications in graphics, these limitations make it difficult to re-pose or reanimate the captured faces. In Zhang et al. [16], one method is presented to compute a single time-varying mesh that closely approximates the depth map sequences while optimizing the vertex motion to be consistent with pixel motion (i.e., optical flow) computed between color frames.

In Figure 3.9, eight examples of mesh models for two subjects are shown. Each mesh has about 23,728 vertices. The sequence for the first subject has 384 mesh models and that for the second has 580 meshes. In Figure 3.10, eight examples of the mesh sequence for the third subject are shown. This sequence consists of 339 meshes and each mesh has 23,728 vertices. Notice that, in Figure 3.9(i, j) and Figure 3.10(i, j), close-up views of some of the face meshes are shown, which demonstrates the ability to capture fine features such as the wrinkles on a frowning forehead and near a squinting eye. These subtle shape details are extremely important for conveying realistic expressions because the human visual system is well tuned to recognize human faces. The uniqueness of the system is its capability to capture not only these shape details, but also how these shape details change over time. In the end, the system is an automatic, dense, and markerless facial motion capture system.

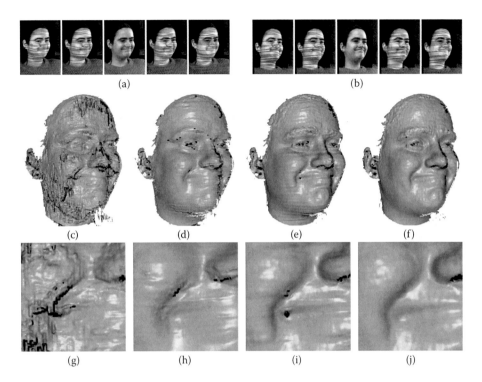

FIGURE 3.8

Comparison of four different stereo matching algorithms. (a, b) Five consecutive frames from a pair of stereo videos. The third frames are nonpattern frames. (c) Reconstructed face at the third frame using traditional stereo matching with a 15×15 window. The result is noisy due to the lack of color variation on the face. (d) Reconstructed face at the second frame using stereo matching with a 15×15 window. The result is much better because the projected stripes provide texture. However, certain face details are smoothed out due to the need for a large spatial window. (e) Reconstructed face at the third frame using local spacetime stereo matching with a $9 \times 5 \times 5$ window. Even though the third frame has little intensity variation, spacetime stereo recovers more detailed shapes by considering neighboring frames together. However, it also yields noticeable striping artifacts due to the parameterization of the depth map. (f) Reconstructed face at the third frame using our new global spacetime stereo matching with a $9 \times 5 \times 5$ window. The new method removes most of the striping artifacts while preserving the shape details. (g–j) Close-up comparison of the four algorithms around the nose and the corner of the mouth.

3.5 Discussion

In this chapter, spacetime stereo methods are presented to capture time-varying 3D surfaces. We discuss limitations of the methods and suggest future works. The spacetime stereo matching techniques are based on window warping using a locally linear disparity variation model, which fails near discontinuity boundaries. For example, rapid lip movements during speech result in temporal depth discontinuities in the scanned sequence. It

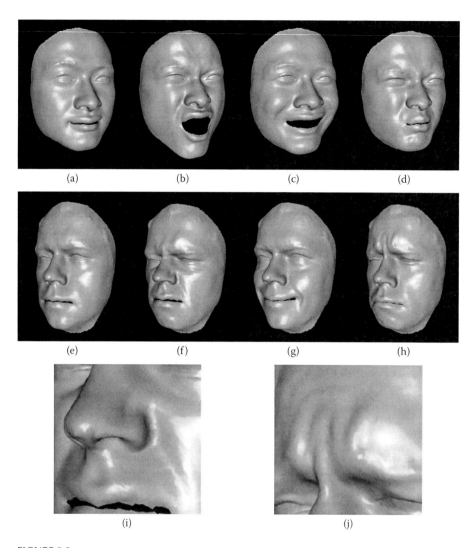

FIGURE 3.9
(a–h) Eight examples of the template tracking results for two subjects. (i) A close-up view of the skin deformation near the nose in the "disgusted" expression shown in (f). (j) A close-up view of the wrinkles on the "frowning" forehead of (h).

is desirable to improve spacetime stereo algorithms for better performance near spatial and temporal discontinuities. Specifically, the window should be adaptively selected based on local spacetime shape variation. In the extreme case of large shape variation over time due to fast motion, the algorithm should automatically switch to one-shot stereo. Kanade and Okutomi [8] proposed an adaptive window selection theory for one-shot stereo assuming a front-parallel disparity model; adaptive window selection for a locally linear disparity model in spacetime remains an important topic for future research.

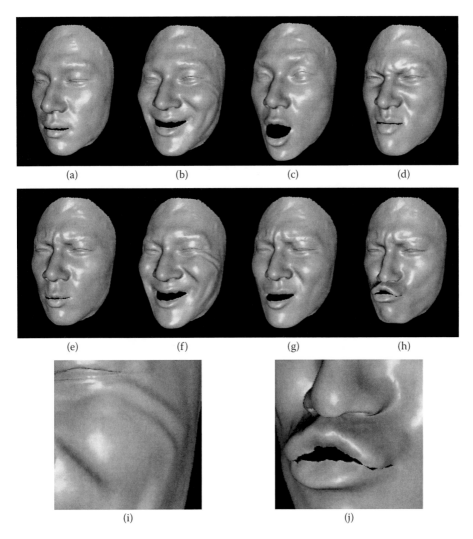

FIGURE 3.10

(a–h) Eight examples of the template tracking results for the third subject. (i) A close-up view of the wrinkles near the left eye in (f). (j) A close-up view of the pouting lips in (h).

Stereo matching algorithms involve various parameters, such as window sizes in spacetime stereo and regularization weight in Markov random field (MRF) stereo. In practice, different data sets require different parameters for optimal performance. Learning parameters automatically is therefore of great importance for these algorithms to be deployed in automated vision systems for real applications. The problem of learning optimal parameters for MRF stereo matching is studied in Zhang and Seitz [15]. How to apply this mechanism to spacetime stereo and other visual reconstruction algorithms is a very interesting problem for future research.

Although the accuracy of spacetime stereo is good enough for several applications (e.g., face modeling and animation), a current limitation is speed. It takes 2–3 minutes to compute a disparity map. Some applications, such as 3D shape recognition of faces, expressions, and poses for natural human computer interfaces, require 3D reconstruction techniques that operate in real time. An important future research avenue is to design new techniques and accelerate existing techniques to fulfill both the speed and accuracy requirements for real-time 3D reconstruction.

References

1. S. Baker and I. Matthews. Lucas–Kanade 20 years on: A unifying framework. *International Journal of Computer Vision* 56 (3): 221–255, March 2004.
2. J.-Y. Bouguet and P. Perona. 3D photography on your desk. *Proceedings International Conference on Computer Vision* 43–50, 1998.
3. K. L. Boyer and A. C. Kak. Color-encoded structured light for rapid active ranging. *IEEE Trans. on Pattern Analysis and Machine Intelligence* 9 (1): 14–28, 1987.
4. B. Curless and M. Levoy. Better optical triangulation through spacetime analysis. *Proceedings International Conference on Computer Vision* 987–994, June 1995.
5. O. Faugeras. *Three-dimensional computer vision.* Cambridge, MA: MIT Press, 1993.
6. P. S. Huang, C. P. Zhang, and F. P. Chiang. High speed 3-D shape measurement based on digital fringe projection. *Optical Engineering,* 42 (1): 163–168, 2003.
7. T. Kanade, A. Gruss, and L. Carley. A very fast VLSI range finder. *Proceedings International Conference on Robotics and Automation* 39: 1322–1329, 1991.
8. T. Kanade and M. Okutomi. A stereo matching algorithm with an adaptive window: Theory and experiment. *IEEE Transactions on Pattern Analysis and Machine Intelligence* 16 (9): 920–932, 1994.
9. S. K. Nayar, M. Watanabe, and M. Noguchi. Real-time focus range sensor. *IEEE Transactions on Pattern Analysis and Machine Intelligence,* 18 (12): 1186–1198, 1996.
10. J. Nocedal and S. J. Wright. *Numerical optimization.* New York: Springer, 1999.
11. M. Proesmans, L. Van Gool, and A. Oosterlinck. One-shot active 3D shape acquisition. *Proceedings International Conference on Pattern Recognition* 336–340, 1996.
12. M. Proesmans, L. Van Gool, and A. Oosterlinck. One-shot active 3d shape acquisition. *International Conference on Pattern Recognition* 336–340, 1996.
13. K. Pulli, H. Abi-Rached, T. Duchamp, L. Shapiro, and W. Stuetzle. Acquisition and visualization of colored 3D objects. *Proceedings International Conference on Pattern Recognition* 11–15, 1998.
14. L. Zhang, B. Curless, and S. M. Seitz. Spacetime stereo: Shape recovery for dynamic scenes. *Proceedings IEEE Conference on Computer Vision and Pattern Recognition* 367–374, 2003.
15. L. Zhang and S. M. Seitz. Parameter estimation for MRF stereo. In *IEEE Computer Society Conference on Computer Vision and Pattern Recognition,* 2005.
16. L. Zhang, N. Snavely, B. Curless, and S. M. Seitz. Spacetime faces: High-resolution capture for modeling and animation. *ACM Annual Conference on Computer Graphics* 548–558, August 2004.

4

Stereo Particle Imaging Velocimetry Techniques: Technical Basis, System Setup, and Application

Hui Hu

CONTENTS

4.1 Introduction

Particle image velocimetry (PIV) [1] is an imaging-based flow diagnostic technique that relies on seeding fluid flows with tiny tracer particles and observing the motions of the tracer particles to derive fluid velocities. For PIV measurements, a sheet of laser light is usually used to illuminate the region of interest. The tracer particles scatter the laser light as they move through it. Photographic film or digital cameras are used to record the positions of the tracer particles at two different times separated by a prescribed

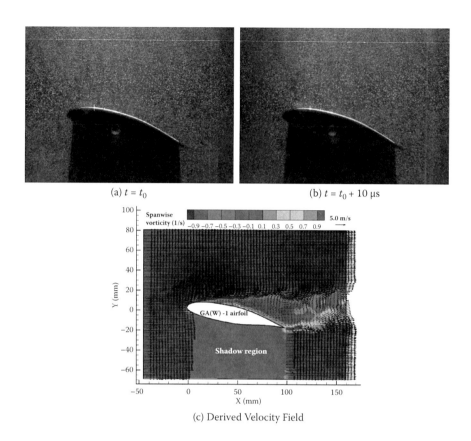

(a) $t = t_0$ (b) $t = t_0 + 10\ \mu s$

(c) Derived Velocity Field

FIGURE 4.1
(See color insert.) A pair of PIV images and the corresponding velocity distribution. (Hu, H. and Yang, Z. 2008. *ASME Journal of Fluid Engineering* 130 (5): 051101. With permission.)

time interval. The displacements of individual tracer particles—or, more often, groups of tracer particles—are determined by a well-developed computer-intensive PIV image processing procedure. The displacements over a known time interval provide the particle velocity vectors. The velocity of the working fluid is deduced based on the assumption that the tracer particles move with the same velocity as local working fluids. Figure 4.1 shows a typical PIV image pair and the corresponding flow velocity distribution derived from the images as the result of the PIV measurements.

A "classical" PIV technique, as shown in Figure 4.1, is a two-dimensional (2D) measuring technique. It is only capable of measuring two components of flow velocity vectors in the plane of the illuminating laser sheet. The out-of-plane component of velocity vectors is lost, while the in-plane components are affected by an unrecoverable error due to the perspective transformation [2]. Recent advances in the PIV technique have been directed toward obtaining all three components of fluid velocity vectors in a plane or in a volume simultaneously to allow wider applications of the PIV technique to study more complex flow phenomena. Several advanced PIV techniques have been developed successfully in recent years, including the holographic PIV (HPIV) technique [3,4], three-dimensional (3D) particle-tracking velocimetry (3D-PTV) [5], tomographic PIV [6], and the stereo PIV (SPIV) to be described in the present study.

Holographic PIV [3,4] utilizes holography techniques for image recording, which enables the determination of all three components of velocity vectors throughout a volume of fluid flow. Of the existing PIV techniques, HPIV is capable of the highest measurement precision and spatial resolution [7]. However, HPIV is also the most complex PIV technique and requires significant investment in equipment and optical alignments as well as the development of advanced data processing techniques. Continuous efforts are still required to make HPIV a practical PIV technique for various complex engineering applications [8–10].

Three-dimensional PTV [5] and tomographic PIV [6] techniques typically use three or more cameras to record the positions of the tracer particles in the measurement volume from different observation directions. Through 3D image reconstruction, the locations of the tracer particles in the measurement volume are determined. By using particle-tracking (for 3D-PTV) or 3D correction-based image processing algorithms, the 3D displacements of the tracer particles in the measurement volume are determined. It should be noted that the positions of almost all the tracer particles in the measurement volume are required to be recorded by each image recording camera for the 3D-PTV or tomographic PIV measurements. Thus, it becomes very difficult, if not impossible, to distinguish the positions of the tracer particles if the image density of the tracer particle in the measurement volume becomes too high. Therefore, the measurement results of 3D-PTV and tomographic PIV systems usually suffer from poor spatial resolution to elucidate the small-scale vortex and flow structures in the fluid flows.

The stereo PIV technique is the most straightforward and easily accomplished method to achieve simultaneous measurements of all three components of velocity vectors in the laser illuminating plane. It uses two cameras at different view axes or with an offset distance to do stereoscopic image recording. In the view reconstruction, the corresponding image segments in the image planes of the two cameras are matched to reconstruct all three components of the velocity vectors in the measurement plane. Stereo PIV measurements can have much higher in-plane spatial resolution compared to the 3D-PTV and tomographic PIV methods. It can provide thousands of flow velocity vectors in the measurement plane.

In the sections that follow, the technical basis of stereo PIV technique is introduced briefly at first. Three basic optical arrangements commonly used for this type of image recording are compared and the advantages and disadvantages of each optical arrangement are briefly described. A general in situ calibration procedure is described to determine the mapping functions between the 3D physical space in the objective fluid flow and the image planes of the cameras used for stereoscopic image recording. The flow chart to reconstruct the three components of the velocity vectors in the measurement plane of physical space from the 2D displacement vectors detected by the two cameras is also given. Finally, the feasibility and implementation of the stereo PIV technique are demonstrated by performing simultaneous measurements of all three components of flow velocity vectors in an airflow exhausted from a lobed nozzle/mixer to reveal the unique 3D flow features in the lobed jet mixing flow.

4.2 Technical Basis of the Stereo PIV Technique

The technical basis of stereo PIV measurements is how to reconstruct all three components of displacement vectors in the measurement plane of physical space from the two projected planar displacement vectors detected by the two cameras. Figure 4.2 shows the schematic on how and why the three components of displacement vectors in the measurement plane of the physical space can be reconstructed from the two projected planar vectors in the image planes detected by the two cameras. The origin, O, is a known point within the measurement plane (i.e., laser illuminating plane), which is visible by both of the cameras, with the physical coordinate of (x_0, y_0, z_0). Assuming an orthogonal, global coordinate system, the X- and Y-axes are aligned with the laser sheet, while the Z-axis is normal to the laser sheet plane.

The two cameras view the area of interest within the light sheet plane from the points $L_1 = (x_1, y_1, z_1)$ and $L_2 = (x_2, y_2, z_2)$. Based on the pinhole lens assumption, the measured displacements of three-dimensional displacement (dx, dy, dz) in the physical space are (dx_1, dy_1) and (dx_2, dy_2) in the image planes of the two cameras. Given the ensemble averaging nature of the PIV measurements and the fact that the illuminating laser sheet thickness is several orders smaller than the observation distance, the displacement vectors can be assumed to be within a zero thickness plane in the physical space. The angles enclosed by the viewing ray and the light sheet normal direction (parallel to the Z-axis direction) are α_1 and α_2 for the respective viewing directions projected on the X–Z plane. Correspondingly, β_1 and β_2 define the angles within the Y–Z plane. Because the viewing distance typically is much greater than the displacement vector, the angle differences along the displacement vector, $\delta\alpha$ and $\delta\beta$, are ignored. Following the work of Brucker [11], the three components of the displacement vector can be reconstructed using the following equations:

$$dx = \frac{dx_2 \tan \alpha_1 - dx_1 \tan \alpha_2}{\tan \alpha_1 - \tan \alpha_2} \tag{4.1}$$

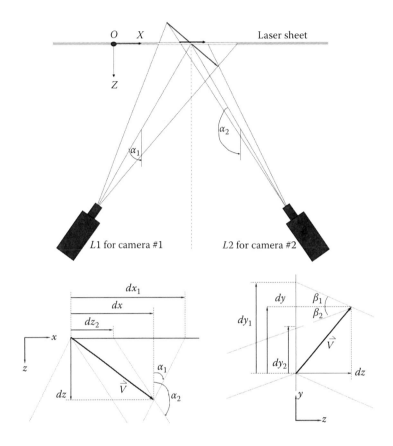

FIGURE 4.2
The schematic for the reconstruction of three components of the displacement vector for stereo PIV measurements.

$$dy = \frac{dy_2 \tan \beta_1 - dy_1 \tan \beta_2}{\tan \beta_1 - \tan \beta_2} \tag{4.2}$$

$$dz = \frac{dx_2 - dx_1}{\tan \alpha_1 - \tan \alpha_2} \tag{4.3}$$

or

$$dz = \frac{dy_2 - dy_1}{\tan \beta_1 - \tan \beta_2} \tag{4.4}$$

These equations can be applied to any stereo imaging geometry. However, the numerators may approach zero as the viewing axes become collinear in either of their two-dimensional projections. For example, the two cameras arranged in the same vertical position as the field of view make the angles β_1, β_2 and their tangents (i.e., $\tan\beta_1$ and $\tan\beta_2$) very small. Clearly, the dz can only be estimated with higher accuracy by using Equation (4.4), while dy has to be rewritten using Equation (4.3), which does not include $\tan\beta_1$ and $\tan\beta_2$ in the nominator [7]:

$$
\begin{aligned}
dy &= \frac{dy_1 + dy_2}{2} + \frac{dz}{2}(\tan\beta_2 - \tan\beta_1) \\
&= \frac{dy_1 + dy_2}{2} + \frac{dx_2 - dx_1}{2}\left(\frac{\tan\beta_2 - \tan\beta_1}{\tan\alpha_1 - \tan\alpha_2}\right)
\end{aligned}
\tag{4.5}
$$

4.2.1 Stereoscopic Image Recording

As in conventional 2D PIV measurements, a laser sheet is usually used to illuminate the flow field in the region of interest for stereo PIV measurements. The tracer particles seeded in the objective fluid flow will scatter the laser light when they pass through the illuminated plane. For stereo PIV measurements, the images of the tracer particles are recorded stereoscopically by using two image-recording cameras. As shown in Figure 4.3, three basic approaches are usually used for the stereo image recording in stereo PIV measurements: the lens translation method, general angle displacement method, and angle displacement arrangement with Scheimpflug condition.

4.2.1.1 Lens Translation Method

In the lens translation method, the two cameras used for stereoscopic image recording are placed side by side with the image planes and camera lens principal plane parallel to the measurement plane (laser sheet), as shown schematically in Figure 4.3(a). Such implementation is the most straightforward approach used for stereoscopic image recording. Because of the parallel geometry, the ratio of the image distance to object distance (i.e., the magnification factor) is constant across the entire acquired PIV images. Therefore, the 2D displacement vectors in the image planes of the two cameras are readily combined to reconstruct the three components of displacement vectors in the measurement plane without the additional manipulations necessary in a variable magnification.

However, it should be noted that when the lens translation method is applied to a liquid flow, the change of the refractive index at the liquid–air interface will cause a variable magnification over the image field [2]. For such cases, the merit of the paralleling arrangement described before will not be valid anymore. Furthermore, the lens translation method has the drawback of the

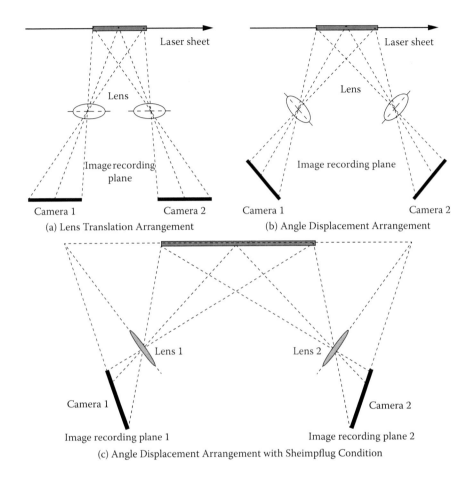

FIGURE 4.3
The camera configurations for stereoscopic PIV image recording.

limited common area in the flow field viewed by both cameras. A large angle
between the cameras is usually desirable in order to have good measurement
accuracy for the measurements of the out-of-plane component of the displacement vectors [12,13]. As a result, the camera image sensors are usually moved
close to the outer edge of the image circle. Since the outer edge of the image
area is used for stereo image recording, the image quality may decrease and
the lens aberrations may become serious in the recorded images [14].

4.2.1.2 General Angle Displacement Method

For the general angle displacement approach, as shown in Figure 4.3(b), the
two cameras view the same region of interest with an angle displacement.
The image planes of the two cameras and the principal planes of the camera
lenses are rotated with respect to the laser illuminating plane. Compared with

the lens translation arrangement described before, the general angle displacement approach can have a much larger overlapped view region, which means a much larger measurement window for the stereo PIV measurements. It can also provide a better image quality since the image sensors are placed near the lens optical axis. However, the main disadvantage of such an arrangement is that the planes with the best focus are parallel to the image planes of the cameras—not in the measurement plane illuminated by the laser sheet. As a result, in order to have the entire measurement window in focus, smaller lens apertures (i.e., larger f-number) are usually needed to increase the depth of focus in the particle image recording, which will result in lower PIV image intensity. Ensuring enough signal-noise-ratio (SNR) in the acquired PIV images will require more powerful lasers for illumination and increase the diffraction limited image of the tracer particles when using small lens apertures (i.e., large f-number). This will impose additional requirements on the illumination system for stereo PIV measurements or cause larger measurement uncertainties.

4.2.1.3 Angle Displacement Arrangement with the Scheimpflug Condition

As described in Prasad and Jensen [15], if the tilted image sensor plane and the lens principal plane can intersect with the measurement plane at a common line to satisfy the Scheimpflug condition [16], the measurement plane in the physical space can be focused onto the image plane of the camera perfectly. As shown in Figure 4.3(c), an angle displacement arrangement with the Scheimpflug condition utilizes the Scheimpflug principle for stereoscopic image recording. This can overcome the disadvantage of conventional angle displacement configuration and provide the best focus for the entire measurement window in the stereo image recording.

The angle displacement arrangement with the Scheimpflug condition approach allows one to keep the best focus plane in the laser illuminating plane while having the cameras view the measurement plane from an off-axial angle; this allows the lens apertures of the cameras to be operated with the same aperture settings as a conventional 2D PIV system. It also enables the image sensors to be near the optical axis of the camera lenses so that the quality of the acquired PIV images can be much better compared to the lens translation approach. Due to its great advantage in stereoscopic image recording, the angle displacement arrangement with the Scheimpflug condition approach is the most commonly used approach in stereo PIV measurements.

4.3 Determination of Mapping Functions between Physical and Image Planes

As described earlier, while angle displacement approaches have the advantage of providing a larger measurement window for stereo PIV measurements,

FIGURE 4.4
The perspective effect of the angle displacement arrangement.

perspective distortions will be introduced in the recorded images since the image planes of the cameras are tilted with respect to the illuminating laser plane (i.e., the measurement plane) for stereoscopic image recording. As shown in Figure 4.4, a rectangular grid in the measurement plane will become trapezoid in the acquired images of the two cameras. A calibration procedure is needed to determine the mapping function between the measurement plane in the measurement plane of the physical space and the image planes of the cameras in order to correct the perspective distortions for quantitative stereo PIV measurements. So far, two approaches have been suggested to determine the mapping functions between the image planes of the cameras and the measurement plane in the physical space. They are usually called the mechanical registration approach and the mathematical registration approach [17].

For the mechanical registration approach [2,15], ray tracing is performed to establish the relationship between the image planes of the cameras and the measurement plane in the physical space. The point sources are located on a square grid in the measurement plane of the physical space and the positions of their images on the two image planes are determined by using a ray tracing technology based on a linearly optical assumption. It should be noted that the mechanical registration method is applicable only for the flow measurements with relatively simple configurations. For example, when the mechanical registration method is used, the air–liquid interface must be parallel to the measurement plane; liquid prisms are usually needed at the interface to compensate for the aberrations due to the different refractive indexes of the air and the liquid flow [15]. It would be very difficult, if not impossible, to use the mechanical registration method when complex nonlinear distortions (such as curved air–liquid interfaces) exist between the measurement plane and the image planes of the cameras.

Compared with the mechanical registration approach described before, the mathematical registration method is more general and can efficiently establish the relationship between the image planes of the cameras and the measurement plane in the physical space. This is usually conducted by performing a general in situ calibration procedure based on the acquisition of one or several images of a calibration target plate—say, a Cartesian grid of small dots—at the center of the illuminating laser sheet (i.e., measurement plane [7]) or across the depth of the laser sheet [12,13,18]. The images of the acquired calibration target plate are used to determine the magnification matrices of the image recording cameras, and the mapping functions between the image planes of the two cameras and the measurement plane are determined mathematically [17]. This can account for various distortions between the measurement plane and the image planes of the cameras.

So far, a number of mathematical models have been suggested to represent the relationship between the measurement plane in physical space and the image planes of the two cameras used for stereoscopic image recording. The three most commonly used models will be compared in the following context.

4.3.1 Parallel Projection Model

Based on the parallel projection assumption of perfect lenses, Lawson and Wu [12,13] suggested using a linear function to express the relationship between the measurement plane in the physical space and the image planes of the two cameras used for stereoscopic image recording, which is expressed as

$$X^{(c)} = M^{(c)}(z\sin\alpha^{(c)} + x\cos\alpha^{(c)}) \tag{4.6}$$

$$Y^{(c)} = M^{(c)}y \tag{4.7}$$

where $c = 1$ and 2 for the left and right cameras, respectively. While $X^{(c)}$ and $Y^{(c)}$ are in the image planes of the cameras, x, y, z are in the physical space. $M^{(c)}$ is the magification factor, which is given by

$$M^{(c)} = \frac{d^{(c)}}{d_0 - z\cos\alpha^{(c)} - x\sin\alpha^{(c)}} \tag{4.8}$$

Differentiating Equations (4.7) and (4.8) in terms of x, y, and z leads to particle image displacement ($\Delta X^{(c)}$ and $\Delta Y^{(c)}$) in the image planes equal to

$$\Delta X^{(c)} = a^{(c)}\Delta x + b^{(c)}\Delta z \tag{4.9}$$

$$\Delta Y^{(c)} = c^{(c)}\Delta x + d^{(c)}\Delta y + e^{(c)}\Delta z \tag{4.10}$$

The displacement of the tracer particle in the physical space ($\Delta x, \Delta y, \Delta z$) can be solved from Equation (4.9) and Equation (4.10), which gives

$$\Delta x = \frac{b^{(2)}\Delta X^{(1)} - b^{(1)}\Delta X^{(2)}}{a^{(1)}b^{(2)} - a^{(2)}b^{(1)}} \tag{4.11}$$

$$\Delta y = \frac{c^{(1)}\Delta Y^{(1)} + c^{(2)}\Delta Y^{(2)}}{2} \tag{4.12}$$

$$\Delta z = \frac{a^{(2)}\Delta X^{(1)} - a^{(1)}\Delta X^{(2)}}{a^{(2)}b^{(1)} - a^{(1)}b^{(2)}} \tag{4.13}$$

where the parameters $a^{(c)}$, $b^{(c)}$, and $c^{(c)}$ in the equations are determined from the geometrical parameters of the system setup or calculated from the calibration images obtained by an in situ calibration procedure. When the displacements in the two image planes of the cameras ($\Delta X^{(1)}$, $\Delta Y^{(1)}$, $\Delta X^{(2)}$, and $\Delta Y^{(2)}$) are known, the displacement of the tracer particles in the measurement plane (Δx, Δy, Δz) can be constructed by using Equation (4.11) to Equation (4.13).

It should be noted that the linear model described here is based on the parallel projection assumption, which is valid only for perfect optics and cannot account for nonlinear distortions such as the effects of imperfect lenses or other optical elements used for stereoscopic image recording. This may cause significant errors in stereo PIV measurements for cases where the nonlinear distortions cannot be ignored.

4.3.2 Second-Order Mapping Approach

Willert [7] suggested a more robust approach to determine the mapping function between the measurement plane in the physical space and the image planes of the two cameras that can account for the nonlinear optical distortions in the stereoscopic image recording. In this approach, second-order functions are used to express the relationships between the measurement plane and the image planes of the cameras, which are written as

$$x = \frac{a_{11}X_{(c)} + a_{12}Y_{(c)} + a_{13} + a_{14}X_{(c)}^2 + a_{15}Y_{(c)}^2 + a_{16}X_{(c)}Y_{(c)}}{a_{31}X_{(c)} + a_{32}Y_{(c)} + a_{33} + a_{34}X_{(c)}^2 + a_{35}Y_{(c)}^2 + a_{36}X_{(c)}Y_{(c)}} \tag{4.14}$$

$$y = \frac{a_{21}X_{(c)} + a_{22}Y_{(c)} + a_{23} + a_{24}X_{(c)}^2 + a_{25}Y_{(c)}^2 + a_{26}X_{(c)}Y_{(c)}}{a_{31}X_{(c)} + a_{32}Y_{(c)} + a_{33} + a_{34}X_{(c)}^2 + a_{35}Y_{(c)}^2 + a_{36}X_{(c)}Y_{(c)}} \tag{4.15}$$

where $c = 1$ and 2 for the left and right cameras, respectively, and the x–y plane is in the measurement plane. $X_{(c)}$ and $Y_{(c)}$ are in the image planes of the cameras. While $a_{33} = 1$ for these equations, the other 17 coefficients are determined by an in situ calibration procedure. As described in Willert [7], by using this approach, only one calibration image is needed with the calibration target plate located at the central plane of the illuminating laser sheet. However, it should be noted that Equations (4.14) and (4.15) can only be used for the coordinate mapping between the measurement plane and the image planes of the cameras. Since variable z, which is in the direction normal to the measurement plane in the physical space, does not appear in the equations, the relationships between the displacement vectors of the tracer particles in the physical space (Δx, Δy, Δz) and the 2D displacements in the image planes (ΔX_L, ΔY_L) and (ΔX_R, ΔY_R) cannot be obtained directly by using the equations. As a result, it is necessary to rely on Equation (4.1) to Equation (4.5), which are based on the pinhole lens assumption, to reconstruct the three components of the displacement vectors in the physical space; this limits the applications of this approach for stereo PIV measurements.

4.3.3 General Multidimensional Polynomial Function Mapping Approach

A more general multidimensional polynomial function mapping approach, suggested by Soloff et al. [18], can compensate various optical distortions such as inaccurate optical alignment, lens nonlinearity, refraction by optical windows, or fluid interfaces involved in the stereo PIV measurements.

It is assumed that the general relationship between the physical space in the objective fluid flow (x, y, z) and the image planes of the cameras (X^1, Y^1) and (X^2, Y^2) can be described by a general mapping function:

$$X^{(c)} = F^{(c)}(x_i) \tag{4.16}$$

where $c = 1,2$ for the left and right cameras, respectively, for stereoscopic image recording, and $i = 1,2,3$ for the x, y, and z directions in the physical space of the objective fluid flow.

Between the time interval of $t = t_0$ and $t = t_0 + \Delta t$ for stereo PIV image acquisition, tracer particles at the original positions of x_i move to new positions of $x_i + \Delta x_i$. The displacements of their images in the image planes can be expressed as

$$\Delta X^{(c)} = F^{(c)}(x_i + \Delta x_i) - F^{(c)}(x_i) \tag{4.17}$$

Performing the Taylor series expansion in this equation and volume averaging over the interrogation cells, the first-order relationships between the displacements in the image planes of the cameras $\Delta X_i^{(c)}$ and the displacement in the physical space (Δx_j) should be

$$\overline{\Delta X_i^{(c)}} \cong F_{i,j}^{(c)}(x)\overline{\Delta x_j} \qquad i = 1,2 \qquad j = 1,2,3 \tag{4.18}$$

Equation (4.18) provides a set of equations for each camera used for stereo PIV measurements. Writing out both sides of the equations in full and augmenting their yields, it will be

$$\begin{pmatrix} \overline{\Delta X_1^{(1)}} \\ \overline{\Delta X_2^{(1)}} \\ \overline{\Delta X_1^{(2)}} \\ \overline{\Delta X_2^{(2)}} \end{pmatrix} = \begin{pmatrix} F_{1,1}^{(1)} & F_{1,2}^{(1)} & F_{1,3}^{(1)} \\ F_{2,1}^{(1)} & F_{2,2}^{(1)} & F_{2,3}^{(1)} \\ F_{1,1}^{(2)} & F_{1,2}^{(2)} & F_{1,3}^{(2)} \\ F_{2,1}^{(2)} & F_{2,2}^{(2)} & F_{2,3}^{(2)} \end{pmatrix} \begin{pmatrix} \overline{\Delta x_1} \\ \overline{\Delta x_2} \\ \overline{\Delta x_3} \end{pmatrix} \qquad (4.19)$$

$$F_{i,j}^{(c)} = \frac{\partial F_i^{(c)}}{\partial x_j} \qquad c = 1,2 \qquad i = 1,2 \qquad j = 1,2,3 \qquad (4.20)$$

or

$$\overline{\Delta X_i^{(c)}} \cong (\nabla F)\overline{\Delta x} \qquad (4.21)$$

In Equation (4.19), the four terms $\overline{\Delta X_1^{(1)}}$, $\overline{\Delta X_2^{(1)}}$, $\overline{\Delta X_1^{(2)}}$, $\overline{\Delta X_2^{(2)}}$ on the left side are the displacements in the image planes for the left and right cameras. They can be determined by using an image processing procedure. The 12 terms in the transformation matrix (∇F) are the derivatives of the general mapping function.

The three-dimensional displacement $\overline{\Delta x_1}$, $\overline{\Delta x_2}$, and $\overline{\Delta x_3}$ in the physical space can be obtained by solving the set of four equations given previously. It should be noted that Equation (4.19) is overdetermined (i.e., there are only three unknown variables with four equations). A least squares method can be used to solve the equations to determine the displacement vectors of the tracer particles in physical space (Δx_1, Δx_2, and Δx_3).

In order to determine the general mapping function between the physical space and the image planes of the cameras used for stereoscopic image recording, a target plate with arrays of dots at a prescribed gap interval, as shown in Figure 4.5, is usually used for the in situ calibration. The front surface of the target plate is aligned with the center plane of the laser sheet, and the images of the calibration target plate are acquired with the target plate translated at several locations across the depth of the laser sheets, as shown schematically in Figure 4.6.

Since any function can be expressed as a multidimensional polynomial function mathematically, the mapping functions between the 3D physical space in the objective fluid flow and the 2D image planes of the two cameras used for stereo image recording can be expressed as multidimensional polynomial functions. For example, the polynomial functions can be chosen to be fourth order in the plane parallel to the illuminating laser sheet plane and second order along the direction normal to the laser sheet plane. This can be expressed as

FIGURE 4.5
The calibration plate used in the present study.

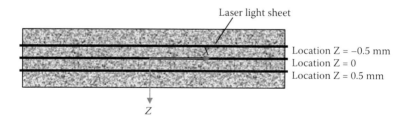

FIGURE 4.6
The locations of the calibration plate for the in situ calibration.

$$F(x, y, z) = a_0 + a_1 x + a_2 y + a_3 z + a_4 x^2 + a_5 xy + a_6 y^2$$
$$+ a_7 xz + a_8 yz + a_9 z^2 + a_{10} x^3 + a_{11} x^2 y + a_{12} xy^2$$
$$+ a_{13} y^3 + a_{14} x^2 z + a_{15} xyz + a_{16} y^2 z + a_{17} xz^2$$
$$+ a_{18} yz^2 + a_{19} x^4 + a_{20} x^3 y + a_{21} x^2 z^2 + a_{22} x^1 y^3$$
$$+ a_{23} y^4 + a_{24} x^3 z + a_{25} x^2 yz + a_{26} xy^2 z + a_{27} y^3 z$$
$$+ a_{28} x^2 z^2 + a_{29} xyz^2 + a_{30} y^2 z^2$$

(4.22)

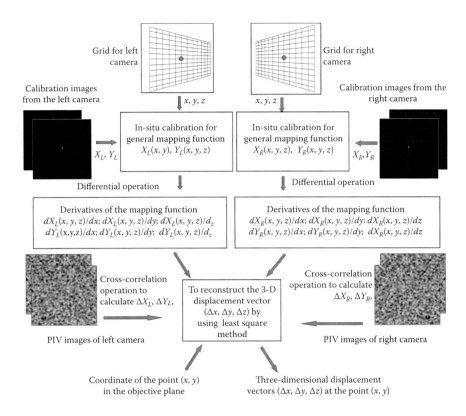

FIGURE 4.7
Illustration of the steps to reconstruct the three components of the displacement vectors in the measurement plane (Dx, Dy, Dz) for stereo PIV measurements.

In this equation, the x- and y-directions are in the plane parallel to the laser sheet plane; the z-direction is normal to the laser sheet plane. The 31 coefficients a_0 to a_{31} in the equation can be determined from the acquired calibration images.

Figure 4.7 shows the flow chart typically used to reconstruct all three components of the displacement vectors of the tracer particles (Δx, Δy, Δz) in the physical space by using the general multidimensional polynomial mapping function described before. As shown in Figure 4.7, in order to obtain the general mapping functions between the 3D physical space and the image planes of the two cameras, the images of the calibration target plate with a grid of dots (as shown in Figure 4.5) are captured at several locations across the thickness of the illuminating laser sheet (as shown in Figure 4.6). The coordinates of the marker grids (i.e., arrays of dots) in the physical space are known, and the coordinate values of the marker grids in the image planes of the left and right cameras can be determined quantitatively by using an image processing procedure.

It should be noted that, as shown in Equation (4.22), there are only 31 unknown coefficients for each image recording camera to determine the relationship between the 3D physical space in the objective fluid flow and the image plane of the camera. The acquired images of the marker grids on the calibration plate at several different locations across the laser sheet will make the number of equations that can be used to determine the 31 coefficients in the several thousands. By using a least squares method [19], the 31 coefficients for each image recording camera can be determined with optimum values.

Once the general mapping functions are determined through the in situ calibration procedure, differentiating the general mapping functions in terms of x, y, and z leads to 12 coefficients of the transformation matrix (∇F) given in Equation (4.19). The derivatives of the mapping functions represent the gradients of the particle image displacements in the image planes of the two cameras due to the displacements of the tracer particles in the physical space.

A sample of the 12 gradients for the two image recording cameras obtained by an in situ calibration procedure is given in Figure 4.8 and Figure 4.9. These gradients are the amount of the particle image displacement, in pixels, in the X- or Y-direction in the image planes caused by a 1.0 mm displacement of the tracer particles along x-, y-, or z-direction in the physical space. For example, at the point of $(0,0,0)$ in the measurement plane of the physical space, the 12 gradient values are

$$L\ dX/dx = 11.1035 \quad L\ dY/dx = 0.0485R\ dX/dx = 11.1029 \quad R\ dY/dx = -0.0104 \quad (4.23)$$

$$L\ dX/dy = 0.0029 \quad L\ dY/dy = 12.1847R\ dX/dy = -0.0278 \quad R\ dY/dy = 12.1586 \quad (4.24)$$

$$L\ dX/dz = 5.1942 \quad L\ dY/dz = -0.0112R\ dX/dz = -4.8357 \quad R\ dY/dz = 0.0311 \quad (4.25)$$

For the values given in Equation (4.23), this indicates that a 1.0 mm displacement at the point of $(0,0,0)$ along the x-direction in the measurement plane of the physical space will cause an image displacement of 11.1035 pixels along the X-direction and 0.0485 pixels along the Y-direction in the image plane of the left camera. An image displacement of 11.1029 pixels along the X-direction and 0.0104 pixels along the Y-direction will be detected in the image plane of the right camera.

Equation (4.24) indicates that a 1.0 mm displacement at the point of $(0,0,0)$ along the y-direction in the measurement plane of the physical space will result in image displacements of 0.0029 pixels along the X-direction and 12.1847 pixels along the Y-direction in the image plane of the left camera. An image displacement of -0.0278 pixels along the X-direction and 12.1586 pixels along the Y-direction will be detected in the image plane of the right camera.

From Equation (4.25), it can be seen that a 1.0 mm displacement at the point of $(0,0,0)$ along the z-direction in the measurement plane of the physical space will lead to an image displacement of 5.1942 pixels along the X-direction and

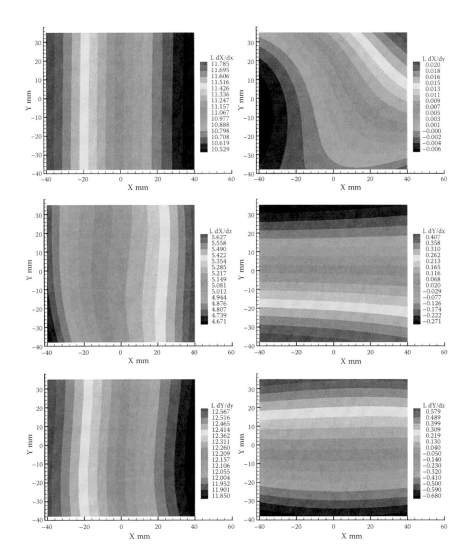

FIGURE 4.8
(See color insert.) The gradients of the left-hand image recording camera for stereo image recording.

−0.0112 pixels along the Y-direction in the image plane of the left camera. An image displacement of −4.8357 pixels along the X-direction and 0.0311 pixels along the Y-direction will be detected in the image plane of the right camera.

Once the general mapping functions are determined through the in situ calibration procedure, the stereo PIV system is ready to be used to conduct quantitative measurements in fluid flows. After the images of the tracer particles in the objective fluid flows are recorded stereoscopically by using the two cameras, the next step for stereo PIV measurements is to reconstruct the

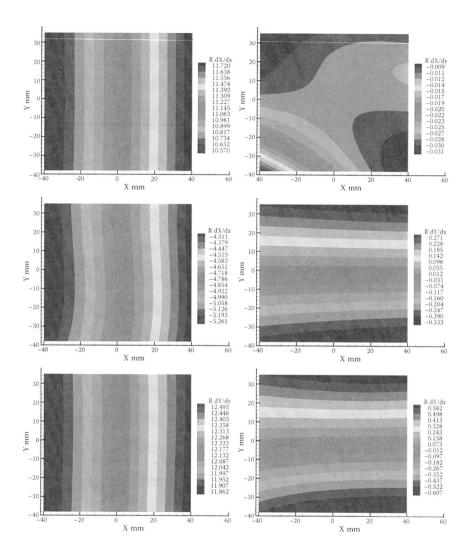

FIGURE 4.9
(See color insert.) The gradients of the right-hand image recording camera for stereo image recording.

three components of the displacement vectors of the tracer particles (Δx, Δy, Δz) based on the 2D particle image displacements (ΔX_L and ΔY_L) and (ΔX_R and ΔY_R) detected by the left and right cameras. The 2D particle image displacements in each image plane can be calculated separately by using a routine PIV image processing procedure, which is the same as that used for conventional 2D PIV measurements.

Figure 4.10 shows a pair of typical PIV raw images along with the 2D particle image displacements (ΔX_L and ΔY_L) and (ΔX_R and ΔY_R) detected by the left and right cameras; these were obtained in an experimental study to characterize a turbulent lobed jet mixing flow [20]. Based on the 2D particle image displacements (ΔX_L and ΔY_L) and (ΔX_R and ΔY_R), along with the distributions of the 12 gradients as shown in Figure 4.8 and Figure 4.9, the three components of the displacement vectors of the tracer particles in the measurement plane (Δx, Δy, Δz) can be reconstructed by solving Equation (4.19) with a least squares method. The reconstructed 3D flow velocity distribution in the measurement plane is given in Figure 4.10(d).

4.4 Using the Stereo PIV Technique to Study a Lobed Jet Mixing Flow

A lobed nozzle/mixer (Figure 4.11), which consists of a splitter plate with convoluted trailing edge, is considered a very promising fluid mechanic device for efficient mixing of two co-flow streams with different velocity, temperature, and/or species [21,22]. Lobed nozzles/mixers have been given a great deal of attention by many researchers in recent years, and they have also been widely used in various engineering applications. For example, for some commercial aeroengines, lobed nozzles have been used to reduce both take-off jet noise and specific fuel consumption (SFC) [23]. In order to reduce the infrared radiation signals of military aircraft, lobed nozzles have also been used to enhance the mixing process of the high-temperature and high-speed gas plume from aeroengines with ambient cold air [24]. More recently, lobed nozzles have also emerged as attractive approaches for enhancing mixing between fuel and air in combustion chambers to improve the efficiency of combustion and reduce the formation of pollutants [25].

In addition to the continuous efforts to optimize the geometry of lobed mixers/nozzles for better mixing performance and to widen the applications of lobed mixers/nozzles, extensive studies about the mechanism of why the lobed mixers/nozzles can substantially enhance fluid mixing have also been conducted in recent years. Based on pressure, temperature, and velocity measurements of the flow field downstream of a lobed nozzle, Paterson [26] revealed the existence of large-scale stream-wise vortices in lobed mixing flows induced by the special geometry of the lobed nozzles/mixers. These vortices were suggested to be responsible for the enhanced mixing.

Werle et al. [27] and Eckerle et al. [28] found that the stream-wise vortices in lobed mixing flows follow a three-step process by which these vortices form, intensify, and then break down and suggested that the high turbulence resulting from the vortex breakdown would improve the overall mixing process. Elliott et al. [29] suggested that there are three primary contributors to the mixing processes in lobed mixing flows. The first is the span-wise

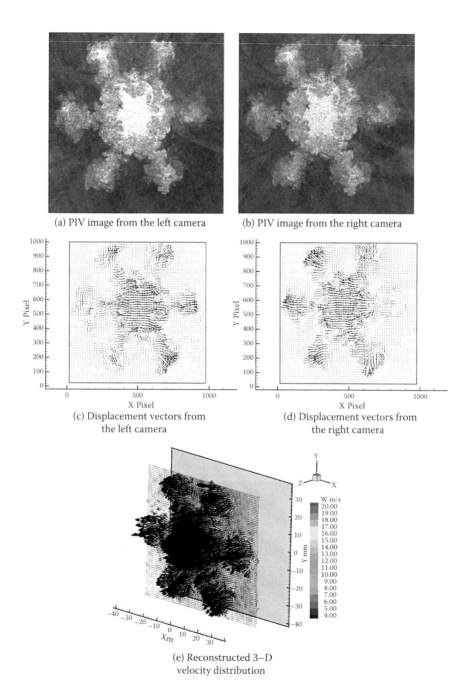

(a) PIV image from the left camera

(b) PIV image from the right camera

(c) Displacement vectors from
the left camera

(d) Displacement vectors from
the right camera

(e) Reconstructed 3–D
velocity distribution

FIGURE 4.10
Stereo PIV measurement results in a lobed jet mixing flow.

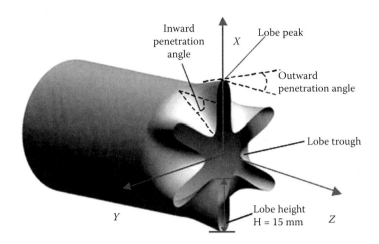

FIGURE 4.11
Schematic of the lobe nozzle/mixer used in the present study.

vortices, which occur in any free shear layers due to the Kelvin–Helmholtz instability. The second is the increased interfacial contact area due to the convoluted trailing edge of the lobed mixer. The last element is the stream-wise vortices produced by the special geometry of the lobed mixer.

Although the existence of unsteady vortices and turbulent structures in lobed mixing flows has been revealed in previous studies by qualitative flow visualization, the quantitative, instantaneous, whole-field velocity and vorticity distributions in lobed mixing flows were not obtained until the work of Hu et al. [30]. These researchers used both planar laser induced fluorescence (PLIF) and conventional 2D PIV techniques to study lobed jet mixing flows. Based on the directly perceived PLIF flow visualization images and quantitative PIV velocity, vorticity, and turbulence intensity distributions, the evolution and interaction of various vortical and turbulent structures in the lobed jet flows were discussed. It should be noted that the conventional 2D PIV system used in Hu et al. [30] is only capable of obtaining two components of velocity vectors in the planes of illuminating laser sheets. The out-of-plane velocity component is lost while the in-plane components may be affected by an unrecoverable error due to perspective transformation [2].

Meanwhile, for the highly 3D turbulent flows like lobed jet mixing flows, conventional 2D PIV measurement results may not be able to reveal the 3D features of the complex lobed jet mixing flows successfully. A high-resolution stereo PIV system, which can provide all three components of velocity vectors in a measurement plane simultaneously, is used in the present study to quantify the turbulent jet mixing flow exhausted from a lobed nozzle/mixer.

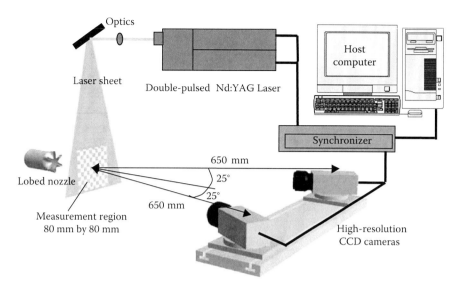

FIGURE 4.12
Experimental setup used for the stereo PIV measurements.

4.4.1 Experimental Setup for Stereo PIV Measurements

Figure 4.12 shows the schematic of the experimental setup used in the present study to achieve stereo PIV measurements. An air jet flow is exhausted from a circular nozzle/mixer at the speed of $U_0 = 20.0$ m/s. The jet flow is seeded with ~1 μm oil droplets by using a droplet generator. Illumination is provided by a double-pulsed Nd:YAG laser (NewWave Gemini 200) adjusted on the second harmonic and emitting two pulses of 200 mJ at the wavelength of 532 nm with a repetition rate of 10 Hz. The laser beam is shaped to a sheet by a set of mirrors with spherical and cylindrical lenses. The thickness of the laser sheet in the measurement region is about 2.0 mm for the present study. Two high-resolution digital cameras used to perform stereoscopic PIV image recording are arranged in an angular displacement configuration. With the installation of tilt-axis mounts, the lenses and camera bodies are adjusted to satisfy the Scheimpflug condition. The cameras and the double-pulsed Nd:YAG lasers are connected to a workstation (host computer) via a synchronizer, which controls the timing of the laser sheet illumination and the charged coupled device camera data acquisition.

A general in situ calibration procedure (described before) is used in the present study to determine the mapping functions between the image planes of the two cameras and the measurement plane in the fluid flow. The mapping function used is taken to be a multidimensional polynomial, which is fourth order for the X- and Y-directions parallel to the laser sheet plane and second order for the Z-direction normal to the laser sheet plane, as expressed in Equation (4.22). The 2D displacements in each image plane of the two cameras are obtained by a frame-to-frame cross-correlation technique involving successive frames of patterns of particle images in an interrogation window of 32 pixels × 32 pixels. An effective overlap of 50% of the interrogation windows is employed in the PIV image processing. Following the flow chart shown in Figure 4.7, by using the mapping functions obtained by the in situ calibration procedure and the 2D displacements in the two image planes detected by the two cameras, the 3D flow velocity vectors in the laser illuminating plane are reconstructed (shown in Figure 4.10).

4.4.2 Measurement Results and Discussion

Figure 4.13 and Figure 4.14 give the stereo PIV measurement results in two typical cross planes of the lobed jet mixing flow, which include typical instantaneous velocity fields, simultaneous stream-wise vorticity distributions, ensemble-averaged velocity, and stream-wise vorticity fields. The ensemble-averaged velocity and stream-wise vorticity fields given in the

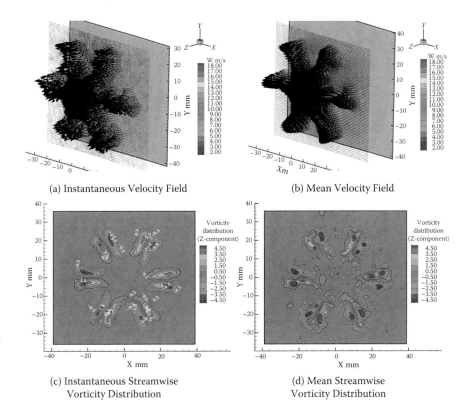

(a) Instantaneous Velocity Field

(b) Mean Velocity Field

(c) Instantaneous Streamwise
Vorticity Distribution

(d) Mean Streamwise
Vorticity Distribution

FIGURE 4.13
(See color insert.) Stereo PIV measurement results in the Z/D = 0.25 (Z/H = 0.67) cross plane.

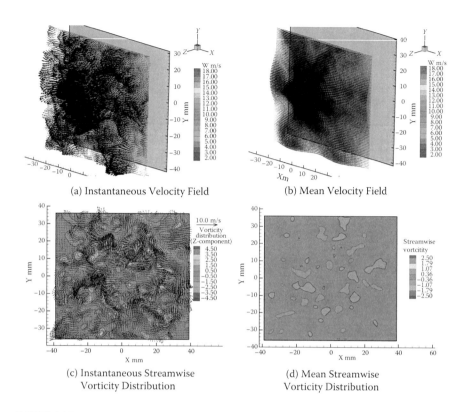

(a) Instantaneous Velocity Field

(b) Mean Velocity Field

(c) Instantaneous Streamwise
Vorticity Distribution

(d) Mean Streamwise
Vorticity Distribution

FIGURE 4.14
(See color insert.) Stereo PIV measurement results in the Z/D = 3.0 (Z/H = 8.0) cross plane.

figures are calculated based on 500 frames of instantaneous stereo PIV measurement results.

As shown clearly in Figure 4.13, the high-speed core jet flow is found to have the same geometry as the lobed nozzle in the Z/D = 0.25 (Z/H = 0.67) cross plane (almost at the exit of the lobed nozzle). The "signature" of the lobed nozzle in the form of a six-lobe structure can be seen clearly from both the instantaneous and ensemble-averaged velocity fields. The existence of very strong secondary streams in the lobed jet flow is revealed clearly in the velocity vector plots. The core jet flow ejects radially outward in the lobe peaks and ambient flows inject inward in the lobe troughs. Both the ejection of the core jet flow and the injection of the ambient flows are generally following the outward and inward contours of the lobed nozzle, which results in the generation of six pairs of counter-rotating stream-wise vortices in the lobed jet flow. The maximum radial ejection velocity of the core jet flow in the ensemble-averaged velocity field is found to be about 5.0 m/s, which is almost equal to the value of $U_0 \bullet \sin(\theta_{out})$. A big high-speed region can also be seen clearly from the ensemble-averaged velocity field; this represents the high-speed core jet flow in the center of the lobed nozzle.

The existence of the six pairs of large-scale stream-wise vortices due to the special geometry of the lobed nozzle can be seen clearly and quantitatively from the stream-wise vorticity distributions shown in Figure 4.13(c) and Figure 4.13(d). The size of these large-scale stream-wise vortices is found to be on the order of the lobe height. Compared with those in the instantaneous stream-wise vorticity field (Figure 4.13c), the contours of the large-scale stream-wise vortices in the ensemble-averaged stream-wise vorticity field (Figure 4.13d) are found to be much smoother. However, they have almost the same distribution pattern and magnitude as their instantaneous counterparts. The similarity between the instantaneous and ensemble-averaged stream-wise vortices suggests that the generation of the large-scale stream-wise vortices at the exit of the lobed nozzle is quite steady.

As revealed from the stereo PIV measurement results given in Figure 4.14, the lobed jet mixing flow is found to become so turbulent that the "signature" of the lobed nozzle can no longer be identified easily from the instantaneous velocity fields as the downstream distance increases to $Z/D = 3.0$ ($Z/H = 8.0$). The flow field is fully filled with many small-scale vortices and turbulent structures. The ensemble-averaged velocity field in this cross plane shows that the distinct high-speed region in the center of the lobed jet flow has dissipated so seriously that iso-velocity contours of the high-speed core jet flow have become small concentric circles. The ensemble-averaged secondary streams in this cross plane become so weak (the maximum secondary stream velocity is less than 0.8 m/s) that they cannot be identified easily from the ensemble-averaged velocity vector plot.

From the instantaneous stream-wise vorticity distribution in the $Z/D = 3.0$ ($Z/H = 8.0$) cross plane given in Figure 4.14(c), it can be seen that there are many small-scale stream-wise vortices in the lobed jet mixing flow that almost fully fill the measurement window. However, the maximum vorticity value of these instantaneous small-scale stream-wise vortices is found to be at the same level of those in the upstream cross planes. Due to serious dissipation caused by the intensive mixing between the core jet flow and ambient flow, almost no apparent stream-wise vortices can be identified from the ensemble-averaged stream-wise vorticity distribution (Figure 4.14d) in the $Z/D = 3.0$ ($Z/H = 8.0$) cross plane anymore.

Based on the stereo PIV measurement results at 12 cross planes, the 3D flow field in the near downstream region of the lobed jet mixing flow is reconstructed. Figure 4.15 shows the ensemble-averaged velocity vector distributions in the lobed jet mixing flow viewed from upstream and downstream positions. The corresponding velocity iso-surfaces of the reconstructed 3D flow field are also given in Figure 4.15. The velocity magnitudes of the iso-surfaces are 4.0, 8.0, 12.0, and 16.0 m/s, respectively. It can be seen clearly that the high-speed core jet flow has the same geometry as the lobed nozzle (i.e., six-lobe structure) at the exit of the lobed nozzle/mixer. Due to the "stirring effect" of the large-scale stream-wise vortices generated by the lobed nozzle,

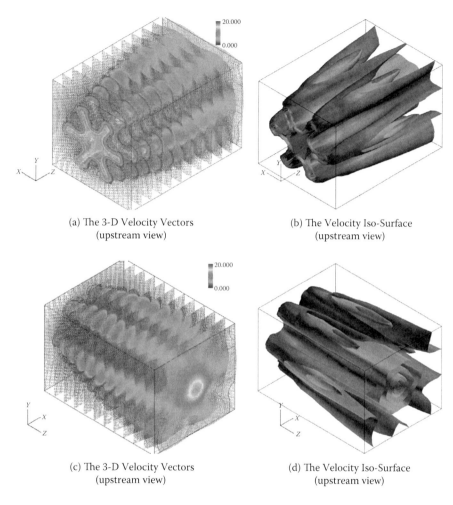

(a) The 3-D Velocity Vectors
(upstream view)

(b) The Velocity Iso-Surface
(upstream view)

(c) The 3-D Velocity Vectors
(upstream view)

(d) The Velocity Iso-Surface
(upstream view)

FIGURE 4.15
(See color insert.) Reconstructed three-dimensional flow fields of the lobed jet mixing flow.

the six-lobe structure of the core jet flow is rounded up rapidly. At $Z/D = 3.0$ ($Z/H = 8.0$) downstream, the iso-surfaces are found to become concentric cylinders, which are very similar to those in a circular jet flow.

The unique flow structures in the lobed jet mixing flow are revealed clearly and quantitatively from the stereo PIV measurement results. By elucidating the underlying physics, our understanding about the important physical process pertinent to the mixing enhancement in lobed mixing flow can be improved significantly. This will enable us to explore and optimize design paradigms for the development of novel mixers/nozzles for various engineering applications.

4.5 Summary

The technical basis of the stereo PIV technique, which is capable of achieving simultaneous measurements of all three components of flow velocity vectors in fluid flows, has been described in the present study. Three basic optical arrangements most commonly used for stereo image recording were introduced along with a brief description about the advantages and disadvantages of each optical arrangement. A general in situ calibration procedure to determine the mapping functions between the 3D physical space in the objective fluid flow and the image planes of the cameras used for stereoscopic image recording was described in detail. The study also gave the flow chart and steps needed to reconstruct the three components of the displacement vectors in the measurement plane of the physical space by using the 2D displacement vectors in the image planes detected by the two cameras and the multidimensional polynomial mapping functions.

The feasibility and implementation of the stereo PIV technique were demonstrated by achieving simultaneous measurements of all three components of flow velocity vectors in an air jet flow exhaust from a lobed nozzle/mixer. The evolution of the large-scale stream-wise vortices and the unique 3D flow structures in the lobed jet mixing flow were revealed clearly and quantitatively from the stereo PIV measurement results. Based on these results, the underlying physics related to the enhanced mixing processes in lobed mixing flows can be elucidated more clearly and quantitatively.

References

1. Adrian, R. J. 1991. Particle-image technique for experimental fluid mechanics. *Annual Review Fluid Mechanics* 261–304.
2. Prasad, A. K. and Adrian, R. J. 1993. Stereoscopic particle image velocimetry applied to fluid flows. *Experiments in Flows* 15:49–60.
3. Barnhart, D. H., Adrian, R. J., and Papen, G. C. 1994. Phase-conjugate holographic system for high-resolution particle image velocimetry. *Applied Optics* 33:7159–7170.
4. Zhang, J., Tao, B., and Katz, J. 1997. Turbulent flow measurement in a square duct with hybrid holographic PIV. *Experiments in Fluids* 23:373–381.
5. Virant, M. and Dracos, T. 1997. 3D PTV and its application on Lagrangian motion. *Measurement Science and Technology* 8:1539.
6. Elsinga, G. E., Scarano, F., Wieneke, B., and van Oudheusden, B. W. 2006. Tomographic particle image velocimetry. *Experiments in Fluids* 41:933–947.
7. Willert, C. 1997. Stereoscopic digital particle image velocimetry for application in wind tunnel flows. *Measurement Science and Technology* 8:1465–1479.

8. Meng, H., Pan, G., Pu, Y., and Woodward, S. H. 2004. Holographic particle image velocimetry: From film to digital recording. *Measurement Science and Technology* 15:673–685.

9. Svizher, A. and Cohen, J. 2006. Holographic particle image velocimetry system for measurement of hairpin vortices in air channel flow. *Experiments in Fluids* 40:708–722.

10. Katz, J. and Sheng, J. 2010. Applications of holography in fluid mechanics and particle dynamics. *Annual Review Fluid Mechanics* 42:531–555.

11. Brucker, C. 1996. 3-D PIV via spatial correlation in a color-coded light sheet. *Experiments in Fluids* 21:312–314.

12. Lawson, N. J. and Wu, J. 1997. Three-dimensional particle image velocimetry: Error analysis of stereoscopic techniques. *Measurement Science Technology* 8:894–900.

13. Lawson, N. J. and Wu, J. 1997. Three-dimensional particle image velocimetry: Experimental error analysis of a digital angular stereoscopic system. *Measurement Science Technology* 8:1455–1464.

14. Bjorkquist, D. C. 1998. Design and calibration of a stereoscopic PIV system. *Proceedings of the Ninth International Symposium on Application of Laser Techniques in Fluid Mechanics,* Lisbon, Portugal, 1998.

15. Prasad, A. K. and Jensen, K. 1995. Scheimpflug stereocamera for particle image velocimetry in liquid flows. *Applied Optics* 34:7092–7099.

16. Larmore, L. 1965. *Introduction to photographic principles.* New York: Dover Publications, Inc.

17. Hill, D. F., Sharp, K.V., and Adrian, R. J. 1999. The implementation of distortion compensated stereoscopic PIV. *Proceedings of 3rd International Workshop on PIV,* Santa Barbara, CA, Sept. 16–18, 1999.

18. Soloff, S. M., Adrian, R. J., and Liu, Z. C. 1997. Distortion compensation for generalized stereoscopic particle imaging velocimetry. *Measurement Science Technology* 8:1441–1454.

19. Watanabe, Z., Natori, M., and Okkuni, Z. 1989. Fortran 77 software for numerical computation. Maruzen Publication, ISBN4-621-03424-3 C3055.

20. Hu, H., Saga, T., Kobayashi, T., and Taniguchi, N. 2002. Mixing process in a lobed jet flow. *AIAA Journal* 40 (7):1339–1345.

21. McCormick, D.C. and Bennett, J. C., Jr. 1994. Vortical and turbulent structure of a lobed mixer free shear layer. *AIAA Journal* 32 (9):1852–1859.

22. Belovich, V. M. and Samimy, M. 1997. Mixing process in a coaxial geometry with a central lobed mixing nozzle. *AIAA Journal* 35 (5):838–841.

23. Presz, W. M., Jr., Reynolds, G., and McCormick, D., 1994. Thrust augmentation using mixer-ejector-diffuser systems. AIAA paper 94-0020, 1994.

24. Hu, H., Saga, T., Kobayashi, T., Taniguchi, N., and Wu, S. 1999. Research on the rectangular lobed exhaust ejector/mixer systems. *Transactions of Japan Society of Aeronautics & Space Science* 41 (134):187–194.

25. Smith, L. L., Majamak, A. J., Lam, I. T., Delabroy, O., Karagozian, A. R., Marble, F. E., and Smith, O. I. 1997. Mixing enhancement in a lobed injector. *Physics of Fluids* 9 (3):667–678.

26. Paterson, R. W. 1982. Turbofan forced mixer nozzle internal flowfield. NASA CR-3492.

27. Werle, M. J., Paterson, R. W., and Presz, W. M., Jr. 1987. Flow structure in a periodic axial vortex array. AIAA paper 87-6l0.

28. Eckerle, W. A., Sheibani, H., and Awad, J. 1993. Experimental measurement of the vortex development downstream of a lobed forced mixer. *Journal of Engineering for Gas Turbine and Power* 14:63–71.

29. Elliott, J. K., Manning, T. A., Qiu, Y. J., Greitzer, C. S., Tan, C. S., and Tillman, T. G. 1992. Computational and experimental studies of flow in multi-lobed forced mixers. AIAA 92-3568.

30. Hu, H., Saga, T., Kobayashi, T., and Taniguchi, N. 2000. Research on the vortical and turbulent structures in the lobed jet flow by using LIF and PIV. *Measurement Science and Technology* 11 (6):698–711.

5

Basic Concepts

Sergio Fernandez and Joaquim Salvi

CONTENTS

Shape reconstruction using coded structured light is considered one of the most reliable techniques to recover object surfaces. With a calibrated projector–camera pair, a light pattern is projected onto the scene and imaged by the camera. Correspondences between projected and recovered patterns are found and used to extract three-dimensional (3D) surface information. Among structured light (SL) techniques, the combination of dense acquisition and real time constitutes an active field of research. To achieve density and real time, most of the work present in the literature is based on the projection of a single one-shot fringe pattern, where depth is extracted analyzing phase deviation of the imaged pattern. However, the algorithms employed to unwrap the phase are computationally slow and can fail in the presence of depth discontinuities and occlusions. This chapter presents an up-to-date review and a new classification of the existing techniques. Moreover, a proposal for a new one-shot dense pattern that combines De Bruijn and fringe pattern projection to obtain an absolute, accurate, and computationally fast 3D reconstruction is presented. Finally, the proposed technique is compared to some of the already existing methods, obtaining both qualitative and quantitative results. The advantages and drawbacks of the proposed technique are discussed.

5.1 Introduction

Three-dimensional measurement constitutes an important topic in computer vision, with different applications such as range sensing, industrial inspection of manufactured parts, reverse engineering (digitization of complex, free-form surfaces), object recognition, 3D map building, biometrics, clothing design, and others. The developed solutions are traditionally categorized into contact and noncontact techniques. Contact techniques, used for a long time in reverse engineering and industrial inspections, present slow performance and high cost due to the necessity of using mechanically calibrated passive arms [68].

On the other hand, noncontact techniques (both active and passive) achieve higher accuracy without the necessity of touching the object, which is highly recommended in many applications. In passive approaches, the scene is first imaged by two or more calibrated cameras and correspondences between the images are found to extract the 3D shape. This implies that the density of the reconstruction is directly related to the texture of the object, obtaining poor results in the presence of textureless surfaces [4, 32]. Methods based on structured light (active techniques) came to cope with this issue, substituting one of the cameras by an active device (a projector) which projects a structured-light pattern onto the scene. This active device is modeled as an inverse camera and can be calibrated correspondingly [55]. The projected pattern imposes the illusion of texture onto an object, increasing the number of correspondences [56], thus being able to obtain dense reconstructions even for textureless surfaces.

This chapter analyzes the different coding strategies used in active structured light, focusing on the improvements presented in the last years. The classification of the different SL approaches presented in the work of Salvi, Batlle, and Mouaddib [57] is considered to this end. Furthermore, a new proposal of one-shot dense reconstruction is presented that combines De Bruijn and fringe pattern projection to obtain an absolute, accurate, and computationally fast 3D reconstruction. This new technique is compared to some SL representative techniques, providing both quantitative and qualitative results. Finally, in the conclusions we analyze the main positives and drawbacks of the technique.

5.2 Classification

Coded structured light (CSL) is based on the projection of one pattern or a sequence of patterns that univocally determines the code word of a projecting pixel (or feature) within a nonperiodic region. Coded structured light has produced many works during the last decades and some recopilatory works can be found in the literature. This is the case of the surveys presented by Batlle, Mouaddib, and Salvi [4] and Salvi et al. [58], which analyzed the different CSL techniques existing in temporal and spatial multiplexing domains from 1998 until 2004, respectively. Regarding frequency multiplexing, Su [62] reviewed the Fourier transform (FT) techniques proposed until 2001. However, there is no previous work comparing the three approaches. Therefore, a classification extracting and analyzing attributes common in all the approaches is missing. This is overcome in the present survey, which also incorporates the most recent contributions done in CSL in the last years.

Table 5.1 shows a new classification of the existing pattern projection techniques. The main distinction has been done regarding the sparse or dense 3D reconstruction achieved. Patterns providing sparse reconstruction present a digital profile with the same value for the region represented by the

TABLE 5.1

Proposed Classification Embracing Every Group of CSL

			Shots	Cameras	Axis	Pixel Depth	Coding Strategy	Subpixel Acc.	Color
SPARSE	Spatial multiplexing	De Bruijn							
		Boyer and Kak 1987	1	1	1	C	A	Y	N
		Salvi et al. 1998	1	1	1	C	A	Y	Y
		Monks et al. 1992	1	1	1	C	A	Y	N
		Pages et al. 2004	1	1	1	C	A	Y	N
		Nonformal							
		Forster 2007	1	1	1	C	A	Y	N
		Fechteler and Eisert 2008	1	1	1	C	A	Y	N
		Tehrani 2008	1	2	1	C	A	N	Y
		Maruyama and Abe 1993	1	1	2	B	A	N	Y
		Kawasaki et al. 2008	1	1	2	C	A	N	Y
		Ko 1995	1	1	2	G	A	N	Y
		Koninckx and Van Gool 2006	1	1	2	C	P	Y	Y
		M-array							
		Griffin et al. 1992	1	1	2	C	A	Y	Y
		Morano et al. 1998	1	1	2	C	A	Y	Y
		Pages et al. 2006	1	1	2	C	A	Y	N
		Albitar et al. 2007	1	1	2	B	A	N	Y

Time multiplexing									
Binary codes									
Posdamer and Altschuler	1982	>2	1	1	B	A	N	Y	
Ishii et al.	2007	>2	1	1	B	A	N	Y	
Sun	2006	>2	2	1	B	A	Y	Y	
N-ary codes									
Shifting codes									
Caspi et al.	1998	>2	1	1	C	A	N	N	
Zhang et al.	2008	>2	1	1	C	A	Y	N	
Sansoni et al.	2000	>2	1	1	G	A	Y	N	
Guhring	2001	>2	1	1	G	A	Y	Y	
Srinivasan et al.	1985	>2	1	1	G	A	Y	Y	
Ono et al.	2004	>2	1	1	G	P	Y	Y	
Wust et al.	1991	1	1	1	C	P	Y	N	
Guan et al.	2004	1	1	1	G	P	Y	Y	
Single phase shifting (SPS)									
Gushov and Solodkin	1991	>2	1	1	G	A	Y	Y	
Pribanic et al.	2009	>2	1	1	G	A	Y	Y	
Multiple phase shifting (MPS)									

DENSE

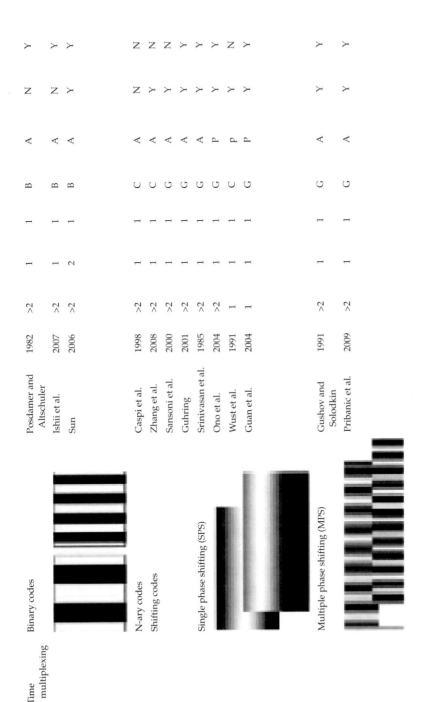

TABLE 5.1 (*Continued*)

Proposed Classification Embracing Every Group of CSL

				Shots	Cameras	Axis	Pixel Depth	Coding Strategy	Subpixel Acc.	Color
DENSE	Frequency multiplexing	Single coding frequency	Takeda and Mutoh	1983	1	1	1	G	P	Y	Y
			Cobelli et al.	2009	1	1	1	G	P	Y	Y
			Li et al.	1990	2	1	1	G	P	Y	Y
			Hu and He	2009	2	2	1	G	P	Y	Y
			Chen et al.	2007	1	1	1	C	P	Y	N
			Yue	2006	1	1	1	G	P	Y	Y
			Chen et al.	2005	2	1	1	G	P	Y	Y
			Berryman et al.	2008	1	1	1	G	P	Y	Y
			Gdeisat et al.	2006	1	1	1	G	P	Y	Y
			Zhang et al.	2008	1	1	1	G	P	Y	Y
			Lin and Su	1995	2	1	1	G	P	Y	Y
			Huang et al.	2005	>2	1	1	G	P	Y	Y
			Jia et al.	2007	2	1	1	G	P	Y	Y
			Wu and Peng	2006	1	1	1	G	P	Y	Y
			Fernandez et al.	2000	1	1	1	C	A	Y	N
	Spatial multiplexing	Grading	Carrhill and Hummel	1985	1	1	1	G	A	Y	N
			Tajima and Iwakawa	1990	1	1	1	C	A	Y	N

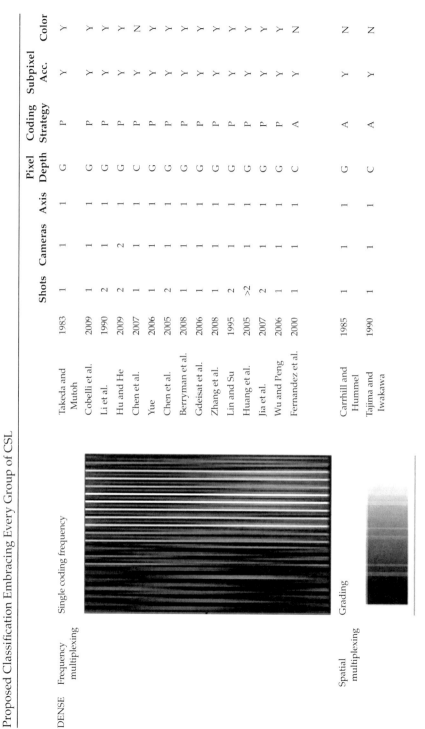

same code word. The size of this region largely determines the density of the reconstructed object. On the other hand, dense reconstruction is achieved by projecting a sequence of digital patterns superposed over time to obtain full pixel coverage or with a smooth profile pattern, where every pixel has a unique code word within the nonperiodicity region. Both approaches achieve dense reconstruction. A posteriori subclassification is done regarding spatial, time, and frequency multiplexing. Columns on the right indicate the value of some intrinsic attributes common to all the patterns. These attributes are

- *Number of projected patterns* determines whether the method is valid or not for measuring moving objects.
- *Number of cameras* uses stereo vision (two or more cameras) coupled to a noncalibrated pattern used only to get texture on the surface pattern, or a unique camera coupled to a calibrated projector.
- *Axis codification* codes the pattern along one or two axes.
- *Pixel depth* refers to the color and luminance level of the projected pattern (B, G, and C stand for binary, grayscale, and color, respectively).
- *Coding strategy* refers to the periodicity of the set of patterns projected on the surface (A stands for absolute and P stands for periodic).
- *Subpixel accuracy* determines whether the features are found considering subpixel precision, thus providing better reconstruction results (yes or no).
- *Color* determines whether the technique can cope with colored objects (yes or no).

5.3 Sparse Reconstruction Methods

In sparse reconstruction methods the pattern presents a digital profile, and spatial or temporal multiplexing is employed to image the scene. Spatial multiplexing techniques code the pattern using the surroundings of a given feature, while temporal multiplexing creates the code word by the successive projection of patterns onto the object. In addition, some methods combine spatial and temporal information to take advantage of both techniques.

5.3.1 Spatial Multiplexing

Spatial multiplexing groups all techniques where the code word of a specific position is extracted from surrounding points. Intensity or color variations are used to create the code word. Three different coding strategies can be distinguished within this group: De Bruijn patterns, nonformal coding, and M-arrays.

5.3.1.1 De Bruijn-Based Techniques

De Bruijn sequences are a set of pseudorandom values with specific properties between them. A k-ary De Bruijn sequence of order n is a circular sequence d_0, d_1, d_{n^k-1} (length n^k) containing each substring of length k exactly once (window property of k). De Bruijn sequences can be constructed by taking a Hamiltonian or Eulerian path of an n-dimensional De Bruijn graph (see Fredricksen [18] for more details). This algorithm allows us to create univocal stripe sequences in the pattern and is able to extract the position by looking at the color of the stripes placed in the same window.

Several proposals can be found using De Bruijn sequences, with both striped and multislit patterns. First proposals of De Bruijn-based striped patterns are found in the method developed by Boyer and Kak [7]. In this approach, RGB (red, green, blue) space was used to code the sequence of stripes. Being c_i^k the color of the stripe i in the subpattern k, the distance between two subpatterns k and l is given by

$$d = \sum_{i=1}^{N} \delta_i \tag{5.1}$$

where

$$\delta_i = \begin{cases} 0 & \text{if } c_i^k = c_i^l \\ 1 & \text{otherwise} \end{cases} \tag{5.2}$$

The pattern proposed by Boyer and Kak [7] contains more than 300 stripes colored by three different colors. Color detection was done with a stripe indexing algorithm preceded by a Hamming filtering. However, no color calibration was pursued to suppress the effect of different albedo, leading to some errors due to leakage from the blue to the green channel.

A different approach was followed by Monks, Carter, and Shadle [41], where a multislit-based De Bruijn sequence was projected. Six colors were used to color the slits, separated by black gaps. The slit colors were chosen so that every subsequence of three colors appeared only once. Colors were chosen in the hue channel (hue, saturation, and intensity [HSI] space), despite the fact that projection was performed in RGB and transformed back to HSI once the image was captured by the camera. Full saturation and full intensity were chosen in the SI (saturation/intensity) channels. A previous color calibration step was performed by the authors in order to determine the transfer function of the optical system. Once the system was calibrated, captured colors were corrected before applying fringe detection. A minimum cost matching algorithm was used in the decoding step in order to find the most probable matching between projected and recovered patterns, considering that some slits might be imaged partly occluded or badly segmented [58].

To simplify the peak detection process, Salvi et al. [56] created a grid of horizontal and vertical colored slits. Every crossing of the two slits was

FIGURE 5.1
(See color insert.) Pattern proposed by Pages et al. RGB pattern and luminance channel. (Pages, J. et al. *17th International Conference on Pattern Recognition, ICPR 2004* 4:284–287, 2004. With permission.)

extracted by simple peak intensity detection. Hue channel was again used (in HSI space) to encode the colors. Three colors were assigned for the horizontal lines and another three for the vertical lines, using a De Bruijn third-order sequence. The decoding step was done back in HSI space, showing negligible errors scanning planar surfaces under scene light control. However, some problems were encountered due to the sensitivity of the hue channel under different albedo of the illuminated object.

Some years later, Pages et al. [46] and Pages, Salvi, and Forest [47] proposed an alternative approach to traditional striped- or multislit-based patterns. They combined a striped pattern in the hue channel with a multislit pattern in the intensity channel (see Figure 5.1), which defined dark and bright areas within the same color stripe. Therefore, the high resolution of classical striped patterns and the accuracy of multislit patterns were combined. The half-illuminated stripes were colored according to a De Bruijn sequence for a subpattern of n stripes, while bright slits were colored equally within the same subpattern. In the experiments, a 128-striped pattern with four colors and a window property of three encoded stripes was applied. Using this codification, their approach doubled the resolution of traditional De Bruijn stripe-based techniques.

5.3.1.2 Nonformal Coding

Nonformal coding comprises all the techniques having nonorthodox codification, in the sense that the pattern is designed to fulfill some particular requirements. Both one-axis and two-axes encoding are suitable for these methods. One-axis coding methods are based on striped or multislit patterns. This is the case of Forster's [17] and Fechteler and Eisert's [15] proposals, which created color-based patterns in which two adjacent colors must differ in at least two color channels in the receptor device (red, green, and blue). This condition is not usually accomplished in De Bruijn sequences. Forster used a striped pattern, while Fechteler and Eisert employed a multislit pattern. In Fechteler and Eisert, a parabola was fitted in every RGB channel (or combination of channels for nonpure RGB colors, the option selected by Forster). Optionally, surface color was acquired by projecting an extra white pattern. Tehrani, Saghaeian, and Mohajerani [65] applied the idea of color slits to reconstruct images taken from two camera views, using 10 hue values to create the slit pattern (the difference between colors was maximal for adjacent slits).

There are also some proposals based on two-axes encoding. For instance, Maruyama and Abe [39] proposed a pattern of randomly cut black slits on a white background. In this approach, coding information was held in the length of the slits and their position within the pattern. Every recorded segment had its own length, which could be similar for several segments. The code word corresponding to a segment was determined by its own length and the lengths of its six adjacent segments. The main drawback of this method is that the length of segments is affected by the projector–object and object–camera distances, as well as by the camera optics, therefore reducing the reliability of the system.

Another solution based on stripe lengths was recently developed by Kawasaki et al. [32], who established a pattern of horizontal and vertical lines. In this work, the uniqueness of a specific location was coded in the spacing between horizontal lines (in blue); vertical lines (in red) were equally spaced. A peak detection algorithm was applied to locate the crossing points (dots) in the recovered image, and a posteriori comparison with distances to neighboring dots determined their positions in the projected pattern.

Ito and Ishii [29] did not use stripes or slits for coding; instead, they used a set of square cells (like a checkerboard) that had one out of three possible intensity values. Every node (intersection between four cells of the checkerboard) was associated with the intensity values of the forming cells. In order to differentiate nodes with the same subcode, epipolar constraints between the camera and the projector were employed.

The idea of using epipolar constraints was also applied in the work presented by Koninckx and Van Gool [35]. They proposed an adaptive system where green diagonal lines ("coding lines") were superimposed on a grid of vertical black lines ("base pattern"). If a coding line was not coincident with an epipolar line, intersections created with the base pattern would all have lain on different epipolar lines on the camera image. This determines a unique point in the projected pattern being able to perform the matching and the triangulation. A greater inclination of diagonal lines gave a higher density of the reconstruction, but a lower noise resistance. Therefore, the density of reconstruction could be chosen depending on how noisy the environment was, giving an adaptive robustness versus accuracy.

5.3.1.3 *M-arrays*

First presented by Etzion [14], M-arrays (perfect maps) are random arrays of dimensions $r \times v$ in which a submatrix of dimensions $n \times m$ appears only once in the whole pattern. Perfect maps are constructed theoretically with dimensions $rv = 2^{nm}$, but for real applications, the zero submatrix is not considered. This gives a total of $rv = 2^{nm} - 1$ unique submatrices in the pattern and a window property of $n \times m$. M-arrays represent in a two-dimensional space what De Bruijn patterns are in a one-dimensional space (see references

14 and 38 for more details). Choosing an appropriate window property will determine the robustness of the pattern against pattern occlusions and object shadows for a given application. Morita, Yajima, and Sakata [43] proposed a two-projection-based technique where an encoded matrix of black dots on a white background was projected; in the second projection some black dots were removed according to a binary-encoded M-array.

There are different approaches to represent nonbinary M-arrays, which are classified regarding the approach used to code the M-array: colored dots (color based) or geometric features like circles and stripes (feature based). For instance, Griffin, Narasimhan, and Yee [21] generated an array of 18 × 66 features using an alphabet of four words—1, 2, 3, 4—comparing color- and feature-based approaches. As the second approach is not color dependent, better results were obtained in the presence of colored objects. Morano et al. [42] used a brute force (not De Bruijn based) algorithm to generate the pattern. An iterative algorithm, adding one new code word and checking it against all the previous ones, was performed. If all the distances between values were at least equal to the specified minimum Hamming distance, the new word was accepted and the next iteration was followed, until the pattern was created. The directions in which the pattern was created are indicated in Figure 5.2.

This algorithm was used a posteriori by Pages et al. [45] to design a 20 × 20 M-array-based pattern with an alphabet of three symbols and a window property of 3 × 3. A color approach was used for the dots' codification, using red, green, and blue in order to separate them in the camera sensor. The decoding algorithm analyzed the four neighbors of every dot. Once this was done, a comparison between all possible combinations of eight neighbors was performed in order to locate the recorded dot univocally in the projected pattern and perform the triangulation.

A different approach was followed by Albitar, Graebling, and Doignon [2], who used a 3 × 3 window property and three different symbols (black circle, circumference, and stripe) to represent the code word. As no color codification was employed, this solution presented robustness against colored objects. In the detection step, orientation of the projected pattern was extracted from the direction of the projected stripes. Once this was done, location of the symbols in the projected pattern was accomplished. Albitar et al. employed this method to create a 3D scan for medical imaging purposes (scanning of parts of the body), stating that this one-shot technique was robust against occlusions (up to a certain limit) and suitable for moving scenarios.

5.3.2 Time Multiplexing

Time multiplexing methods are based on the code word created by the successive projection of patterns onto the object surface. Therefore, the code word associated with a position in the image is not completely formed until all patterns have been projected. Usually, the first projected pattern corresponds to the most significant bit, following a coarse-to-fine paradigm. Accuracy

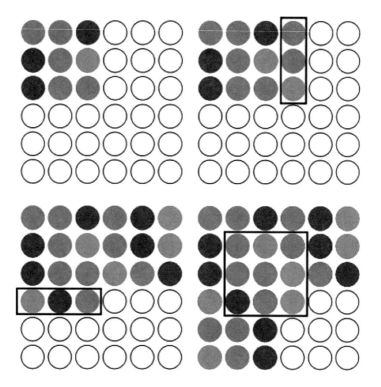

FIGURE 5.2
(See color insert.) Code generation direction followed by Morano et al. with colored spots representation. (Morano, R. A. et al. *IEEE Transactions on Pattern Analysis and Machine Intelligence* 20 (3): 322–327, 1998. With permission.)

directly depends on the number of projections, as every pattern introduces finer resolution in the image. In addition, code word bases tend to be small, providing higher resistance against noise. There are several approaches in sparse time multiplexing, which are explored here.

5.3.2.1 Temporal Binary Codes

These codes were first proposed by Posdamer and Altschuler [48] in 1982. A sequence of patterns with black and white stripes was projected onto the object. The number of stripes increased by two in every pattern, following a coarse-to-fine strategy. Therefore, the length of the code word was given by 2^m bits, where m was the total number of projected patterns. An edge detection algorithm was employed to localize the transition between two consecutive stripes (black/white or vice versa). Moreover, Hamming distance between the code words of two adjacent points could be maximized to reduce errors in the detection step, as was proposed by Minou, Kanade, and Sakai [40].

5.3.2.2 Temporal n-ary Codes

Based on the use of n-ary codes, Caspi, Kiryati, and Shamir [9] proposed a color-based pattern where n^m stripes were coded in RGB space. The parameters to set were the number of colors to be used (N), the number of patterns to be projected (M), and the noise immunity factor alpha (α). For the calibration step, Caspi et al. proposed a reflectivity model given by the following equation:

$$\underbrace{\begin{bmatrix} R \\ G \\ B \end{bmatrix}}_{\vec{C}} = \underbrace{\begin{bmatrix} a_{rr} & a_{rg} & a_{rb} \\ a_{gr} & a_{gg} & a_{gb} \\ a_{br} & a_{bg} & a_{bb} \end{bmatrix}}_{A} \underbrace{\begin{bmatrix} k_r & 0 & 0 \\ 0 & k_g & 0 \\ 0 & 0 & k_b \end{bmatrix}}_{K} \vec{P}\underbrace{\left\{\begin{matrix} r \\ g \\ b \end{matrix}\right\}}_{\vec{c}} + \underbrace{\begin{bmatrix} R_0 \\ G_0 \\ B_0 \end{bmatrix}}_{\vec{C}_0} \tag{5.3}$$

where

$\vec{c}$ is the projected instruction for a given color

$\vec{P}$ is the nonlinear transformation from projected instruction to the projected intensities for every RGB channel

A is the projector–camera coupling matrix

K is the reflectance matrix (constant reflectance in every RGB channel is assumed)

$\vec{C}_0$ is the reading of the camera under ambient light

5.3.2.3 Temporal Hybrid Codes

In order to reduce the number of projections, Ishii et al. [28] proposed a system where temporal and spatial coding were combined. The level of spatial or temporal dependence was given by the speed and accuracy requirements. For a given pixel $p(x, y)$ at time t of the projected pattern, the value was determined by using the following equation:

$$I(x, y, t) = G\left(\hat{I}\frac{x}{m} + t°(\mathrm{mod}\,n), y\right) \tag{5.4}$$

where

$$G(k, y) = G\left(\hat{I}\frac{2^k y}{I_y} + \frac{1}{2}°(\mathrm{mod}\,2)\right) \tag{5.5}$$

G is a binary image obtained from a camera at time t

n is the space code size

m is the light pattern width in the x direction

I_y is the image size in the y direction

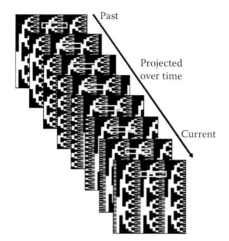

	2	3	4	5	6	7	0
2	3	4	5	6	7	0	1
3	4	5	6	7	0	1	2
4	5	6	7	0	1	2	3
5	6	7	0	1	2	3	4
6	7	0	1	2	3	4	5
7	0	1	2	3	4	5	6
0	1	2	3	4	5	6	7

FIGURE 5.3
Spatiotemporal algorithm proposed by Ishii et al. (Ishii, I. et al. *IEEE/RSJ International Conference on Intelligent Robots and Systems (IROS)*, 925–930, 2007. With permission.)

There were n selectable code values for a pixel at time t, depending on the importance of temporal encoding or spatial encoding. As shown in Figure 5.3, combination of temporal and spatial information can be done from total temporal encoding (represented by $p = 1$) to total spatial encoding (given by $p = 8$). The parameter p is called the space coding weighter, as it provides an idea of how temporal or spatial the codification is.

5.4 Dense Reconstruction Methods

This group of techniques provides 3D reconstruction of all the pixels captured by the image device. It is constituted by discrete or continuous shifting patterns, frequency patterns, and spatial grading, showing continuous variations on intensity or color throughout one axis or two axes. Among these methods, the use of periodic and absolute patterns can be found. Periodic patterns are used in time multiplexing shifting methods and in frequency multiplexing. Additionally, absolute patterns are based on spatial grading.

5.4.1 Time Multiplexing

The same concept of time multiplexing in sparse reconstruction techniques is applied for dense reconstruction approaches. Dense time multiplexing is represented by shifting techniques with discrete and continuous patterns.

5.4.1.1 Discrete Shifting Methods

There are some discrete implementations that use the shifting of patterns to obtain dense reconstructions. This is the case of Sansoni, Carocci, and Rodella [59], Guhring [23], and Zhang, Curless, and Seitz [72]. The proposals of Sansoni et al. and Guhring projected a set of black and white striped patterns (as in binary codes). Afterward, the work of Sansoni et al. projected four shifted versions of the last pattern, while Guhring's proposal projected shifted versions of a slit-based pattern covering every pixel in the image. Binary patterns provided an absolute location of the information given by shifted patterns, avoiding ambiguities in the decoding step. Using a different strategy, Zhang et al. employed color to project De Bruijn sequences, which are smoothed and shifted versions of the same pattern. The smoothing process provided subpixel accuracy to this method. In order to avoid errors due to occlusions and discontinuities, multipass dynamic programming (a variance of the dynamic programming proposed by Chen et al. [10]) was employed to match observed to projected patterns.

5.4.1.2 Continuous Phase-Shifting Methods

When projecting a sinusoidal grating onto a surface, every point along a line parallel to the coding axis can be characterized by a unique phase value. Any nonflat 3D shape will cause a deformation in the recorded pattern with respect to the projected one, which is recorded as a phase deviation. This phase deviation provides information about the illuminated shape. By matching the recovered image with the projected pattern, the object shape is recovered. The pattern must be shifted and projected several times in order to extract the phase deviation (this is not the case with frequency multiplexing approaches). Due to the grayscale nature of the projected patterns, they present advantages like resistance to ambient light and resistance to reflection variation. Depending on the number of frequencies used to create the pattern, we can distinguish between simple and multiple phase-shifting (PS) methods.

5.4.1.2.1 Single Phase Shifting

Single phase-shifting (SPS) techniques use only one frequency to create the sequence of patterns. In order to recover phase deviation, the pattern is projected several times; every projection is shifted from the previous projection by a factor of $2\pi/N$, with N the total number of projections, as shown in the following equation (superindex P indicates the projected pattern):

$$I_n^p\left(y^p\right) = A^p + B^p \cos\left(2\pi f_\phi y^p - 2\pi n/N\right) \tag{5.6}$$

where A^p and B^p are the projection constants and (x^p, y^p) are the projection coordinates, $n = 0, 1, \dots N$. The received intensity values from the object surface, once the set of patterns is projected, is

$$I_n(x,y) = \alpha(x,y)\left[A + B\cos\left(2\pi f_\phi y^p + \phi(x,y) - 2\pi n/N\right)\right] \tag{5.7}$$

As can be observed from Equation (5.7), it suffers of intensity and phase deviation; it is necessary to cancel the effect of a different albedo ($\alpha(x, y)$) to extract the phase correctly. This is shown in the following equation:

$$\phi(x,y) = \arctan\left[\frac{\sum_{n=1}^{N} I_n(x,y)\sin\left(2\pi n/N\right)}{\sum_{n=1}^{N} I_n(x,y)\cos\left(2\pi n/N\right)}\right] \tag{5.8}$$

From a minimum of three projected shifted patterns it is possible to create a relative phase map and to reconstruct the phase deviation caused by the object shape. However, the arctangent function returns values in the range $(-\pi, \pi]$ and therefore a phase unwrapping procedure is necessary to work with a nonambiguous phase value out of the wrapped phase. This is the reason why these patterns provide effective dense reconstruction only under the restriction of smoothed surfaces.

Phase-shifting methods have been used in a variety of applications during the last years. For instance, Ono et al. [44] created the so-called correlation image sensor (CIS), a device that generates temporal correlations between light intensity and three external reference signals on each pixel using phase shifting and a space–temporal unwrapping. Some approaches using phase shifting have also been developed from the work proposed by Srinivasan, Liu, and Halious [61].

One of the drawbacks of phase-shifting methods is the necessity to project several patterns in time, which is more than the theoretic minimum of three patterns considered for real conditions. A solution to reduce the total time required in the projection step is to multiplex the patterns either in color space or in frequency. Following this idea, Wust and Capson [70] proposed a method that projected three overlapping sinusoidal patterns shifted 90° and coded in red, green, and blue. Therefore, in this way the camera recorded phase deviation of every pattern in a different color channel and a normal phase extraction algorithm, like the one shown in the following equation:

$$\Phi(x,y) = \arctan\left(\frac{I_r - I_g}{I_g - I_b}\right) \tag{5.9}$$

where $\Phi(x, y)$ is the phase of a given pixel, and I_r, I_g, and I_b are the red, green, and blue intensities, respectively.

A different approach was proposed by Guan, Hassebrook, and Lau [22], where the patterns were combined in frequency using the orthogonal

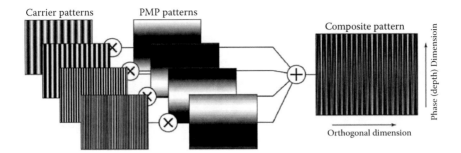

FIGURE 5.4
Composite pattern formed by the multiplexation of modulated phase shifting profilometry (PMP) patterns using the algorithm of Guan et al. (Guan, C. et al. *Optics Express* 11 (5): 406–417, 2003. WIth permission.)

dimension, as shown in Figure 5.4. Basically, traditional band pass filtering was performed on the recorded pattern, as is theoretically done in communications for frequency multiplexing. This step filters noise without suppressing the information hold in the surroundings of the carriers. In particular, Guan et al. [22] used a maximally flat magnitude Butterworth filter. Once this step was done, a normal phase extraction was performed over the obtained patterns. This method provided higher signal-to-noise ratio than color multiplexing approaches and it was not dependent on the surface color. However, some errors arose in the presence of a different albedo and abrupt shape variations.

5.4.1.2.2 Multiple Phase Shifting

The use of more than one frequency in phase shifting comes to cope with the uncertainty created in the extracted wrapped phase. As stated in the remainder theorem [52], an absolute phase map can be computed from two different relative phase maps with frequencies that are relative prime numbers. This principle was used by Gushov and Solodkin [24] for interferometry, where an interferometer able to deal with vibrations or relief parameters was constructed. More recently, Pribanic, Dapo, and Salvi [49] presented a technique based on multiple phase shifting (MPS) where only two patterns were used to create the relative phase maps. Two sinusoidal patterns were shifted and projected in time, in order to recover phase deviation (see Figure 5.5). From these sets of images it was possible to obtain two relative phase maps, using normal phase-shifting decoding algorithms (as shown in Equation 5.8). With this, the absolute phase map was recovered. This map can be directly compared to the ideal phase-shifting map, providing correspondences for the triangulation step. The algorithm was tested for different pairs of frequencies over a flat surface. Finally, the reconstruction of a footprint and a face were pursued, providing small 3D reconstruction errors.

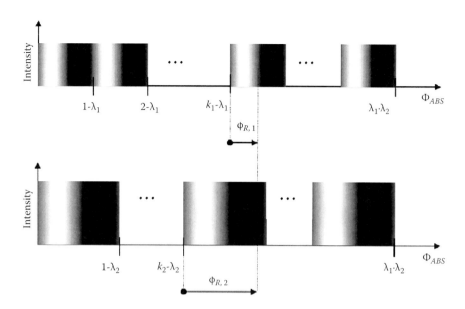

FIGURE 5.5

Pair of projected sinusoidal patterns with two different frequencies (k is the number of periods).

5.4.2 Frequency Multiplexing

Frequency multiplexing methods group all the techniques where phase decoding is performed in the frequency domain rather than in the spatial domain. There are different approaches depending on the frequency analysis performed on the image. Traditionally, Fourier methods have been used for this purpose, although wavelet-based methods have also been studied.

5.4.2.1 Fourier Transform

Fourier transform (FT) was introduced to solve the necessity of having a phase-shifting-based method for moving scenarios. FT was first proposed by Takeda and Mutoh [64], who extracted depth from one single projected pattern. A sinusoidal grating was projected onto the object, and the reflected deformed pattern was recorded. The projected signal for a sinusoidal grating was represented by

$$I_n^p\left(y^p\right) = A^p + B^p \cos\left(2\pi f_\phi y^p\right) \tag{5.10}$$

Once reflected onto the object, the phase component was modified by the shape of the object, thus giving an intensity value expressed as

$$I(x,y) = \alpha(x,y)\Big[A + B\cos\left(2\pi f_\phi y^p + \phi(x,y)\right)\Big]_p \tag{5.11}$$

The phase component must be isolated to extract shape information. This was achieved performing a frequency filtering in the Fourier domain. The background component was suppressed and a translation in frequency was done to bring the carrier component (which holds the phase information) to the zero frequency axis. When the following sequence of equations is applied, the phase can be extracted from the signal. First, the input signal was rewritten as

$$I(x, y) = a(x, y) + c(x, y)e^{2\pi i f_0 y^p} + c^*(x, y)e^{-2\pi i f_0 y^p} \tag{5.12}$$

where

$$c(x, y) = \frac{1}{2}b(x, y)e^{i\phi(x, y)} \tag{5.13}$$

where $c^*(x, y)$ is the complex value of constant $c(x, y)$. Finally, the phase component was extracted from the imaginary part of the following equation:

$$\log[c(x, y)] = \log\left[\left(\frac{1}{2}\right)b(x, y)\right] + i\phi \tag{5.14}$$

The obtained phase component range is $(-\pi, \pi]$, and it is necessary to apply an unwrapping algorithm in order to obtain a continuous phase related to the object. Once the phase was unwrapped, the relative depth information was extracted using

$$h(x, y) = L \cdot \frac{\Delta\phi(x, y)}{\left(\Delta\phi(x, y) - 2\pi f_0 d\right)} \tag{5.15}$$

where L is the distance to the reference plane and d is the distance between the camera and the projector devices.

FT has been widely used in industrial applications. For instance, Cobelli et al. [13] used FT for global measurement of water waves. In their work, two sources of noise were considered in the filtering step. The first one was related to illumination inhomogeneities of background variations over the field of view, which remains present as an additive variation. The second one was due to the local surface reflectivity. As this reflection varies much more slowly than the sinusoidal modulation impinged on the surface, it can also be treated as background noise. Thus, both sources of noise can be suppressed using the background component filtering procedure proposed by Takeda and Mutoh [64]. Due to the periodic nature of the projected pattern, this method was constrained by the maximum reconstructible slope given by

$$\left|\frac{\partial h(x,y)}{\partial x}\right|_{max} < \frac{L}{3d} \qquad (5.16)$$

In order to increase this slope limitation, Li, Su, and Guo [36] proposed the so-called π-phase-shifting FT. Two sinusoidal patterns were projected using this method; the second one was a half-period shifted version of the first one. This solution multiplies by three the detectable range in depth slope. This principle was used by Hu and He [25] to scan moving objects that had uniform velocity (as in an assembly line). In their work, two scan line cameras were used, and one single pattern was projected. The distance between the two cameras corresponded to half the period of the grating. As the velocity of the object was known, matching two scannings of the same point at different instants of time could be done. This procedure avoids the projecting of two patterns and takes advantage of the uniform motion present in assembly lines.

There are some proposals that combine both π-phase-shifting FT patterns in one single projected pattern using color or frequency multiplexing. For instance, Chen et al. [11] used color space to project a bicolor sinusoidal fringe pattern consisting of the sum of π-phase-shifting FT patterns represented by blue and green patterns. Another approach was considered by Yue, Su, and Liu [71]. In this work the same principle used by Guan et al. for phase shifting was developed for FT. Appropriate carrier frequencies were chosen regarding the characteristics of the projector and camera used, assuming that the Nyquist sampling theorem was satisfied. These frequencies were kept away from zero frequency as much as possible. When the results are analyzed, the standard deviation error is slightly lower than for normal FT, while accuracy remains unaltered.

In case of scanning coarse objects where discontinuities and speckle-like structures can appear, two-dimensional (2D) FT filtering must be used [62], as it permits better separation of the desired information from noise. This is due to the fact that noise is normally in two dimensions distributed in a fringe pattern, having a spectra scattered in a 2D frequency domain. For instance, Hung and more recently Lin and Su [37] proposed a method for 2D FT scanning where the filtering step, aimed to prevent frequency spreading, was performed using a 2D Hanning window. However, some other filters that have similar characteristics can also be used. This is the case of Chen et al. [12], who applied a Gaussian filter. Two-dimensional FT filtering has been used by Berryman et al. [6] to create a low-cost automated system to measure the three-dimensional shape of the human back, obtaining an accuracy of ±1 mm.

Spatial phase detection (SPD) constitutes an alternative to FT that was initially proposed by Toyooka and Iwaasa [67]. The analysis of the received signal (Equation 5.17) is done using the sine and cosine functions, as can be observed in Equations (5.18) and (5.20).

$$I(x,y) = \alpha(x,y)\left[a + B\cos\left(2\pi fy^p + \phi(x,y)\right)\right] \tag{5.17}$$

$$I_c(x,y) = \alpha(x,y)\left[A + B\cos\left(2\pi fy^p + \phi(x,y)\right)\right]\cdot\cos\left(2\pi fy^p\right) \tag{5.18}$$

$$= \alpha(x,y)\cdot A\cos\left(2\pi fy^p\right) + \frac{1}{2}\alpha(x,y)\cdot B\cos\left(4\pi fy^p + \frac{1}{2}\alpha(x,y)\cdot B\cos\left(\phi(x,y)\right)\right) \tag{5.19}$$

$$I_s(x,y) = \alpha(x,y)\left[A + B\cos\left(2\pi fy^p + \phi(x,y)\right)\right]\cdot\sin\left(2\pi fy^p\right) \tag{5.20}$$

$$= \alpha(x,y)\cdot A\sin\left(2\pi fy^p\right) + \frac{1}{2}\alpha(x,y)\cdot B\sin\left(4\pi fy^p + \frac{1}{2}\alpha(x,y)\cdot B\sin\left(\phi(x,y)\right)\right) \tag{5.21}$$

Now, $\phi(x, y)$ varies more slowly than any term containing f, so only the last term in each new function is a low-frequency term. This part of the function can then be extracted by low-pass filtering. Regarding Euler's formula for the sine and cosine functions and the principles of Fourier transform applied on sinusoidal functions [50], this step provides similar results to obtaining the real and the imaginary components of the Fourier transform applied to the incoming signal. Therefore, the last step is to extract the phase component from these components, which is obtained by applying the arctangent function:

$$\phi(x,y) = \arctan\left[\frac{r(x,y) * I_s(x,y)}{r(x,y) * I_c(x,y)}\right] \tag{5.22}$$

where $r(x, y)$ represents a low-pass filter and $*$ denotes convolution. It is important to note that Toyooka and Iwaasa use integration to extract the phase terms, whereas other authors using related spatial domain methods apply different low-pass filters [5]. As in FT, this method suffers from leakage distortion when working with fringe patterns, as no local analysis is performed to avoid spreading errors due to discontinuities and different albedo.

5.4.2.2 Window Fourier Transform

The task of suppressing the zero component and avoiding the frequency overlapping between background and data (the leakage distortion problem) has also been studied using other frequency-based approaches.

This is the case of the windowed Fourier transform (WFT; also called the Gabor transform), which splits the signal into segments before the analysis in frequency domain is performed. The received signal is filtered applying the WFT analysis transform shown in Equations (5.23) and (5.25).

$$Sf(u,v,\xi,\eta) = \int_{-\infty}^{\infty}\int_{-\infty}^{\infty} f(x,y)\cdot g(x-u,y-v)\cdot\exp(-j\xi x - j\eta y)\,dx\,dy \text{ is } (x,y) \quad (5.23)$$

(ξ, η) are the translation and frequency coordinates, respectively, and $g(x, y)$ is the windowing function

When $g(x, y)$ is a Gaussian window, the WFT is called a Gabor transform; that is,

$$g(x,y) = \frac{1}{\sqrt{\pi\sigma_x\sigma_y}}\cdot\exp\left(-\frac{x^2}{2\sigma_x^2}-\frac{y^2}{2\sigma_y^2}\right) \quad (5.24)$$

where σ_x and σ_y are the standard deviations of the Gaussian function in x and y, respectively.

Equation (5.23) provides the four-dimensional (4D) coefficients $Sf(u, v, \xi, \eta)$ corresponding to the 2D input image. The windowing permits the WFT to provide frequency information of a limited region around each pixel. The Gaussian window is often chosen as it provides the smallest Heisenberg box [34]. Once the 4D coefficients are computed, the phase can be extracted. There are two main techniques for phase extraction in WFT: windowed Fourier filtering (WFF) and windowed Fourier ridge (WFR). In WFF the 4D coefficients are first filtered, suppressing the small coefficients (in terms of its amplitude) that correspond to noise effects. The inverse WFT is then applied to obtain a smooth image:

$$\overline{f(x,y)} = \int_{-\infty}^{\infty}\int_{-\infty}^{\infty}\int_{-\eta_1}^{\eta_h}\int_{-\xi_1}^{\xi_h} \overline{Sf(u,v,\xi,\eta)}\cdot g_{u,v,\xi,\eta}(x,y)\,d\xi\,d\eta\,du\,dv \quad (5.25)$$

where

$$\overline{Sf(u,v,\xi,\eta)} = \begin{cases} Sf(u,v,\xi,\eta) \text{ if } |Sf(u,v,\xi,\eta)| > \text{threshold} \\ 0 \text{ if } |Sf(u,v,\xi,\eta)| < \text{ threshold} \end{cases} \quad (5.26)$$

The estimated frequencies $\omega_x(x, y)$ and $\omega_y(x, y)$ and corresponding phase distribution are obtained from the angle given by the filtered WFF, as explained in Kemao [34]. In WFR, however, the estimated frequencies are extracted from the maximum of the spectrum amplitude:

$$\left[\omega_x(u,v),\omega_y(u,v)\right] = \arg\max_{\xi,\eta}\left|Sf(u,v,\xi,\eta)\right| \quad (5.27)$$

The phase can be directly obtained from the angle of the spectrum for those frequency values selected by the WFR (phase from ridges) or integrating the frequencies (phase by integration). Phases from ridges represent a better solution than phases from integration (despite some phase correction that may need to be applied [34]), as phases from integration errors are accumulated and lead to large phase deviations.

Using WFT, Chen et al. [11] proposed its use (Gabor transform) to eliminate the zero spectrum. However, as was demonstrated by Gdeisat, Burton, and Lalor [19], Chen and colleagues' technique was not able to eliminate the zero spectrum in fringe patterns that have large bandwidths or in cases where the existence of large levels of speckle noise corrupts the fringe patterns. This is mainly caused by an erroneous selection of the width and shape of the window for the Fourier analysis. The window size must be small enough to reduce the errors introduced by boundaries, holes, and background illumination; at the same time, it must be big enough to hold some periods and hence allow the detection of the main frequency to perform an optimal filtering. This problem is studied in the work of Fernandez et al. [16], where an automatic algorithm for the optimal selection of the window size using adapted mother wavelets is proposed. However, in applications where the frequency varies considerably during the analysis (in space or in time), this trade-off is difficult to achieve and noise arises due to an incorrect frequency detection.

5.4.2.3 Wavelet Transform

Wavelet transform (WT) was proposed to solve the aforementioned trade-off. In WT the window size increases when the frequency to analyze decreases, and vice versa. This allows removal of the background illumination and prevents the propagation of errors produced during the analysis, which remain confined in the corrupted regions alone [19]. Additionally, the leakage effects are reduced, avoiding large errors at the edges of the extracted phase maps. The continuous wavelet transform (CWT) is a subfamily of WT that performs the transformation in the continuous domain. Moreover, it is common to use CWT with complex wavelets for the analysis of the fringe patterns [1]. The one-dimensional (1D)-CWT algorithm analyzes the fringe pattern on a row-by-row basis, whereas the 2D-CWT algorithm is an extension of the analysis to the two-dimensional space. In 2D analysis, a 4D transform is obtained from WT. (The daughter wavelets are obtained by translation, dilation, and rotation of the previously selected mother wavelet).

Once this is performed, phase extraction is pursued using the phase from ridges or the phase by integration algorithms, also named phase estimation and frequency estimation (similarly to WFT). As in WFT, it has been proven that the phase from ridges provides better results than the phase from integration, due to the accumulative effect in the phase from the integration algorithm [1]. The work done by Gdeisat et al. [19] applied a two-dimensional wavelet function to the recovered image, based on the phase from ridge extraction. Rotation and scale were considered jointly with x and y coordinates resulting in a four-dimensional wavelet transform. To apply the transformation, the mother wavelet $\psi(x, y)$ must satisfy the admissibility condition. Under this condition Gdeisat et al. used a differential of Gaussian as the mother wavelet; Zhang, Chen, and Tang [73] employed a 2D complex Morlet wavelet. Four subimages were created at one iteration of the wavelet decomposition algorithm corresponding to the low and high frequencies in both axes. The phase component was extracted from the ridge information present in the corresponding high-frequency subimage. The task of choosing appropriate values for rotation and scale parameters determined the results of filtering and phase extraction.

Related to this, a novel method for choosing the adaptive level of discrete wavelet decomposition has been proposed by Zhang et al. [73]. They have achieved higher accuracy in the principal frequency estimation and low-frequency energy suppression against traditional zero suppression algorithms used in FT. More recently, Salvi, Fernandez, and Pribanic [54] proposed a colored one-shot fringe pattern based on 2D wavelet analysis. This work proposes a color-based coding strategy to face the problem of phase unwrapping. Using a combination of color fringes at some specific frequencies with a unique relation between them, it is possible to obtain the absolute phase without any uncertainties in the recovered depth. However, there are some problems related to the relationship between the window size and the frequency of the fringes. In WT, the window size increases when the horizontal or vertical fringe frequencies decrease.

This can be troublesome for the analysis of some fringe patterns where the carrier frequency is extremely low or high, as was pointed out by Kemao et al. [33]. Moreover, in computational applications, a dyadic net is used to generate the set of wavelet functions. That is, the size of the wavelet is modified by the factor 2^j. This can lead to some problems in applications like fringe pattern analysis, where the change in the spatial fringe frequencies throughout the image is not high enough to produce a relative variance of 2^j in the size of the optimal wavelet.

5.4.2.3.1 The Problem of Phase Unwrapping

Phase unwrapping represents a crucial step in frequency multiplexing techniques. In absence of noise, if all phase variations between neighboring pixels are less than π, the phase unwrapping procedure can be reduced to add the corresponding multiple of 2π when a discontinuity appears. Unfortunately, noise, local shadows, undersampling, fringe discontinuities, and irregular surface brightness make the unwrapping procedure much more difficult to solve. Plenty of approaches have been presented [3, 69]. For instance, phase unwrapping based on modulation follows an iterative algorithm, starting from the pixel with higher intensity value and comparing it to the pixels inside a 3×3 surrounding square region. The comparison step is done one by one, queuing the affected pixels from maximum to minimum intensity.

This method can also be applied when dealing with moving objects, substituting the searching area to a $3 \times 3 \times 3$ voxel.

Additionally, Wu and Peng [69] presented a phase unwrapping algorithm based on region growing. The phase was unwrapped from the smoothest area to the surroundings according to a linear estimation. In order to decrease the error, a quality map was used to guide the unwrapping. The map can be defined in different ways as far as it provides quality information. For instance, second-order partial derivative can be used to determine the pixels to unwrap—that is, those pixels having this value lower than a specified threshold. Statistical methods can also be used to consider the variance within a mask for every pixel. Finally, Gorthi and Lolla [20] projected an extra color-coded pattern, which can be univocally identified once the image is captured, thus giving rough information about the required phase to add or subtract in the unwrapping step. A further explanation of different unwrapping methods used in profilometry can be found in Judge and Bryanston-Cross [31].

5.4.2.3.2 Alternatives to Sinusoidal Grating

Not all frequency transform methods use sinusoidal fringes for the projected pattern. As Huang et al. [27] stated, structured-light techniques based on sinusoidal phase-shifting methods have the advantage of pixel-level resolution, large dynamic range, and few errors due to defocusing. However, the arctangent computation makes them relatively slow. As an alternative, they used three 120° phase-shifted trapezoidal fringe patterns. The phase deviation was extracted from the so-called intensity ratio image:

$$r(x,y) = \frac{I_{med}(x,y) - I_{min}(x,y)}{I_{max}(x,y) - I_{min}(x,y)} \tag{5.28}$$

where $I_{min}(x, y)$, $I_{med}(x, y)$, and $I_{max}(x, y)$ are the minimum, median, and maximum intensities of the three patterns for the image point (x, y). Image defocus does not cause major errors when using a sinusoidal pattern, as it is still sinusoidal when the image is defocused. However, errors caused by blurring have to be taken into account when dealing with trapezoidal patterns. Modeling these errors as a Gaussian filtering, Huang and colleagues' experiments yielded defocusing errors not bigger than 0.6%. More recently, another approach using triangular patterns has been proposed by Jia, Kofman, and English [30]. This approach used only two triangular patterns shifted half the period, making it more feasible to be implemented in real-time applications. Ronchi grating has also been used in pattern projection as an alternative to sinusoidal grating. This is the case of Lin and Su [37], who proposed an algorithm where only one pattern was needed. Phase information was obtained taking the imaginary part of

$$\Delta\Phi(x,y) = \log\left[\hat{I}(x,y)\hat{I}_0^*(x,y)\right] \tag{5.29}$$

where $\hat{I}(x,y)$ and $\hat{I}_0(x,y)$ are the recorded illuminance from the setup and the reference plane, respectively. A Ronchi grating was also used by Spagnolo et al. [60] in real applications, in order to recover 3D reconstructions of artwork surfaces.

5.4.3 Spatial Multiplexing (Grading Methods)

Grading methods refer to all techniques containing the entire code word for a given position only in the pixel value. Therefore, the resolution can be as high as the pixel resolution of the projector device is. However, these methods suffer from high sensitivity to noise and low sensitivity to surface changes, due to the short distances between the code words of adjacent pixels. This is the reason why some authors use methods introducing temporal redundancy, projecting the same pattern several times. As a drawback, note that restriction to static scenarios is imposed when more than one pattern is projected. There are two main techniques based on grading methods: grayscale-based patterns and color-based patterns. Regarding grayscale-based methods, Carrihill and Hummel [8] proposed a linear grayscale wedge spread going from white to black, along the vertical axis. The authors achieved a mean error of 1 cm, due to the high sensitivity to noise and nonlinearity of the projector device. In color-based patterns, the pixel is coded using color instead of grayscale values. As a drawback, color calibration is required. Tajima and Iwakawa [63] presented a rainbow pattern codified in the vertical axis. In order to project this spectrum, a nematic liquid crystal was used to diffract white light. Two images were projected to suppress the effect of colored surfaces.

5.5 One-Shot Dense Reconstruction Using De Bruijn Coding and WFT

As can be observed, the main contributions in SL done during the last years rely on the fields of real-time response (that is, one-shot pattern projection) and dense projection. Some one-shot techniques obtain dense reconstruction; this is the case of many frequency multiplexing-based techniques and spatial grading. However, both groups present some problems in 3D reconstruction. The former fails under presence of discontinuities and slopes of the 3D surface, suffering from errors due to the periodicity of the projected fringes in the unwrapping step. Additionally, the latter presents low-accuracy results due to the high sensitivity to noise and low sensitivity to surface changes. Moreover, as can be extracted from the work of Salvi et al. [57], the best results in terms of accuracy are obtained by De Bruijn-based techniques for one-shot pattern projection.

However, De Bruijn codes are present only in sparse reconstruction algorithms. The new method presented in this section attempts to solve this problem. The proposed algorithm uses a De Bruijn codification integrated with a fringe pattern projection and consequent frequency analysis. Therefore, it is possible to obtain the accuracy provided by a classical De Bruijn stripe pattern and the density of the reconstruction provided in fringe pattern analysis. To this end, a set of colored fringes is projected on the object. The color of these fringes is established following a De Bruijn codification. The analysis of the recovered image is done in two different ways: one for De Bruijn decoding and another for windowed Fourier transform. A general scheme of the proposed algorithm is shown in Figure 5.6.

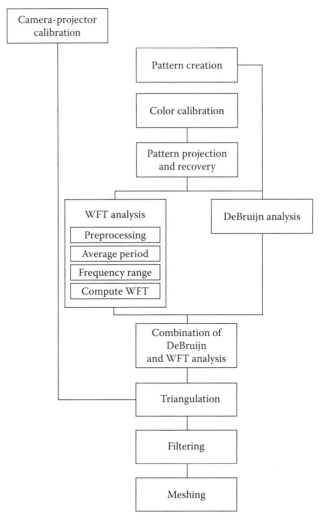

FIGURE 5.6
Diagram of the proposed algorithm.

5.5.1 Camera–Projector Calibration

Many camera calibration techniques can be found in the literature [2–5]. However, projector calibration has not been so widely studied, mainly due to the more limited field of application, which has dealt with custom application-oriented solutions. As Berryman et al. [6] stated, there are three different ways of modeling a projector. Among them, the plane structured-light model, which approximates the projector as the inverse of the pinhole camera model, is the most common. The establishment of 2D–3D point correspondences required for the projector calibration is not straightforward and usually requires specific projector calibrating patterns and nonautomatic image processing algorithms. Our proposal makes use of three different patterns: the printed checkerboard calibrating pattern (CCP), the projector calibrating pattern (PCP), and the projected CCP onto the PCP. For comparison, we have manufactured two PCPs. The first consists of a white plane in which a checkerboard pattern has been printed at one of the corners of the plane (PCP1). The second consists of a white plane in which a checkered row of points has been printed all along the perimeter of the plane (PCP2). Both patterns are shown in Figure 5.7.

First, the camera is calibrated using the printed CCP and Bouguet's implementation of Zhang's technique [2]. The camera is modeled according to the pinhole camera model, where a given 2D image point, $\hat{m} = [u, v, 1]$, and a 3D object point, $\hat{M} = [X, Y, Z, 1]$, are related by $s\hat{m} = A \cdot [R\ t]\hat{M}$, where s is an arbitrary scale factor and A and $[R\ t]$ are the camera intrinsic and extrinsic matrices, respectively. Also, lens distortion is modeled considering both linear and nonlinear distortion up to six levels of radial and tangential distortion.

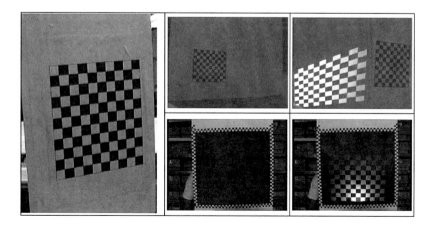

FIGURE 5.7
The left column shows the printed CCP. The middle column shows the PCP1 (top) and PCP2 (bottom). The right column shows one view of the projected CCP on PCP1 and PCP2, respectively.

Second, we place the world coordinate system at the camera center and we face the projector calibration. The projector projects another CCP (similar to the one used for camera calibration) on the white area of the PCP. Then, the PCP is shown to the camera making sure that it is placed at many different positions and orientations, as done in camera calibration, and an image I_i is taken for every PCP pose i.

Third, for every I_i, the PCP printed pattern is segmented and the PCP corner points in pixels extracted. Since the metric distance among these corner points on the PCP is known, the homography between the 2D corner points on the image and the 2D corner points on the PCP is computed. A proper algorithm considering both linear and nonlinear distortion is applied.

Fourth, for every I_i, the CCP pattern projected on the white area of the PCP is segmented and the CCP corner points on the image extracted. Using the previous homography, the projection of these corner points on the PCP is computed, obtaining a set of 3D corner points in metrics. Their 2D correspondences in pixels are known since they are given by the corners on the CCP image projected by the projector device. The same pinhole model described for camera calibration is applied for the calibration of the projector. Therefore, Zhang's method is performed using the 2D–3D correspondences for all I_i, $i = 1\dots n$, thus obtaining the optimized intrinsic and extrinsic parameters for the projector. Note that radial and tangential distortion parameters are computed considering nondistorted 2D points and distorted projected 3D points inversely to those used for camera calibration.

5.5.2 Pattern Creation

The first step in any SL technique is focused on the creation of the pattern to be projected on the measuring surface. The sequence of colors used in the fringes is selected using a De Bruijn-based sequence generator. As was mentioned in previous sections, a k-ary De Bruijn sequence of order n is a circular sequence $d_0, d_1, \dots, d_{n^k-1}$ (length n^k) containing each substring of length k exactly once (window property of k). In our approach we set $n = 3$ as we work only with red, green, and blue colors. Moreover, we set the pattern to have 64 fringes—a convention given the pixel resolution of the projector and the camera. Therefore, $n^k \geq 64$, so we set the window property to $k = 4$. An algorithm performing the sequence generation provides us an arbitrary De Bruijn circular sequence $d_0, d_1, \dots, d_{80}$.

The next step is to generate the colored fringes according to the color sequence. To this end, an HSI matrix is created and converted to RGB space. The intensity channel matrix is created from a sinusoidal vector with values from 0 to 255 and $n = 64$ periods in total for the image height. This vector defines every column in the intensity channel matrix. The hue channel matrix is created using the previously computed De Bruijn sequence. For all the pixels corresponding to the same period in the intensity channel matrix, this matrix is assigned to the same value of the De Bruijn sequence, starting

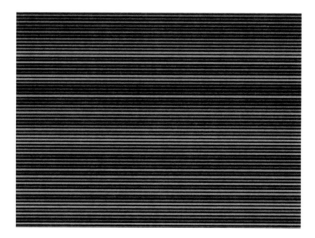

FIGURE 5.8
(See color insert.) Pattern of the proposed method; $m = 64$ sinusoidal fringes are coded in color using a De Bruijn code generator algorithm.

from the first element. Finally, the saturation channel matrix is set to the maximum value for all the pixels. The proposed pattern containing $m = 64$ fringes can be observed in Figure 5.8.

5.5.3 Color Calibration

Once the deformed pattern is captured by the camera, the first task is to split the three color channels obtained from the camera and perform a color enhancement to reduce the effect of albedo and cross talk in every color channel. To cope with this, a previous color calibration has been pursued. Different proposals of color calibration can be found in the literature [9, 41, 47, 51]. The most exhaustive work for SL was proposed by Caspi et al. [9], who developed a precise color calibration algorithm based on linearizing the projector–camera matrix and the surface reflectance matrix specific to every scene point projected into a camera pixel.

A simpler version of this method has been performed in our work. The proposed algorithm uses least-squares to linearize the combination matrix corresponding to the projector–camera pair and the surface reflectance matrices, in terms of response to color intensity, for each pixel in the received image and each color channel (red, green, and blue). For every pixel and every color channel, the projected intensity is increased linearly and the corresponding captured color is stored.

Figure 5.9 shows the projected and captured color values. As can be observed, the red channel suffers the higher cross talk effects, mainly when projecting green onto the image. This is compensated by the transformation matrix. A linear regression is computed that yields a matrix estimation of the projected color values for every received value. Having the set of three

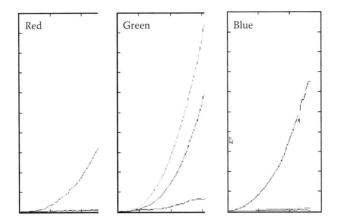

FIGURE 5.9
(See color insert.) Received color intensities for projected increasing values of red, green, and blue, respectively.

received color values—R_0, G_0, B_0—the estimated projected values R, G, B are given by Equation (5.30). It is important to note that this calibration has been done under the assumption that all objects have a reflectance similar to the flat-white Lambertian plane used for calibration, as only Lambertian white surface objects were analyzed.

$$\begin{bmatrix} R_0 \\ G_0 \\ B_0 \end{bmatrix} = \begin{bmatrix} a_{rr} & a_{rg} & a_{rb} \\ a_{gr} & a_{gg} & a_{gb} \\ a_{br} & a_{bg} & a_{bb} \end{bmatrix} \begin{Bmatrix} R \\ G \\ B \end{Bmatrix} \tag{5.30}$$

The matrix of Equation (5.30) represents the whole system (projector–camera) and aims to subtract the effect of cross talk between color channels. However, as it approximates the system as a linear transformation between projected and received images, some errors will persist due to nonlinearities. This is the reason why De Bruijn vocabulary was minimized to three colors, in order to maximize the Hamming distance between different colors in the hue channel.

5.5.4 Pattern Projection and Recovery

Once the deformed pattern is recovered by the camera and the color calibration has been pursued, an image processing step is applied. First, the color calibrated (corrected) RGB image is transformed to the HSV space. Afterward, a mask is created regarding the information held in the V plane. A closure morphological operation is applied, followed by a binarization. Those pixels exceeding the value given by the Otsu thresholding algorithm are selected as active pixels for the mask. Finally, the mask is applied separately to the corrected RGB image and to the V matrix. The masked V matrix is used in the WFT analysis, whereas the masked RGB image is the input of the De Bruijn detection algorithm.

5.5.5 Windowed Fourier Transform Analysis

A dense 3D reconstruction of the imaged scene (subjected to the robustness of the phase unwrapping) can be obtained extracting the phase of the recovered image. There are five different techniques used traditionally for phase extraction: phase measurement profilometry (PMP), SPD, FT, WFT, and WT. Among them, only those based on frequency analysis (FT, WFT, and WT) project one single shot and thus are able to work with moving objects. Regarding these frequency-based techniques, the main differences among them are related to the section of the imaged pattern that is considered in the frequency analysis. In the state of the art, FT performs a global analysis, which is appropriate for stationary signals with poor spatial localization. However, this is not the case in CSL, which is by nature limited in space and thus nonstationary. This fact led to the use of the other two frequency-based transforms (WFT and WT), which analyze local information in the imaged pattern.

WFT and WT are constituted by two main steps: windowing the imaged pattern in local patches and computing the transform at every local patch. The crucial point in these techniques relies on the necessity of selecting an optimal window size, which constitutes a trade-off between resolution in space and resolution in frequency. To this end, theoretical and practical comparisons were performed in the work of Fernandez et al. [16]. As stated in this work, the main difference between both techniques is the way the window size is set, depending on whether it has a fixed or a variable value. WT performs better with signals that have a wide range of frequencies with shorter correlation times for the higher frequencies than for the lower frequencies. This is the case in natural scenes, where low-frequency components usually last for longer durations than high-frequency components.

However, in fringe patterns, their periodicity and spatial extension do not depend on the selected frequency. Nevertheless, they mostly present spatial-harmonic components around the selected frequency. This is the reason why, despite many authors' claim of the goodness of WT [1, 19], some recent works state the best suitability of WFT [26, 34]. Another point to consider is the resistance to noise. It has been demonstrated [26] that, for noiseless fringe patterns, the frequency components can be accurately recovered in either small or large windows, regardless of the frequency value. However, in the presence of higher noise on the imaged fringe pattern, an optimal selection of the window size is crucial for filtering the noise while preserving the main frequency components.

Under these circumstances, the fixed window size of WFT performs better than the variable window size of WT. This is mainly due to the dyadic net used in practical applications of WT. This net changes the window size for adjacent levels of dilation geometrically (by two) and is excessive for some applications where the main frequency stands close to a fixed value (as in fringe pattern analysis). This is the reason why the WFT technique was selected for the analysis of the recovered pattern.

Another point to consider is the importance of selecting a window with good localization in both frequency and space, in order to perform an optimal analysis of the fringes. This point was also studied in the work of Fernandez et al. [16]. As a result, the Morlet wavelet adapted to WFT presented a better localization in the frequency domain than the others and was thus more suitable for demodulating fringe patterns with slow phase variations and relatively low signal-to-noise ratios. It must be mentioned that the adaptation of the Morlet wavelet to the use in WFT includes a normalization in frequency and size as, to preserve the value of energy provided by the WFT algorithm, a change in the window size must be compensated with an increment of the modulus of the signal. Finally, the ability to adapt the size of the wave envelope relative to the wave period must be considered. In wavelet analysis, this parameter is used to create a set of complex mother wavelets within the same wavelet family. In WFT, this is equivalent to changing the size of the window, as the preset frequency does not change with this size. Further information can be found in Fernandez et al. [16].

5.5.5.1 *Preprocessing the Image*

The preprocessing step consists of salt-and-pepper filtering and histogram equalization. This reduces the noise present in the captured image and enhances the image contrast for a later frequency component extraction, respectively. Finally, a DC filter is applied to extract the DC component of the image. This step delivers an enhanced image where the fringes are perceived more clearly.

5.5.5.2 *Setting the Average Period and the Standard Deviation*

This step represents the main idea of the automatic selection of the window. The algorithm extracts an approximated value of the number of periods existing in every line along the coding axis of the image. To do so, a local maximum extraction is performed for both maximum and the minimum values in every line along the coding axis. The algorithm avoids false positives by suppressing those local maxima that are not followed by a local minimum. Once the number of periods is extracted for every image column, an average of the global period, the corresponding frequency, and its variance are computed. This variance represents the uncertainty in the estimated frequency and is crucial to perform a global analysis of the image.

Regarding this point, a discussion about whether the selection of global or local variance for patches in the image is required needs to take place.

In principle, a local selection seems to be more appropriate, as it can distinguish frequencies of different patches. However, it requires more computation because the WFT must be applied in every patch. Using a global WFT and the appropriate range for the analytic frequencies, a trade-off to delete noisy frequencies and to preserve the ones related to the real shape must be set. This reduces the total number of WFTs to one, thus reducing the computational time.

5.5.5.3 Setting the Range of Frequencies and the Window

The selection of the appropriate range of frequencies is done according to the variance and the average values of the period. For instance, considering the range $[f_m - 3 \cdot std(f), f_m + 3 \cdot std(f)]$ in both x and y axes, the 95% of detected frequencies are analyzed, according to the central limit theorem [53]. The frequencies outbounding this range are considered outliers. In practice, this range can be reduced to $[fm - 2 \cdot std(f), fm + 2 \cdot std(f)]$ (90% of the frequencies are represented) without a significant loss in accuracy.

Another variable to consider is the window size related to the number of periods of the sinusoidal signal. In contrast to the mother wavelets in WT, WFT does not require the number of periods to be linked to the sinusoidal oscillation of the signal. In WT, the number of periods determines a mother wavelet within the same wavelet family and usually goes from one up to three or four periods, allowing information about the frequency to be held without losing local information. In WFT, though, the number of periods can be directly set from the definition of the signal. Our algorithm has tested from one up to three periods, determining the optimal value by the ridge extraction algorithm (WFR).

5.5.5.4 Computing the WFT

Once all the parameters are defined, the set of signals that have different sinusoidal frequencies and windows are convolved with the enhanced image. As a result, a 4D matrix is obtained (having dimensions of x and y axes, window size, and frequency). The WFR algorithm is then applied to compute the most likely values of window (w_x, w_y) and the corresponding phase value, delivering the wrapped phase in the interval $[-\pi, \pi]$.

5.5.6 De Bruijn

The corrected RGB image consists of a set of sinusoidal deformed color fringes coded following a certain De Bruijn structure. The aim of this step is to extract the color lines associated with every deformed color fringe. As mentioned, in the state of the art, multislit and striped 1D De Bruijn-based patterns can be found in the literature. When a multislit pattern is projected and reflected on the scene, the lines get smoothed, resulting in a Gaussian shape for every line in the recovered pattern. This scenario is similar to the one resulting after the projection of sinusoidal fringes, where the fringes are also sinusoidal (Gaussian like) when they are imaged by the camera. Therefore, our recovered pattern can be seen as a slit-based pattern in terms of the decoding process.

Following this idea, the localization of the maxima in every line must be found. This is done with subpixel accuracy, assuring that the total number of maxima found at every column does not exceed the number of fringes present in the pattern. The positions and color values of the maxima are used in the matching algorithm, which solves the correspondences through dynamic programming. Afterward, correspondences in every column are compared with the surrounding columns in order to detect errors in the matching procedure (due to an erroneous color extraction or a wrong matching process). The correction is done assuming color smoothness for single pixels in the image regarding the value of the surrounding pixels.

5.5.7 Combination of De Bruijn Pattern and Wrapped Phase Pattern

Having, on one hand, the wrapped phase of the recovered pattern and on the other hand the correspondences of the De Bruijn lines for every fringe in the pattern, the next step is to merge them into a correspondence map containing all the pixels present in the recovered pattern. This is done assuming that the subpixel accuracy provided by the De Bruijn slit-based decoding algorithm is better than the precision provided by the WFT. Therefore, the maxima of the fringes found in the De Bruijn decoding are used as ground truth where the fringe pattern is fitted in. Errors in the fringe pattern may cause some fringes to be wider or thinner than the distance between the corresponding maxima. This is caused by the 2D nature of the WFT algorithm, which may include some frequencies of adjacent positions in the Fourier transform, leading to an erroneous phase value for that position.

This effect is corrected by shrinking or expanding the wrapped phase according to the De Bruijn correspondences for the maxima. A nonlinear fourth-order regression line is used to this end. This process is done for every column in the image, obtaining the modified wrapped-phase map. Finally, an interpolation is done in the correspondence map in order to find the correspondences of every pixel to its maximum. That is, for every column, the modified wrapped-phase map between two maximum positions goes from $-\pi$ to π; therefore, a direct correlation is set between these values and the position of the projected and the recovered color intensities.

5.5.8 Triangulation

Every pair of (x, y) projector–camera coordinates given by the matching step is an input in the triangulation module, which also makes use of the extrinsic and intrinsic parameters provided by the calibration module. The output is a cloud of points in (x, y, z) representing the shape of the reconstructed object.

5.5.9 Filtering

A post-triangulation filtering step is necessary due to some erroneous match-ings that originate outliers in the 3D cloud of points. Two different filtering steps are applied regarding the constraints given by the statistical 3D posi-tions and the bilaterally based filter:

- *Three-dimensional statistical filtering:* In the 3D space, the outliers are characterized by their extremely different 3D coordinates regarding the surrounding points. Therefore, pixels that have 3D coordinates different from the 95% of the coordinates of all the points are consid-ered for suppression. This is done in two steps for all the points in the 3D cloud. First, the distance to the centroid of the cloud of points is computed for every pixel. Afterward, those pixels with a distance to the centroid greater than two times the standard deviation of the cloud of points are considered outliers.

- *Bilateral filtering:* Still, there can be some misaligned points after applying the statistical filtering. In this case it would be propitious to apply some anisotropic filtering that filters the data while pre-serving the slopes. Of course, 3D points must be modified only when they present isolated coordinates that do not correspond to the shape of a reconstructed 3D object. To this end, an extension to 3D data of the 2D bilateral filter proposed by Tomati and Manduchi [66] was implemented. The bilateral filter is a nonrecursive aniso-tropic filter whose aim is to smooth the cloud of points (up to a given value), while preserving the discontinuities, by means of a nonlinear combination of nearby image values. Equations (5.31) and (5.32) are the distance mask and the height mask for a given set of points X, Y, Z around the selected 3D point; the corresponding filtered height value is computed. Afterward, points with a height value different from the output of more than a given threshold are considered points to be filtered out, and height is substituted by the filtered value.

$$G(x, y) = \exp\left(-\left((x - x_c)^2 + (y - y_c)^2\right)\Big/\left(2 * \sigma_1^2\right)\right) \tag{5.31}$$

$$H(z) = \exp\left(-(z - z_c)^2\Big/\left(2 * \sigma_2^2\right)\right) \tag{5.32}$$

5.5.10 Meshing

Finally, a meshing step is applied to obtain a surface from the 3D cloud of points. To do this, a 2D bidimensional Delaunay meshing algorithm is applied to the 3D coordinates with respect to the camera in order to avoid duplicities in the depth value, as this cannot occur from the camera point of view.

5.6 Results

In the field of structured light, the performance of the different techniques has been compared in many different ways over the years, as can be found in the literature. The main difficulty is to set some fixed parameters to evaluate the goodness of a proposed method with respect to the others. In this chapter we proposed a new technique for one-shot dense structured light. Therefore, a comparison with some other techniques requires compulsory testing not only of the main architecture premises of this proposal (number of shots and density of the reconstruction), but also of some parameters common to all SL approaches. However, it would be too much in terms of time and computational requirements to test every single technique and compare it to the others; therefore, six representative techniques corresponding to the main groups existing in SL were selected. Table 5.2 shows these techniques and briefly explains their characteristics.

Three discrete coding techniques and three continuous coding techniques have been chosen for comparison. It is important to mention that all the methods presented here have been implemented directly from the corresponding papers (original code not available), and the parameters have been set in order to obtain optimal reconstruction results. Among sparse spatial multiplexing, one-axis coding was chosen as it presents an easier decoding algorithm than two-axes coding. Among them, the technique of Monks et al. [41] presents a colored slit, pattern-based technique that provides bigger

TABLE 5.2

Selected Methods with Their Main Attributes

Group		Method	Characteristics	Ref.
DC	Spatial m.	Monks et al.	De Bruijn slits pattern; six hue colors (1 pattern)	41
DC	Time m.	Posdamer et al.	Stripes patterns; 7 bits Gray code (24 patterns)	48
DC	Time m. (PS)	Guhring	Time multiplexing + shifting (16 patterns)	23
CC	Time m. (PS)	Pribanic et al.	Multiple phase shifting (18 patterns)	49
CC	Frequency m.	Li et al.	Sinusoidal pattern, π-phase shifting (two patterns)	36
CC	Spatial m.	Carrihill and Hummel	Grading grayscale pattern (one pattern)	8

vocabulary than grayscale approaches as well as easier detection and matching than stripe patterns techniques.

For sparse time multiplexing, the Posdamer algorithm [48] was selected for being a well-known, effective technique in time multiplexing. Dense time multiplexing using shifting codes was proposed by Sansoni et al. [59] and Guhring [23] to obtain dense reconstruction. Between them, Guhring's method was selected because it uses slit shifting, which is easier to segment than the fringe shifting used by Sansoni et al. Also, the technique presented by Pribanic et al. [49] was selected for being the latest dense time multiplexing technique using multiple phase shifting. In frequency multiplexing, the π-phase-shifting FTP method proposed by Li et al. [36] provides higher resistance to slopes than the traditional FTP of Takeda and Mutoh [64], without the necessity to perform wavelet filtering or having to deal with the blurring associated with nonsinusoidal patterns. Chen et al. [11] and Yue et al. [71] use the same π-phase-shifting FTP, multiplexing the patterns into one single projection.

However, the main idea remains unaltered, and therefore the simpler solution proposed by Li et al. is still a good representative to evaluate the performance of these techniques. The grayscale spatial grading proposed by Carrihill and Hummel [8] was chosen against the rainbow pattern implemented by Tajima and Iwakawa [63], which employs a nematic liquid crystal. Finally, the proposed one-shot dense reconstruction algorithm of this chapter was implemented and compared to these techniques.

The setup used for the tests was composed of an LCD video projector (Epson EMP-400W) with a resolution of 1024×768 pixels, a camera (Sony 3 CCD), and a frame grabber (Matrox Meteor-II) digitizing images at 768×576 pixels with 3×8 bits per pixel (RGB). Both camera and video projector were calibrated using the projector camera calibration method developed by Fernandez et al. [55]. The baseline between camera and projector was about 1 m. The results and time estimates were computed using a standard Intel Core2 Duo CPU at 3.00 GHz and 4 GB RAM memory. The algorithms were programmed and run in MATLAB® 7.3.

5.6.1 Quantitative Results

Quantitative results have been analyzed reconstructing a white plane at a distance of about 80 cm from the camera. Principal component analysis (PCA) was applied to obtain the equation of the 3D plane for every technique and for every reconstruction. This technique is used to span the 3D cloud of points onto a 2D plane defined by the two eigenvectors corresponding to the two largest eigenvalues. The results of the experiment are shown in Table 5.3. Observe that the algorithm of Li et al. [36] is conceived to measure deviation of smooth surfaces with respect to the reference plane; therefore, a plane is not conceived to be reconstructed by depth deviation.

Among the techniques obtaining sparse reconstruction, the De Bruijn one-shot projection algorithm developed by Monks et al. [41] presents the

TABLE 5.3

Quantitative Results

Technique	Average (mm)[a]	Standard (mm)[b]	3D Points[c]	Patterns[d]	Ref.
Monks et al.	1.31	1.19	13,899	1	41
Posdamer et al.	1.56	1.40	25,387	14	48
Guhring	1.52	1.33	315,273	24	23
Pribanic et al.	1.12	0.78	255,572	18	49
Li et al.	—	—	—	1	36
Carrihill and Hummel	11.9	5.02	202,714	1	8
Proposed technique	1.18	1.44	357,200	1	

[a] Average deviation of the reconstructing error.
[b] Standard deviation of the reconstructing error.
[c] Number of 3D points reconstructed.
[d] Number of projected patterns.

best results, in terms of average error and standard deviation, against the traditional time multiplexing represented by Posdamer and Altschuler [48]. Dense reconstruction techniques can be divided into one-shot and multiple pattern projection techniques. Among one-shot techniques, the proposed technique defeats the other implemented technique, based on spatial grading. The technique proposed by Carrihill and Hummel [8] obtains the poorest results due to the low variance existing between adjacent pixels in the projected pattern. In contrast, Fourier analysis represented by the proposed technique presents a lower error rate, thanks to the frequency filtering process that is performed in the analysis.

Among multiple pattern projection techniques, the method developed by Pribanic et al. [49] gives the best results in terms of sensitivity to noise, as can be extracted from the values of average error and standard deviation. Regarding the computing time, it can be observed that methods obtaining dense reconstructions (Guhring, Pribanic et al., Li et al., Carrihill and Hummel, and the proposed algorithm) need to compute more 3D points, requiring higher computational time. However, our proposal does not need to compute many images and no unwrapping algorithm is required. This makes our technique faster in terms of computational time. Among methods providing sparse reconstruction, the color calibration step makes Monks and colleagues' algorithm slower than that of Posdamer and Altschuler (also affecting the proposed technique) even though it preserves the same order of magnitude. Still, real-time response is achievable working with the appropriate programming language and firmware.

5.6.2 Qualitative Results

The reconstruction of a real object permits one to analyze the performance of the programmed techniques in terms of accuracy and noise sensitivity. The reconstructed object used to perform the qualitative analysis of the results is a ceramic figure placed at a distance of about 80 cm from the camera. In order to show the results, both 3D cloud of points and surfaces are used. The surface has been generated performing a 2D Delaunay triangulation over (x, y) coordinates.

As can be observed in Figures 5.10, 5.11, 5.12, and 5.13, the best results are obtained with time multiplexing shifting approaches (Guhring [23] and Pribanic et al. [49]). These techniques obtain the best accuracy results and also provide dense reconstruction. Furthermore, both algorithms perform well in the presence of surface slopes, as can be observed in some of the details of the reconstructed object (see, for instance, the ears of the horse). However, more than one projection is necessary to reconstruct the object and this makes them unable to cope with moving scenarios. This is also the case of Posdamer and Altschuler [48], which suffers from some noise in the recovered cloud of points caused by nonlinearities of the camera, which produces some leakage from white to black fringes that can lead to some errors in the position of the recovered edges.

Among one-shot techniques, De Bruijn-based coding presents the best results in terms of accuracy. This is the case of the Monks et al. algorithm [41], which employs De Bruijn color coding to obtain a dynamic sparse reconstruction. Another approach, proposed by Li et al. [36], employs frequency multiplexing (π-phase shifting). This also provides one-shot dense reconstruction. However, high frequencies are lost in the filtering step, causing the loss of some information in the surface details. Moreover, traditional frequency multiplexing approaches can work only on smooth surfaces with slopes not exceeding three times the value given in Equation (5.16).

It is important to mention that the method chosen for phase unwrapping employs a qualitative map to determine the region where the unwrapping should start. Our proposal, also based on frequency analysis, combines it with De Bruijn coding to provide the best performance, in terms of accuracy density of the reconstruction, for one-shot techniques. It obtains results similar to those of Monks et al. [41], but dense reconstruction is achieved. This provides a final 3D shape where details appear much better defined. Moreover, it is robust against slopes in the shape, which is not the case for other frequency-based approaches. Finally, the grading technique proposed by Carrihill and Hummel [8] showed high sensitivity to noise and low sensitivity to changes in depth, caused by the low range existing between adjacent pixels.

5.7 Conclusion

In this chapter, an up-to-date review and a new classification of the different techniques existing in structured light have been proposed, based on the

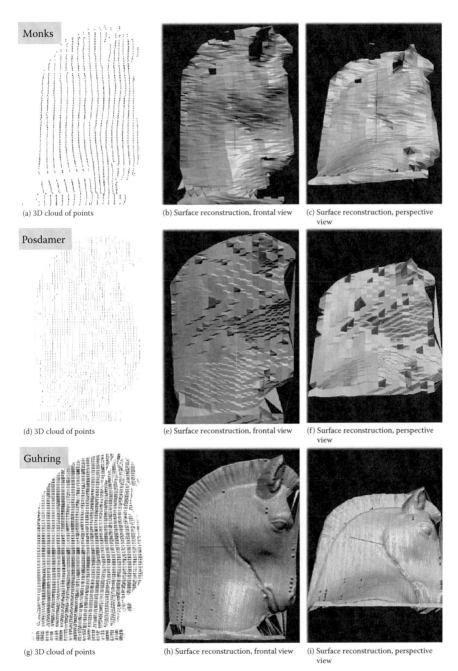

FIGURE 5.10
Results of Monks et al. [41], Posdamer and Altschuler [48], and Guhring [23], respectively. (Monks, T. P. et al. *IEEE 4th International Conference on Image Processing,* 327–330, 1992; Posdamer, J. L. and M. D. Altschuler. *Computer Graphics and Image Processing* 18 (1): 1–17, 1982; Guhring, J. *Videometrics and Optical Methods for 3D Shape Measurement* 4309:220–231, 2001.)

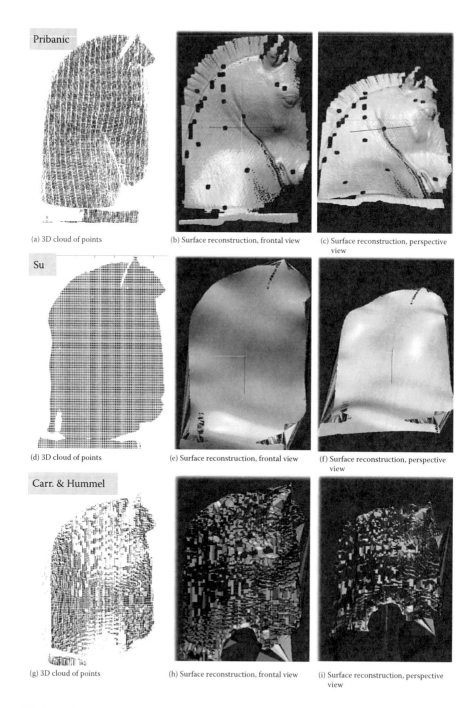

FIGURE 5.11
Results of Pribanic et al. [49], Su et al. [36], and Carrihill and Hummel [8], respectively.

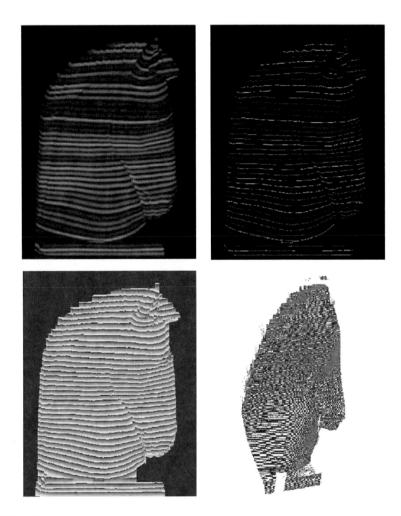

FIGURE 5.12
(See color insert.) Proposed algorithm: input image (top left), extracted color slits (top right), combined slits and WFT fringes (bottom left), and 3D cloud of points (bottom right).

survey of Salvi et al. [57]. The classification was done regarding the sparse or dense 3D reconstruction of the imaged scene. A subclassification regarding the spatial, frequency, or time multiplexing strategies was done. Moreover, a new proposal for one-shot dense 3D reconstruction has been presented that combines the accuracy of De Bruijn spatial multiplexing with the density of reconstruction obtained using frequency multiplexing in fringe projection.

This proposal was implemented jointly with some representative techniques of every group in the classification. Both quantitative and qualitative comparisons were performed, extracting advantages and drawbacks of each technique. The results show that the best results are obtained with time multiplexing shifting approaches, which obtain dense reconstruction

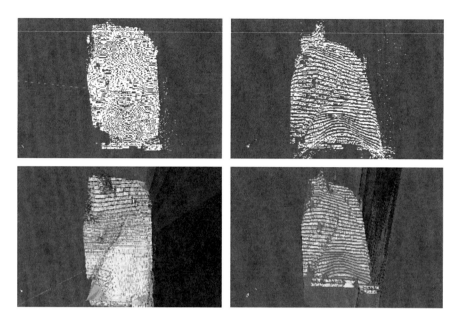

FIGURE 5.13
Different views of the 3D reconstruction using the proposed algorithm. 3D cloud of points (upper row) and 3D mesh (lower row).

and excellent accuracy. However, they are only valid for static scenarios. Among one-shot techniques, our proposed method achieves the best results in terms of accuracy, comparable with other De Bruijn-based spatial coding. Moreover, our proposal achieves dense reconstruction and absolute coding. Additionally, other frequency multiplexing methods provide dense reconstruction for moving scenarios, but present high sensitivity to nonlinearities of the camera, reducing the accuracy and sensitivity to details in the surface, and they can fail in the presence of big slopes.

Among spatial multiplexing approaches able to work in moving scenarios, the use of De Bruijn codes gives good accuracy in the reconstruction, but at the expense of having discrete reconstruction and high sensitivity to changes in color surface or background illumination. Regarding this point, it is important to mention that the background illumination is filtered in the proposed technique by the frequency analysis step.

Summarizing the main contributions done in structured light in the last years, it is important to mention that most of the work has been concerned with frequency multiplexing approaches, trying to increase the robustness in the decoding step and the resistance to slopes under the constraint of moving scenarios [11, 19, 71, 73]. Time multiplexing in phase shifting has arisen also to overcome the problem of slopes in the objects [49]. Hybrid techniques have been the main contribution in both time multiplexing and spatial multiplexing approaches [15, 17, 28, 32, 45], preserving the principles proposed in previous work. Under this scenario, the proposal made in this chapter of merging De Bruijn and frequency-based one-shot patterns achieves a dense reconstruction, with the robustness in the decoding step provided by frequency analysis jointly with the accuracy given by spatial De Bruijn-based patterns. This combination gives us a one-shot absolute dense pattern with the highest accuracy achievable for moving scenarios.

Acknowledgments

This work has been partly supported by the FP7-ICT-2011-7 project PANDORA—Persistent Autonomy through Learning, Adaptation, Observation and Replanning (Ref 288273)—funded by the European Commission and the project RAIMON—Autonomous Underwater Robot for Marine Fish Farms Inspection and Monitoring (Ref CTM2011-29691-C02-02)—funded by the Spanish Ministry of Science and Innovation. S. Fernandez is supported by the Spanish government scholarship FPU.

References

1. A. Z. A. Abid. Fringe pattern analysis using wavelet transforms. PhD thesis, General Engineering Research Institute (GERI). Liverpool John Moores University, Liverpool, UK, 2008.
2. C. Albitar, P. Graebling, and C. Doignon. Design of a monochromatic pattern for a robust structured light coding. *IEEE International Conference Image Processing ICIP* 6: 529–532, 2007.
3. A. Baldi, F. Bertolino, and F. Ginesu. On the performance of some unwrapping algorithms. *Optics and Lasers in Engineering* 37 (4): 313–330, 2002.
4. J. Batlle, E. Mouaddib, and J. Salvi. Recent progress in coded structured light as a technique to solve the correspondence problem: A survey. *Pattern Recognition* 31 (7): 963–982, 1998.
5. F. Berryman, P. Pynsent, and J. Cubillo. A theoretical comparison of three fringe analysis methods for determining the three-dimensional shape of an object in the presence of noise. *Optics and Lasers in Engineering* 39 (1): 35–50, 2003.
6. F. Berryman, P. Pynsent, J. Fairbank, and S. Disney. A new system for measuring three-dimensional back shape in scoliosis. *European Spine Journal* 17 (5): 663–672, 2008.
7. K. L. Boyer and A. C. Kak. Color-encoded structured light for rapid active ranging. *IEEE Transactions on Pattern Analysis and Machine Intelligence* 9 (1): 14–28, 1987.
8. B. Carrihill and R. Hummel. Experiments with the intensity ratio depth sensor. *Computer Vision, Graphics, and Image Processing* 32 (3): 337–358, 1985.

9. D. Caspi, N. Kiryati, and J. Shamir. Range imaging with adaptive color structured light. *IEEE Transactions on Pattern Analysis and Machine Intelligence* 20 (5): 470–480, 1998.
10. C. S. Chen, Y. P. Hung, C. C. Chiang, and J. L. Wu. Range data acquisition using color structured lighting and stereo vision. *Image and Vision Computing* 15 (6): 445–456, 1997.
11. W. Chen, P. Bu, S. Zheng, and X. Su. Study on Fourier transforms profilometry based on bi-color projecting. *Optics and Laser Technology* 39 (4): 821–827, 2007.
12. W. Chen, X. Su, Y. Cao, Q. Zhang, and L. Xiang. Method for eliminating zero spectrum in Fourier transform profilometry. *Optics and Lasers in Engineering* 43 (11): 1267–1276, 2005.
13. P. J. Cobelli, A. Maurel, V. Pagneux, and P. Petitjeans. Global measurement of water waves by Fourier transform profilometry. *Experiments in Fluids* 46 (6): 1037–1047, 2009.
14. T. Etzion. Constructions for perfect maps and pseudorandom arrays. *IEEE Transactions on Information Theory* 34(5, Part 1): 1308–1316, 1988.
15. P. Fechteler and P. Eisert. Adaptive color classification for structured light systems. *IEEE Computer Society Conference on Computer Vision and Pattern Recognition Workshops* 1–7, 2008.
16. S. Fernandez, M. A. Gdeisat, J. Salvi, and D. Burton. Automatic window size selection in windowed Fourier transform for 3D reconstruction using adapted mother wavelets. *Optics Communications* 284 (12): 2797–2807, 2011.
17. F. Forster. A high-resolution and high accuracy real-time 3D sensor based on structured light. *Proceedings 3rd International Symposium on 3D Data Processing, Visualization, and Transmission* 208–215, 2006.
18. H. Fredricksen. A survey of full length nonlinear shift register cycle algorithms. *SIAM Review* 195–221, 1982.
19. M. A. Gdeisat, D. R. Burton, and M. J. Lalor. Eliminating the zero spectrum in Fourier transform profilometry using a two-dimensional continuous wavelet transform. *Optics Communications* 266 (2): 482–489, 2006.
20. S. S. Gorthi and K. R. Lolla. A new approach for simple and rapid shape measurement of objects with surface discontinuities. *Proceedings SPIE* 5856: 184–194, 2005.
21. P. M. Griffin, L. S. Narasimhan, and S. R. Yee. Generation of uniquely encoded light patterns for range data acquisition. *Pattern Recognition* 25 (6): 609–616, 1992.
22. C. Guan, L. Hassebrook, and D. Lau. Composite structured light pattern for three-dimensional video. *Optics Express* 11 (5): 406–417, 2003.
23. J. Gühring. Dense 3-D surface acquisition by structured light using off-the-shelf components. *Videometrics and Optical Methods for 3D Shape Measurement* 4309: 220–231, 2001.
24. V. I. Gushov and Y. N. Solodkin. Automatic processing of fringe patterns in integer interferometers. *Optics and Lasers in Engineering* 14 (4–5): 311–324, 1991.
25. E. Hu and Y. He. Surface profile measurement of moving objects by using an improved π-phase-shifting Fourier transform profilometry. *Optics and Lasers in Engineering* 47 (1): 57–61, 2009.
26. L. Huang, Q. Kemao, B. Pan, and A. K. Asundi. Comparison of Fourier transform, windowed Fourier transform, and wavelet transform methods for phase extraction from a single fringe pattern in fringe projection profilometry. *Optics and Lasers in Engineering* 48 (2): 141–148, 2010.

27. P. S. Huang, S. Zhang, F. P. Chiang, et al. Trapezoidal phase-shifting method for three-dimensional shape measurement. *Optical Engineering* 44:142–152, 2005.

28. I. Ishii, K. Yamamoto, K. Doi, and T. Tsuji. High-speed 3D image acquisition using coded structured light projection. *IEEE/RSJ International Conference on Intelligent Robots and Systems (IROS)*, 925–930, 2007.

29. M. Ito and A. Ishii. A three-level checkerboard pattern (TCP) projection method for curved surface measurement. *Pattern Recognition* 28 (1): 27–40, 1995.

30. P. Jia, J. Kofman, and C. English. Two-step triangular-pattern phase-shifting method for three-dimensional object-shape measurement. *Optical Engineering* 46:083201, 2007.

31. T. R. Judge and P. J. Bryanston-Cross. A review of phase unwrapping techniques in fringe analysis. *Optics and Lasers in Engineering* 21 (4): 199–240, 1994.

32. H. Kawasaki, R. Furukawa, R. Sagawa, and Y. Yagi. Dynamic scene shape reconstruction using a single structured light pattern. *IEEE Conference on Computer Vision and Pattern Recognition, CVPR* 1–8, 2008.

33. Q. Kemao. Windowed Fourier transform for fringe pattern analysis. *Applied Optics* 43 (17): 3472–3473, 2004.

34. Q. Kemao. Two-dimensional windowed Fourier transform for fringe pattern analysis: Principles, applications and implementations. *Optics and Lasers in Engineering* 45 (2): 304–317, 2007.

35. T. P. Koninckx and L. Van Gool. Real-time range acquisition by adaptive structured light. *IEEE Transactions on Pattern Analysis and Machine Intelligence* 28 (3): 432–445, 2006.

36. J. Li, X. Su, and L. Guo. Improved Fourier transform profilometry for the automatic measurement of three-dimensional object shapes. *Optical Engineering* 29 (12): 1439–1444, 1990.

37. J. F. Lin and X. Su. Two-dimensional Fourier transform profilometry for the automatic measurement of three-dimensional object shapes. *Optical Engineering*, 34: 3297–3297, 1995.

38. F. J MacWilliams and N. J. A. Sloane. Pseudo-random sequences and arrays. *Proceedings of the IEEE* 64 (12): 1715–1729, 1976.

39. M. Maruyama and S. Abe. Range sensing by projecting multiple slits with random cuts. *IEEE Transactions on Pattern Analysis and Machine Intelligence* 15 (6): 647–651, 1993.

40. M. Minou, T. Kanade, and T. Sakai. A method of time-coded parallel planes of light for depth measurement. *Transactions of IECE Japan* 64 (8): 521–528, 1981.

41. T. P. Monks, J. N. Carter, and C. H. Shadle. Color-encoded structured light for digitization of real-time 3D data. *IEEE 4th International Conference on Image Processing*, 327–330, 1992.

42. R. A. Morano, C. Ozturk, R. Conn, S. Dubin, S. Zietz, and J. Nissano. Structured light using pseudorandom codes. *IEEE Transactions on Pattern Analysis and Machine Intelligence* 20 (3): 322–327, 1998.

43. H. Morita, K. Yajima, and S. Sakata. Reconstruction of surfaces of 3-D objects by M-array pattern projection method. *Second International Conference on Computer Vision* 468–473, 1988.

44. N. Ono, T. Shimizu, T. Kurihara, and S. Ando. Real-time 3-D imager based on spatiotemporal phase unwrapping. *SICE 2004 Annual Conference* 3: 2544–2547, 2004.

45. J. Pages, C. Collewet, F. Chaumette, J. Salvi, S. Girona, and F. Rennes. An approach to visual serving based on coded light. *IEEE International Conference on Robotics and Automation, ICRA* 6: 4118–4123, 2006.

46. J. Pages, J. Salvi, C. Collewet, and J. Forest. Optimized De Bruijn patterns for one-shot shape acquisition. *Image Vision and Computing* 23: 707–720, 2005.

47. J. Pages, J. Salvi, and J. Forest. A new optimized De Bruijn coding strategy for structured light patterns. *17th International Conference on Pattern Recognition, ICPR 2004* 4: 284–287, 2004.

48. J. L. Posdamer and M. D. Altschuler. Surface measurement by space-encoded projected beam systems. *Computer Graphics and Image Processing* 18 (1): 1–17, 1982.

49. T. Pribanic, H. Dapo, and J. Salvi. Efficient and low-cost 3D structured light system based on a modified number-theoretic approach. *EURASIP Journal on Advances in Signal Processing* 2010: article ID 474389, 11 pp., 2009.

50. J. G. Proakis, D. G. Manolakis, D. G. Manolakis, and J. G. Proakis. *Digital signal processing: Principles, algorithms, and applications,* vol. 3. Englewood Cliffs, NJ: Prentice Hall, 1996.

51. J. Quintana, R. Garcia, and L. Neumann. A novel method for color correction in epiluminescence microscopy. *Computerized Medical Imaging and Graphics,* 35(7–8): 646–652, 2011.

52. P. Ribenboim. *Algebraic numbers,* ed. R. Courant, L. Bers, and J. J. Stoker. New York: John Wiley & Sons, 1972.

53. J. A. Rice. *Mathematical statistics and data analysis.* Belmont, CA: Duxbury Press, 1995.

54. J. Salvi, S. Fernandez, and T. Pribanic. Absolute phase mapping for one-shot dense pattern projection. *IEEE Workshop on Projector–Camera Systems* (in conjunction with IEEE International Conference on Computer Vision and Pattern Recognition) 64–71, 2010.

55. J. Salvi, S. Fernandez, D. Fofi, and J. Batlle. Projector–camera calibration using a planar-based model. Submitted to *Electronics Letters*, 2011.

56. J. Salvi, J. Batlle, and E. Mouaddib. A robust-coded pattern projection for dynamic 3D scene measurement. *Pattern Recognition Letters* 19 (11): 1055–1065, 1998.

57. J. Salvi, S. Fernandez, T. Pribanic, and X. Llado. A state of the art in structured light patterns for surface profilometry. *Pattern Recognition* 43 (8): 2666–2680, 2010.

58. J. Salvi, J. Pages, and J. Batlle. Pattern codification strategies in structured light systems. *Pattern Recognition* 37 (4): 827–849, 2004.

59. G. Sansoni, M. Carocci, and R. Rodella. Calibration and performance evaluation of a 3-D imaging sensor-based on the projection of structured light. *IEEE Transactions on Instrumentation and Measurement* 49 (3): 628–636, 2000.

60. G. S. Spagnolo, G. Guattari, C. Sapia, D. Ambrosini, D. Paoletti, and G. Accardo. Contouring of artwork surface by fringe projection and FFT analysis. *Optics and Lasers in Engineering* 33 (2): 141–156, 2000.

61. V. Srinivasan, H. C. Liu, and M. Halious. Automated phase-measuring profilometry: A phase mapping approach. *Applied Optics* 24: 185–188, 1985.

62. X. Su and W. Chen. Fourier transform profilometry: A review. *Optics and Lasers in Engineering* 35 (5): 263–284, 2001.

63. J. Tajima and M. Iwakawa. 3-D data acquisition by rainbow range finder. *Pattern Recognition, 1990. Proceedings 10th International Conference* 1: 309–313, 1990.

64. M. T. K. Takeda and M. Mutoh. Fourier transform profilometry for the automatic measurement of 3-D object shapes. *Applied Optics* 22: 3977–3982, 1983.

65. M. A. Tehrani, A. Saghaeian, and O. R. Mohajerani. A new approach to 3D modeling using structured light pattern. *Information and Communication Technologies: From Theory to Applications, 2008. ICTTA 2008* 1–5, 2008.

66. C. Tomasi and R. Manduchi. Bilateral filtering for gray and color images. *Sixth International Conference on Computer Vision, 1998* 839–846.

67. Y. Iwaasa and S. Toyooka. Automatic prolometry of 3-D diffuse objects by spatial phase detection. *Applied Optics* 25 (10): 1630–1633, 1986.

68. J. Vanherzeele, P. Guillaume, and S. Vanlanduit. Fourier fringe processing using a regressive Fourier-transform technique. *Optics and Lasers in Engineering* 43 (6): 645–658, 2005.

69. L. S. Wu and Q. Peng. Research and development of fringe projection-based methods in 3D shape reconstruction. *Journal of Zhejiang University-Science A* 7 (6): 1026–1036, 2006.

70. C. Wust and D. W. Capson. Surface profile measurement using color fringe projection. *Machine Vision and Applications* 4 (3): 193–203, 1991.

71. H. M. Yue, X. Y. Su, and Y. Z. Liu. Fourier transform profilometry based on composite structured light pattern. *Optics and Laser Technology* 39 (6): 1170–1175, 2007.

72. L. Zhang, B. Curless, and S. M. Seitz. Rapid shape acquisition using color structured light and multi-pass dynamic programming. *Proceedings of the 1st International Symposium on 3D Data Processing, Visualization, and Transmission (3DPVT)* Padova, Italy, June 19–21, 2002, pages 24–36, 2002.

73. Q. Zhang, W. Chen, and Y. Tang. Method of choosing the adaptive level of discrete wavelet decomposition to eliminate zero component. *Optics Communications* 282 (5): 778–785, 2008.

6

Digital Holography for 3D Metrology

Anand Asundi, Qu Weijuan, Chee Oi Choo, Kapil Dev, and Yan Hao

CONTENTS

6.1 Introduction to Digital Holography

6.1.1 Holography and Digital Holography

Holography is a well established technique for three-dimensional (3D) imaging. It was first introduced by D. Gabor in 1948 [2] in order to improve the resolution of the electron microscope. Light scattered from an object is recorded with a reference beam in a photographic plate to form a hologram. Later, the processed photographic plate is illuminated with a wave identical to the reference wave, and a wave front is created that is identical to that originally produced by the object. As the development of laser, holography is applied in optical metrology widely. The phase of the test object wave could be reconstructed optically but not measured directly until the introduction of digital recording devices for the hologram recording. The digital recorded hologram can be stored in the computer and the numerical reconstruction method used to extract both the amplitude and phase information of the object wave front. By digital holography, the intensity and the phase of electromagnetic wave fields can be measured, stored, transmitted, applied to simulations, and manipulated in the computer.

For a long time, it has been hard to extract the phase from a single off-axis hologram because the phase of the test object is overlapped with the phase of the illuminating wave and the off-axis tilt. Off-axis digital holographic interferometry (double exposure) can easily remove the phase of the illuminating wave and the off-axis tilt by subtraction of the holograms with and without the test object to provide a correct phase measurement. Since the mid-1990s, it has been extended, improved, and applied to many measurement tasks, such as deformation analysis, shape measurement, particle tracking, and refractive index distribution measurement [3] within transparent media due to temperature or concentration variations and bio-imaging application. The drawback of this method is the necessity of recording multiple holograms, which is not suitable in real-time dynamic phase monitoring. In 1999, Cuche, Bevilacqua, and Depeursinge proposed that not only amplitude but also phase can be extracted from a single digital hologram [4]. The introduction of the concept of a digital reference wave that compensates the role of the reference wave in off-axis geometry has successfully removed the off-axis tilt of the optical system. Since then, the numerical reconstruction method for the digital hologram has been thoroughly developed [5–7].

The limited sampling capacity of the electronic camera gives impetus to finding different approaches to achieve microscopic imaging with digital holography and thus revives digital holographic microscopy (DHM) [8]. DHM, by introduction of a microscope objective (MO), gives very satisfactory measurement results in lateral resolution and in vertical resolution. Nevertheless, MO only in the object path introduces a phase difference between the object wave and reference wave. The phase difference will render the phase measurement a failure if a powerful numerical phase compensation procedure is lacking [9, 10].

Numerical phase compensation is based on the computation of a phase mask to be multiplied by the recorded hologram or by the reconstructed wave field in the image plane. In the early days, the phase mask was computed depending on the precisely measured parameters of the optical setup. It was then multiplied by the reconstructed wave front. The correct phase map was obtained by a time-consuming digital iterative adjustment of these parameters to remove the wave front aberration [11–13]. To exclude the necessity of physically measuring the optical setup parameters, a phase mask is

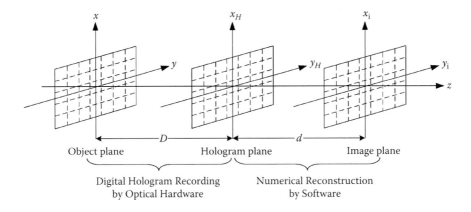

FIGURE 6.1
Digital hologram recording and numerical reconstruction.

computed in a flat portion of the hologram, where the specimen is flat and used for the aberration compensation. This was proposed by Ferraro et al. [14] and then developed by Montfort et al. [15] using a polynomial fitting procedure to adjust the parameters for one dimension at first and then for two dimensions. Numerical compensation, however, makes the reconstruction algorithm complex by the iterative adjustment and extrapolation of the fitted polynomials in different areas. For the special case of the microlens shape (spherical shape), this may result in false compensation [16]. In the case of the compensation in the reconstruction plane, the phase mask has to be adapted when the reconstruction distance is changed.

It is also possible to compensate the phase difference between the object wave and reference wave physically. There are examples, such as the Linnik interferometer [17] and Mach–Zehnder interferometer [18, 19], that can physically solve the problem by introducing the same curvature in the reference wave. Previously, it was difficult to align all the elements precisely to acquire the correct compensation. Now it is easy with the help of real-time phase monitoring software. With the help of the software, we also find that when the two phases between the object wave and the reference wave are matched, the fringe pattern of the hologram is a straight parallel line. A lens or a diverging spherical wave can be used in the reference wave to achieve a quasi-spherical phase compensation [16, 20]. Phase aberration-free DHM can be achieved by a common-path DHM system based on a single cube beam splitter (SCBS) interferometer [21] using an MO [22] or diverging spherical wave [23] to provide magnification to the test specimen. Since the object beam and reference beam share the same optical path, the wave front curvatures can physically compensate each other during the interference.

Digital holography is recording a digitized hologram by using an electronic device (e.g., a charge coupled device [CCD]) and later numerical reconstruction with a computer; both the amplitude and the phase of an optical wave arrive from a coherently illuminated object. Thus, both the hardware and software are needed in digital holography. As shown in Figure 6.1, there are three planes: object plane, hologram plane, and image plane. From the object plane to the hologram plane, the recording of a digital hologram is done by the optical interferometer—namely, optical hardware of digital holography. The CCD or complementary metal oxide semiconductor (CMOS) arrays are used to acquire the hologram and store it as a discrete digital array. From the hologram plane to the image plane, the numerical reconstruction is done by the computer software with a certain algorithm to calculate the diffraction propagation of the waves. In this section, DH from the object plane to the hologram plane—namely, the digital hologram recording process—will be introduced in detail.

As is well known, each optical wave field consists of an amplitude distribution as well as a phase distribution, but all detectors can register only intensities. Consequently, the phase of the wave is missing in the recording process. Interference between two waves can form certain patterns. The interference pattern can be recorded. Its intensity is modulated by phases of the involved interference wave fronts. Interference is the way of phase recording. In conventional holography, the interference pattern is recorded by a photographic plate. In digital holography, the interference pattern is recorded by a digital device. Different from the optical reconstruction of the processed photographic plate, numerical reconstruction will be performed in digital holography. In digital hologram recording, an illuminating wave front modulated by an unknown wave front coming from the object, called object wave O, is added to the reference wave R to give an intensity modulated by their phases. The intensity $I_H(x,y)$ of the sum of two complex fields can be written as

$$I_H(x,y) = |O+R|^2 = |O|^2 + |R|^2 + RO^* + R^*O \tag{6.1}$$

where RO^* and R^*O are the interference terms with R^* and O^* denoting the complex conjugate of the two waves. Thus, if

$$O(x,y) = A_O \exp\left[-j\varphi_O(x,y)\right] \tag{6.2}$$

$$R(x,y) = A_R \exp\left[-j\varphi_R(x,y)\right] \tag{6.3}$$

where A_O and A_R are the constant amplitude of the object wave and reference wave, respectively; $\phi_O(x, y)$ and $\phi_R(x, y)$ are the phase of the object wave and reference wave, respectively. The intensity of the sum is given by

$$I_H(x,y) = A_O^2 + A_R^2 + A_O A_R \cos\left[\varphi_R(x,y) - \varphi_O(x,y)\right] \tag{6.4}$$

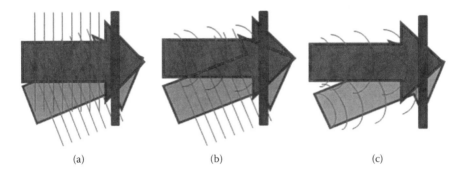

FIGURE 6.2
Interference between two waves: (a) two monochromatic plane waves; (b) one monochromatic diverging spherical wave and one monochromatic plane wave; (c) two monochromatic diverging spherical waves.

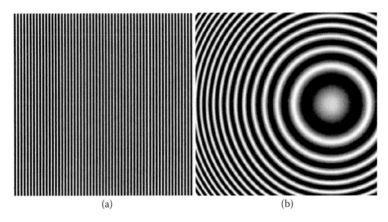

FIGURE 6.3
Interference pattern from two waves: (a) two monochromatic plane waves; (b) one monochromatic diverging spherical wave and one monochromatic plane wave.

It is obvious that the intensity depends on the relative phases of the two involved interference waves. In order to get sufficient information for the reconstruction of the phase of the test object, one needs to specify the detailed characters of the involved waves.

The wave fronts involved in an interferometer depend on the configurations of the optical interferometer. If two monochromatic plane waves interfere with each other, as shown in Figure 6.2(a), the interference pattern will always be straight lines, as shown in Figure 6.3(a). For digital holographic microscopy systems, as an imaging lens combination (condenser lens, MO, and tube lens) is used in the object beam path, the output wave front is a spherical one. If in the reference beam path a plane wave is used, as shown in Figure 6.2(b), the interference pattern will be closed or unclosed circular fringes, as shown in Figure 6.3(b). The spherical phase from the object beam

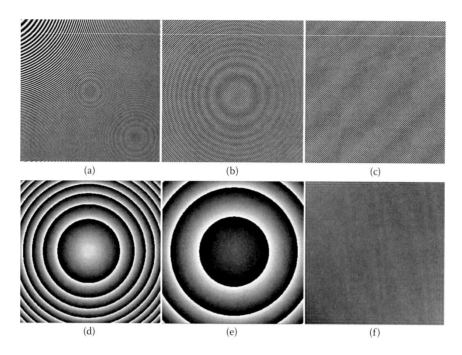

(a)   (b)   (c)

(d)   (e)   (f)

FIGURE 6.4
System phase of holograms in different fringe patterns: (a, b) hologram with circular fringe pattern; (c) hologram with straight fringe pattern; (d) system phase of hologram in (a); (e) system phase of hologram in (b); (f) system phase of hologram in (c).

path plays the important role in the phase modulation of the intensity. If the reference beam path uses a spherical wave, the interference pattern can be straight lines or circular lines depending on the relative positions of the two point sources of the diverging spherical waves. Detailed illustrations are given next.

If two diverging spherical waves are involved in the interference, the fringe patterns of the off-axis digital hologram are shown in Figure 6.4(a)–(c). In fact, the fringe of the interference pattern is decided by the final phase of the system. In order to illustrate this more clearly, the hologram is reconstructed to give the system phase. From the holograms shown in Figure 6.4(a)–(c), the reconstruction of the system phase by using a plane reconstruction reference wave normal to the hologram plane is shown in Figure 6.4(d)–(f). This confirms that different recording conditions give different system phases. In Figure 6.4(d)–(f), the spherical phase curvatures of the system phases are decreasing. In Figure 6.4(f), the system phase is constantly flat. In such a case, there will be no other phase introduced to the phase of the test object if it is involved. This is a very special case in the hologram recording process.

A specific theoretical analysis of the involved spherical wave fronts' interference in DHM is given as follows. We assumed that the reference wave is generated by a point source located at coordinates $(S_{Rx}, S_{Ry}, (z_R^2 - S_{Rx}^2 - S_{Ry}^2)^{1/2})$ and that the illuminating wave is generated by a point source located at coordinates $(S_{Ox}, S_{Oy}, (z_O^2 - S_{Ox}^2 - S_{Oy}^2)^{1/2})$. z_R and z_O are, respectively, the distance between the source points of the reference and illuminating waves and the hologram plane. Using quadratic-phase approximations to the spherical waves involved, the reference wave front in the hologram plane is thus given by

$$R(x, y) = \exp\left\{-j\frac{\pi}{\lambda z_R}\left[(x - S_{Rx})^2 + (y - S_{Ry})^2\right]\right\} \tag{6.5}$$

The illuminating wave is modulated by the phase of the object. In the hologram plane it is given by

$$O(x, y) = A_O \exp\left\{-j\frac{\pi}{\lambda z_O}\left[(x - S_{Ox})^2 + (y - S_{Oy})^2\right]\right\}\exp\left[j\varphi(x,y)\right] \tag{6.6}$$

where A_O is the unit amplitude and $\varphi(x,y)$ is the phase introduced by the test object. The corresponding intensity distribution in the pattern of the interference between the two waves is

$$I_H(x, y) = 1 + |A_O|^2$$

$$+ A_O \exp\left[-j\frac{\pi}{\lambda}\left(\frac{S_{Rx}^2}{z_R} - \frac{S_{Ox}^2}{z_O} + \frac{S_{Ry}^2}{z_R} - \frac{S_{Oy}^2}{z_O}\right)\right]$$

$$\times \exp\left[-j\frac{\pi}{\lambda}\left(\frac{1}{z_R} - \frac{1}{z_O}\right)(x^2 + y^2) + j\frac{2\pi}{\lambda}\left(\frac{S_{Rx}}{z_R} - \frac{S_{Ox}}{z_O}\right)x\right]$$

$$+ j\frac{2\pi}{\lambda}\left(\frac{S_{Ry}}{z_R} - \frac{S_{Oy}}{z_O}\right)y\right]\exp\left[-j\varphi(x,y)\right] \tag{6.7}$$

$$+ A_O \exp\left[j\frac{\pi}{\lambda}\left(\frac{S_{Rx}^2}{z_R} - \frac{S_{Ox}^2}{z_O} + \frac{S_{Ry}^2}{z_R} - \frac{S_{Oy}^2}{z_O}\right)\right]$$

$$\times \exp\left[j\frac{\pi}{\lambda}\left(\frac{1}{z_R} - \frac{1}{z_O}\right)(x^2 + y^2) - j\frac{2\pi}{\lambda}\left(\frac{S_{Rx}}{z_R} - \frac{S_{Ox}}{z_O}\right)x\right]$$

$$- j\frac{2\pi}{\lambda}\left(\frac{S_{Ry}}{z_R} - \frac{S_{Oy}}{z_O}\right)y\right]\exp\left[j\varphi(x,y)\right]$$

The interference term of interest includes combinations of a spherical wave front, tilt in the x direction, tilt in the y direction, and a constant phase. One may not be able clearly to discern them directly from the interference pattern. Its Fourier transform gives the Fourier spectra distribution of

$$I_H^F\left(f_x, f_y\right) = \delta\left(f_x, f_y\right)$$

$$+ j\lambda \frac{z_R z_O}{z_O - z_R} \exp\left[j\pi\lambda \frac{z_R z_O}{z_O - z_R}\left(f_x^2 + f_y^2\right)\right]$$

$$\otimes \delta\left(f_x - \frac{1}{\lambda}\left(\frac{S_{Rx}}{z_R} - \frac{S_{Ox}}{z_O}\right), f_y - \frac{1}{\lambda}\left(\frac{S_{Ry}}{z_R} - \frac{S_{Oy}}{z_O}\right)\right) \otimes \text{FFT}\left\{\exp\left[j\varphi(x, y)\right]\right\} \quad (6.8)$$

$$+ j\lambda \frac{z_R z_O}{z_O - z_R} \exp\left[-j\pi\lambda \frac{z_R z_O}{z_O - z_R}\left(f_x^2 + f_y^2\right)\right]$$

$$\otimes \delta\left(f_x + \frac{1}{\lambda}\left(\frac{S_{Rx}}{z_R} - \frac{S_{Ox}}{z_O}\right), f_y + \frac{1}{\lambda}\left(\frac{S_{Ry}}{z_R} - \frac{S_{Oy}}{z_O}\right)\right) \otimes \text{FFT}\left\{\exp\left[-j\varphi(x, y)\right]\right\}$$

where $\otimes$ denotes the convolution operation.

It is obvious that the spectrum of the interference term of interest (the virtual original object) consists of three parts:

$$j\lambda \frac{z_R z_O}{z_O - z_R} \exp\left[j\pi\lambda \frac{z_R z_O}{z_O - z_R}\left(f_x^2 + f_y^2\right)\right]$$

$$\delta\left(f_x - \frac{1}{\lambda}\left(\frac{S_{Rx}}{z_R} - \frac{S_{Ox}}{z_O}\right), f_y - \frac{1}{\lambda}\left(\frac{S_{Ry}}{z_R} - \frac{S_{Oy}}{z_O}\right)\right)$$

and

$$\text{FFT}\left\{\exp\left[j\varphi(x, y)\right]\right\}$$

These three parts determine the shape of the spectrum. The first term is a spherical factor, which results in the spherical spectrum extending out. The second term is a delta function, indicating the position of the spectrum as

$$\left[\left(\frac{S_{Rx}}{z_R} - \frac{S_{Ox}}{z_O}\right), \left(\frac{S_{Ry}}{z_R} - \frac{S_{Oy}}{z_O}\right)\right]$$

The third term is the information of the test object.

The difference between z_R and z_O results in a different hologram pattern and thus different frequency spectra distribution in the hologram frequency domain. If $z_R > z_O$,

$$\frac{z_R z_O}{z_O - z_R} < 0$$

The spherical wave front of the interference term of interest is a converging one. This means the divergence of the spherical wave front coming from the illuminating wave front or the imaging MO is smaller than that of the reference wave front. In the interference term, the conjugate of the reference wave front is a converging one. Consequently, a converging wave front is left in the DHM system. If $z_R = z_O = z$, then

$$\frac{1}{z_R} - \frac{1}{z_O} = 0$$

The spherical wave front disappears and leaves only the tilt and constant phase. The pattern of the hologram is a set of straight fringes, which is described by the following equation:

$$
\begin{aligned}
I_H(x, y) = & 1 + |A_O|^2 \\
& + A_O \exp\left[-j\frac{\pi}{\lambda}\left(\frac{S_{Rx}^2 - S_{Ox}^2 + S_{Ry}^2 - S_{Oy}^2}{z}\right)\right] \\
& \times \exp\left[j\frac{2\pi}{\lambda}\left(\frac{S_{Rx} - S_{Ox}}{z}\right)x + j\frac{2\pi}{\lambda}\left(\frac{S_{Ry} - S_{Oy}}{z}\right)y\right]\exp\left[-j\varphi(x,y)\right] \\
& + A_O \exp\left[j\frac{\pi}{\lambda}\left(\frac{S_{Rx}^2 - S_{Ox}^2 + S_{Ry}^2 - S_{Oy}^2}{z}\right)\right] \\
& \times \exp\left[-j\frac{2\pi}{\lambda}\left(\frac{S_{Rx} - S_{Ox}}{z}\right)y - j\frac{2\pi}{\lambda}\left(\frac{S_{Ry} - S_{Oy}}{z}\right)y\right]\exp\left[j\varphi(x,y)\right]
\end{aligned}
$$
(6.9)

This means the spherical wave front coming out from the illuminating wave front or the imaging lens combination is totally compensated by the reference wave front during interference. Consequently, a plane wave front should be left in the DHM system. Its Fourier transform gives the Fourier spectra distribution as follows:

$$
\begin{aligned}
I_H^F(f_x, f_y) = & \delta(f_x, f_y) \\
& + \delta\left(f_x - \frac{S_{Rx} - S_{Ox}}{\lambda z}, f_y - \frac{S_{Ry} - S_{Oy}}{\lambda z}\right) \otimes \text{FFT}\left\{\exp\left[j\varphi(x, y)\right]\right\} \\
& + \delta\left(f_x + \frac{S_{Rx} - S_{Ox}}{\lambda z}, f_y + \frac{S_{Ry} - S_{Oy}}{\lambda z}\right) \otimes \text{FFT}\left\{\exp\left[-j\varphi(x, y)\right]\right\}
\end{aligned}
$$
(6.10)

If there is no test object, the spectrum of interest will be a delta function with a sharp point distribution. When a different object is testing, the convolution in Equation (6.10) will make the sharp point distribution a complicated shape.

If $z_R < z_O$, then

$$\frac{z_R z_O}{z_O - z_R} > 0$$

The spherical wave front of the interference term of interest is a diverging one. This means the divergence of the spherical wave front coming from the illuminating wave front or the imaging lens combination is bigger than that of the reference wave front. Consequently, a diverging wave front is left in the DHM system.

In conclusion, when $z_R \neq z_O$, the left spherical wave front can be either diverging or converging, depending on the relative position of the two point sources. This means the spherical wave front coming from the illuminating wave front or the imaging lens combination cannot be physically compensated by the reference wave front during interference. When an individual interference term of interest is considered, a system phase with a spherical curvature is presented. For numerical reconstruction, a collimated reference wave is used in the off-axis digital holographic microscopy setup [11–14, 24] due to the simplicity of the digital replica of such a reference wave. Thus, no other spherical phase will be introduced to the whole interference term of interest. For the wanted phase of the test object, the system's spherical phase must be numerically compensated.

When $z_R = z_O$, the spherical wave front can be removed by the physical matching of the involved object and reference wave front. Thus, the phase directly reconstructed from the hologram is the phase introduced by the test object without any other further numerical process.

From the preceding analysis, it is obvious that the shape of the spectrum can indicate whether the wave front aberration between the object and reference waves can be physically compensated during the hologram recording. The numerical reference wave front should be carefully chosen to ensure that no other phase factor is introduced to the reconstructed object phase. One can monitor the shape of the spectrum to judge whether the spherical phase curvature is totally compensated in the setup alignment process.

As an example, the difference in Figure 6.4(a) and (b) is obvious in the off-axis extent. But it is hard to tell the difference from Figure 6.4(b) and (c) directly due to the almost similar fringe pattern. In this case, the frequency spectra in the spatial frequency domain may provide useful information about the difference between the two digital holograms. Fourier transform of the holograms has been undertaken to give the frequency spectra distribution in the spatial frequency domain as shown in Figure 6.5(a)–(c). In

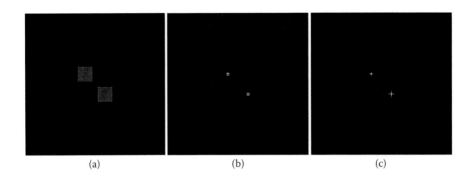

(a)	(b)	(c)

FIGURE 6.5
Fourier spectra of holograms in different fringe patterns: (a) Fourier spectra of hologram in Figure 6.4(a); (b) Fourier spectra of hologram in Figure 6.4(b); (c) Fourier spectra of hologram in Figure 6.4(c).

Figure 6.5(a)–(c), the position of the spectrum is not changed. But the size of the spectrum is decreased from a rectangular shape to a point. This indicates that the status of the physical phase curvature matches between the involved object wave front and reference wave front for the configured interferometer.

Consequently, in the digital hologram recording process, the selection of the configuration of the optical interferometer is important for a successful phase measurement. If the two interference wave fronts match each other, the system phase will not affect the phase measurement of the test object. If the two interference wave fronts do not match, the final system phase must be compensated by using numerical methods to provide the correct phase measurement of the test object. In the following section, numerical reconstruction of the recorded digital hologram will be introduced in detail.

6.1.2 Numerical Reconstruction Algorithms

The digital hologram recording is described by Equation (6.4). Then digital hologram reconstruction is achieved by illumination with a numerical reference wave C. The reconstructed wave front in the hologram plane is then given by

$$CI_H(x,y) = C\left(|O|^2 + |R|^2\right) + CRO^* + CR^*O \qquad (6.11)$$

For in-line recording geometry, the zero order and the two twin images are superposed with one another. It is hard to separate the object information of interest from one single hologram, which limits its application in real-time inspection. For off-axis recording geometry, the separation of the three terms enables further operation to the digital hologram, such as apodization and spatial filtering [25]. After that, the terms to reproduce the original wave front $\psi^H(x,y) = CR^*O$ in the hologram plane are achieved and propagated to the image plane to have a focused image $\psi^I(x,y)$.

Two different numerical reconstruction algorithms are using to calculate the scalar diffraction between ψ^H and ψ^I: the single Fresnel transform (FT) formulation in the spatial domain and the angular spectrum method (ASM) in the spatial frequency domain.

Numerical reconstruction by the Fresnel transform method gives a central reconstruction formula of digital holography as the following:

$$
\psi^I\left(n\Delta x_i, m\Delta y_i\right)
$$

$$
= e^{j\pi d\lambda\left(\frac{n^2}{N^2\Delta x_H^2} + \frac{m^2}{M^2\Delta y_H^2}\right)} \sum_{k=0}^{N-1}\sum_{l=0}^{M-1} \psi^H\left(k\Delta x_H, l\Delta y_H\right)R^* \\ \left(k\Delta x_H, l\Delta y_H\right)e^{\frac{j\pi}{d\lambda}\left(k^2\Delta x_H^2 + l^2\Delta y_H^2\right)}e^{-2j\pi\left(\frac{kn}{N} + \frac{lm}{M}\right)} \tag{6.12}
$$

where

$m, n, k,$ and l are $M \times N$ matrixes denoting the address of the sampling pixel in x and y direction, respectively

d is the hologram reconstruction distance

$\Delta x_i, \Delta y_i, \Delta x_H,$ and Δy_H are the sampling pixel size of the image plane and the hologram plane in x and y directions, respectively, and are related by

$$
\Delta x_i = \frac{d\lambda}{N\Delta x_H}, \Delta y_i = \frac{d\lambda}{M\Delta y_H}
$$

Numerical reconstruction of ψ^H by the angular spectrum method needs a Fourier transform and an inverse Fourier transform:

$$
\begin{cases}
\psi^I\left(n\Delta x_i, m\Delta y_i\right) = \dfrac{\exp(jkd)}{j\lambda d}\,\mathrm{FFT}^{-1}\left\{\mathrm{FFT}\left\{\psi^H\left(k\Delta x_H, l\Delta y_H\right)\right\}\cdot G\left(k\Delta\xi, l\Delta\eta\right)\right\} \\[2mm]
G\left(k\Delta\xi, l\Delta\eta\right) = \exp\left[j\dfrac{2\pi d}{\lambda}\sqrt{1 - (\lambda k\Delta\xi)^2 - (\lambda l\Delta\eta)^2}\right]
\end{cases} \tag{6.13}
$$

where $G\left(k\Delta\xi, l\Delta\eta\right)$ is the optical transfer function in the spatial frequency domain; $\Delta\xi$ and $\Delta\eta$ are the sampling intervals in the spatial frequency domain. The relation between the sampling intervals of the hologram plane and that of the image plane is

$$
\Delta x_i = \frac{1}{N\Delta\xi} = \Delta x_H \text{ and } \Delta y_i = \frac{1}{M\Delta\eta} = \Delta y_H.
$$

Using the single Fresnel transform formulation, the resolution of the reconstructed optical wave field depends not only on the wavelength of the illuminating light but also on the reconstruction distance. And it is always lower than the resolution of the CCD camera. While using the angular spectrum method, one can obtain the reconstructed optical wave field with maximum resolution the same as the pixel size of the CCD camera.

Nevertheless, both the single Fresnel transform formulation and the ASM can give correct reconstruction of the recorded hologram if the distance between the hologram plane and image plane, d, is not too small. If d is such a small value that

$$d \gg \sqrt[3]{\frac{\pi}{4\lambda}\left[(x_i - x_H)^2 + (y_i - y_H)^2\right]_{max}^2} \tag{6.14}$$

the assumption of scalar diffraction, cannot be satisfied, the calculation of the light diffraction propagation between ψ^H and ψ^I will give the wrong results. This means the reconstruction distance of the single Fresnel transform formulation cannot be set as zero. The angular spectrum method can give correct reconstruction of the recorded hologram at any reconstruction distance.

The procedure of digital hologram reconstruction by using the angular spectrum method is shown in Figure 6.6. It is composed of the following steps:

1. Make the digital hologram a Fourier transform to get a superposition of the plane wave front with a different direction.
2. The spectrum of the virtual image is filtered out and moved to the center of the coordinate system.
3. Multiply by the optical diffraction transform function in the frequency domain to calculate the propagation of the light waves.
4. Perform an inverse Fourier transform to the wave front to get its distribution at the spatial domain.

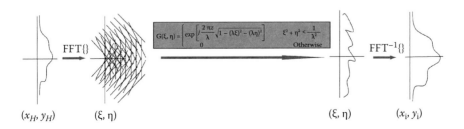

FIGURE 6.6
Digital hologram reconstruction by angular spectrum method.

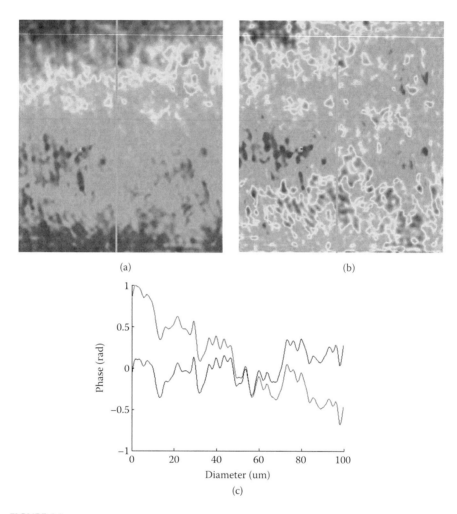

FIGURE 6.7
(See color insert.) (a) Phase with subpixel tilt; (b) phase without subpixel tilt; (c) phase profile comparison along the dark lines in (a) and (b).

It should be noted that in step 2 there are two ways to move the selected spectrum to the center of the coordinate system. One way is to use a plane wave to illuminate the hologram at a certain angle. This angle will offset the off-axis tilt of the selected spectrum. The other way is by direct geometrical movement of the selected spectrum, pixel by pixel. This may cause a problem called subpixel tilt since the discrete pixel always has a certain size. If there is subpixel tilt in the phase, the only way to remove it is a plane wave illumination at a small angle. As for the subpixel tilt shown in Figure 6.7(a), a plane wave is used to illuminate the hologram at a small angle to give the phase reconstruction shown in Figure 6.7(b). The phase profile comparison is given in Figure 6.7(c).

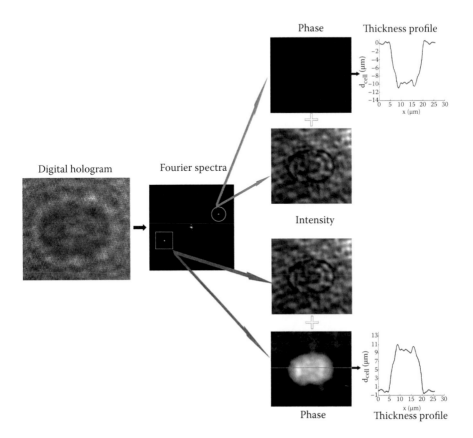

FIGURE 6.8
Digital hologram reconstruction by selecting different spectra: (a) digital hologram of a dividing Vero cell; (b) Fourier spectra of hologram (a); (c) phase from the minus order; (d) intensity from the minus order; (e) phase from the plus order; (f) intensity from the plus order; (g) thickness profile from the phase in (c); (h) thickness profile from the phase in (f).

In step 2, the spectrum of the virtual image is mentioned. One may ask whether there is any difference between the reconstruction results of the spectrum of the virtual image and that of the real image. The reconstruction results of dividing Vero cells are shown in Figure 6.8 to give a clear illustration of the difference. The digital hologram is shown in Figure 6.8(a). Its Fourier spectra are shown in Figure 6.8(b). The reconstruction distance for the spectrum selected by the white circle is $-d$. It should be the virtual image of the test object. The reconstructed phase and intensity are shown in Figure 6.8(c) and (d). The reconstruction distance for the spectrum selected by the white rectangle is d. It should be the real image of the test object. The reconstructed phase and intensity are shown in Figure 6.8(e) and (f). It is obvious that the intensities from the two different spectra are the same as each other and the phases are opposite to each other.

6.1.3 Phase to Profile

Reflection digital holography is widely used for deformation and displacement measurement. The relation between the shape of the test object and the achieved phase is

$$\varphi(x,y) = \frac{4\pi}{\lambda} d(x,y,z) \tag{6.15}$$

where $\varphi(x,y)$ is the achieved phase distribution $d(x,y,z)$ and is the profile of the test object. The depth or height one can measure is

$$-\frac{\lambda}{2} < d(x,y,z) < \frac{\lambda}{2}$$

Transmission digital holography can be applied to refractive index measurement. Light transmitting the test transparent object experiences a change of the optical path length and thereby a phase variation. The interference phase due to refractive index variations is given by

$$\Delta\varphi(x,y) = \frac{2\pi}{\lambda} \int_{-l}^{l} n(x,y) \mathrm{d}z \tag{6.16}$$

where $n(x,y)$ is the wanted refractive index. The light passes the medium in the z-direction and the integration is taken along the propagation direction. Given the thickness of the specimen, the phase distribution can be transferred to refractive index distribution or, given the refractive index of the specimen, the phase distribution can be transferred to thickness. As shown in Figure 6.7(g) and (h), the detected phase of the dividing Vero cell is transferred to a cell thickness with an average refractive index of 1.37.

6.2 Digital Holoscopes

6.2.1 Reflection Digital Holoscope

Three configurations have been developed for the compact reflection digital holoscope. The premise behind these configurations includes the need to make measurements on-site or during processes such as etching or thin-film coating. The goal is to incorporate the system in a noninvasive manner for real-time quantitative measurements with nanometer axial sensitivity.

(a)   (b)

FIGURE 6.9
(a) Product prototypes for compact digital holoscopes showing scanning system on translation stage and portable system that can be placed directly on sample in foreground; (b) prototype of an integrated CDH system.

Figure 6.9(a) shows three of the models. The one mounted on the translation stage enables measurement of large samples using a scanning image system. The items in the foreground are two portable systems that can be placed directly on the specimen to be measured. Finally, the system in Figure 6.9(b) shows a drawing of the new system, which is about 30% smaller than the previous versions with the camera integrated into the system.

6.2.1.1 Calibration (Height Calibration by Comparison with AFM)

A calibration of the system was performed to ensure correct measurements as well as to highlight the accuracy and repeatability of the measurement. One of the concerns here is that there is no specific calibration target that can provide a whole-field calibration of the system. Hence, this study compares the compact digital holoscope (CDH) results of available targets such as the U.S. Air Force (USAF) or internally fabricated microelectromechanical systems (MEMS) step-height target with an atomic force microscope (AFM) scan. The concern here is that the line used for the AFM scan and the line selected from the CDH image may not be exactly the same. Hence, some errors may arise but it gives good ballpark data.

Figure 6.10(a) shows the results of a USAF target measured using the CDH system along with a typical line scan plot from this image. Figure 6.10(b) shows the result from the AFM scan. The AFM scan area plot is also shown, but this one covers a smaller area and took a longer time to scan as compared with the DH result. The results indicate good correlation between the CDH results and the AFM scan. The CDH was then used to test a step-height sample fabricated using the MEMS etching process. The sample was designed

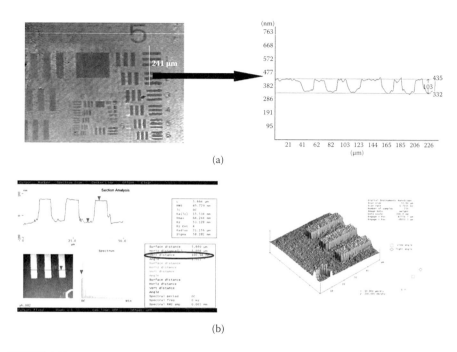

(a)

(b)

FIGURE 6.10
(a) CDH phase image of the USAF target and a line scan plot along the marked line showing the profile; (b) AFM line scan plot as well as the area plot of a small segment.

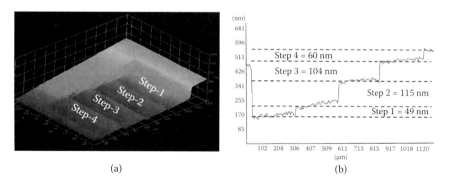

(a)   (b)

FIGURE 6.11
Profiling of MEMS etched target: (a) whole-field profile map; (b) line scan plot.

with etch heights of 50, 100, 100, and 60 nm. The color-coded 3D phase image is shown in Figure 6.11(a) and the line scan profile along a typical section is shown in Figure 6.11(b). The heights match quite well; however, during the etching process, the etch depth is determined by the time of etch rather than any measurement. CDH provides an elegant way to monitor the etch depth during the etching process in real time.

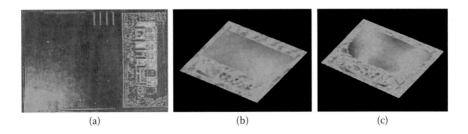

(a) (b) (c)

FIGURE 6.12
(See color insert.) Profile of accelerometer: (a) bare, unmounted; (b) mounted on ceramic substrate; (c) mounted on plastic substrate.

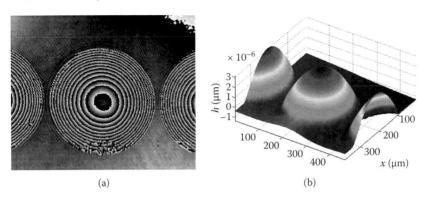

(a) (b)

FIGURE 6.13
(a) Phase of reflection microlens array in pitch 250 mm modulo 2p; (b) height map of the microlens array.

6.2.1.2 MEMS Inspection

A typical example of the use of the CDH for MEMS inspection involved the testing of a MEMS accelerometer mounted onto different substrates holding the related electronics. In the process of bonding the accelerometer to the substrate, deformation or distortions occur that would adversely affect the performance of the accelerometer. Figure 6.12(a) shows the profile of a bare accelerometer and Figure 6.12(b) and (c) the profile for the accelerometer mounted on ceramic substrate and on plastic substrate. The distortion in Figure 6.12(c) as compared to Figure 6.12(b) is quite evident and indicative of the poorer performance of the accelerometer on plastic substrates.

6.2.1.3 Reflective Microlens Array Characterization

A 4 mm × 14 mm × 0.5 mm reflective planoconvex linear microlens array from SUSS micro-optics with a 250 μm pitch was tested by use of the CDH system. Figure 6.13(a) shows the wrapped phase map of a single microlens

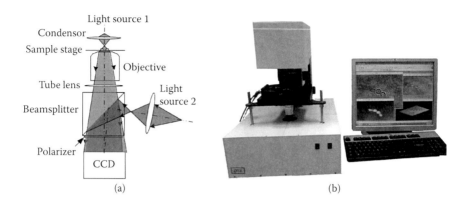

FIGURE 6.14
(a) Schematic of transmission mode DHM setup for short coherence light source; (b) packaged DHM setup for transmission specimen.

and Figure 6.13(b) shows the 3D unwrapped height map of the corresponding microlens.

6.2.2 Transmission Digital Holoscope

A DHM system built in the Michelson interferometer configuration is proposed by using an adjustable lens in the reference beam to perform the quasi-physical spherical phase compensation. It is built for using a light source with a short coherence length. As shown in Figure 6.14(a), an adjustable lens is put in the reference beam path. The coherence length of the light source may limit the position-adjustable capability of the reference wave. In such a case, the adjustable lens can be used to change the phase curvature to fulfill the spherical phase compensation in the hologram recording process.

6.2.2.1 Quantitative Mapping of Domain Inversion in Ferroelectric Crystal

A very interesting electrochromism effect has been observed during the ferroelectric domain inversion on RuO_2-doped $LiNbO_3$ crystals [26, 27]. They have become a new kind of material for investigation in domain inversion engineering. Digital holographic microscopy has been successfully applied to in situ visualization, monitoring, quantitative measurement, and analysis of the domain inversion in ferroelectric crystals. Here, DHM is applied for domain inversion investigation of RuO_2-doped $LiNbO_3$ crystals.

The refractive index and thickness of the ferroelectric crystals under a uniform voltage will change because of the electro-optic (EO) effect and the piezoelectric effect. These changes cause a phase retardation of a normally transmitted wave field. For a crystal wafer with antiparallel domain structure, the refractive index will change according to its polarization. The

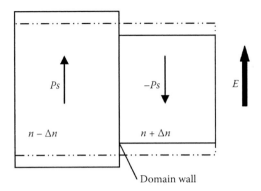

FIGURE 6.15
Schematic of the variation of the ferroelectric crystal under external voltage.

spontaneous polarization of the crystal sample is along its optical axis, the crystallographic c axis. If we applied the external electric field along the optical axis, there is a refractive index variation $\Delta n < 0$ in the part with spontaneous polarization and a refractive index variation $\Delta n > 0$ in the other part with reversed polarization, as shown in Figure 6.15. Additionally, the external electric field causes the thickness variation of the crystal due to the piezoelectric effect. The phase retardation experienced by the normally transmitted wave field when the external electric voltage is applied can be calculated by the following formula [28]:

$$\Delta\varphi = 2\left[\frac{2\pi}{\lambda}\Delta nD + \frac{2\pi}{\lambda}(n_0 - n_w)\Delta D\right] = \frac{2\pi}{\lambda}\left[-\gamma_{13}n_0^3 + 2(n_0 - n_w)k_3\right]U \quad (6.17)$$

where
$\Delta n \propto \gamma_{13}\,U/d$
$n_0 = 2.286$ (when $\lambda = 632.8$ nm) is the refractive index of the original crystal
$n_w = 1.33$ is the refractive index of water
U is the electric voltage
D is the thickness of the crystal along its optical axis
γ_{13} is the linear EO coefficient
k_3 is the ratio between the linear piezoelectric tensor and the stiffness tensor [29]
ΔD is the piezoelectric thickness change

The phase retardation experienced by the object wave field during poling can be calculated.

The sample holder used for the experiments is illustrated by Figure 6.16. The crystal sample, which is a z-cut, 0.5-mm-thick, double-side-polished

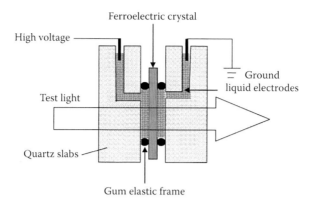

FIGURE 6.16
Schematic of the sample crystal holder.

RuO_2:$LiNbO_3$ substrate, is mounted between two rectangular gum elastic frames, which are clamped among two quartz slabs. The two cavities filled with water act as liquid electrodes. The crystal sample has an area of 36.00 mm² (6.0 mm × 6.0 mm) contacting the liquid electrodes, which provide the homogeneous electric field. A high-voltage supply is used for an electric pulse or continuous external field with 20.0 kV and 2.0 mA maximum output. The switching current caused by the charge redistribution within the crystal is monitored by a microamperemeter.

In the domain inversion, one stable domain nucleus is observed. The phase variation during the whole nucleus growth is obtained. The bottom of the nucleus is a hexagonal shape. As the nucleus grows, the top becomes sharper and sharper, as shown in Figure 6.17. The top view of the domain nucleus in 400 s is shown in Figure 6.18. The hexagonal shape is not very clear due to the nonuniform distribution of the top phase. The phase profiles in different times are given along the white line in Figure 6.18(a) as shown in Figure 6.18(b) and (c). From the comparison, as the domain nucleus grows, the phase retardation increases. The speed of the increment of the phase retardation is different. Along line 3, the phase increment is the fastest. The unequal increment speed indicates that the shape of the domain nucleus will change from a hexagonal shape to a different one during its growth. The reason for this variation is very complex. One reasonable explanation is the redistribution of the poling electrons in the domain inversion. It is related closely to the initial property of the crystal.

6.2.2.2 Characterization of Transmission Microlens Array

Microlens arrays have numerous and diverse applications and are employed in coupling light to optical fibers, increasing the light collection efficiency

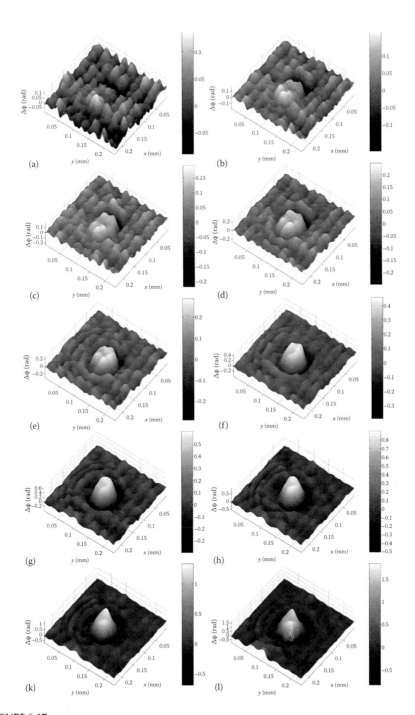

FIGURE 6.17
The growth of the domain nucleus in the crystal. (a) 1 s; (b) 20 s; (c) 30 s; (d) 50 s; (e) 100 s; (f) 150 s; (g) 200 s; (h) 250 s; (k) 300 s; (l) 400 s.

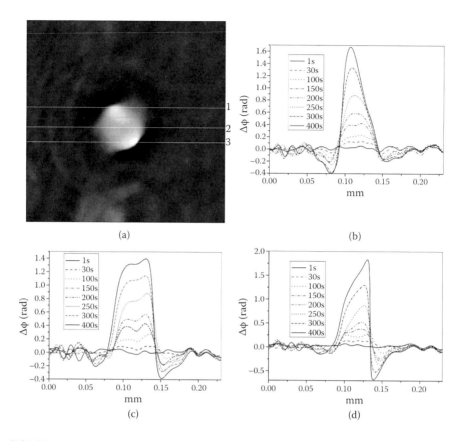

FIGURE 6.18
Evolution of phase profile of domain nucleus: (a) 2D image; (b) along line 1; (c) along line 2; (d) along line 3.

of CCD arrays, compact imaging in photocopiers and mobile-phone cameras, and enabling integral photography in 3D imaging and displays [30]. Precise control of the shape, surface quality, and optical performance of the microlenses is required, as well as the uniformity of these parameters across the array. A common-path DHM system based on a single cube beam splitter (SCBS) interferometer [21] uses an MO [22] to provide magnification to the test specimen (Figure 6.19). Since the object beam and reference beam share the same optical path, the wave front curvatures can physically compensate each other during the interference. This provides a better way for microlens characterization as well as uniformity inspection across a whole array.

A 1 mm × 12.8 mm × 1.2 mm refractive planoconvex linear microlens array from SUSS micro-optics with a 127 μm pitch was tested by the previously described DHM system. With a known refractive index of the lens material, the geometrical thickness of the lens can be deduced from the quantitative

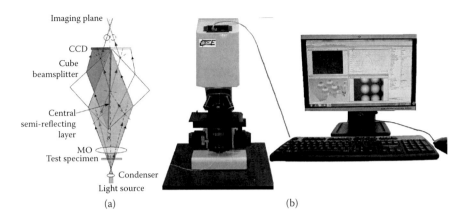

FIGURE 6.19
(a) Schematic of transmission mode common-path DHM used for 3D imaging of linear micro-lens array; (b) packaged CPDHM setup.

phase map as well as the lens shape, height, and radius of curvature (ROC), if the lens has a flat face (e.g., a planoconvex lens). When light is transmitted through the lens, the optical path length will be changed according to the height and refractive index of the lens. By using DHM, the optical path length change can be easily achieved from the phase map of the wave front. Given the refractive index, one can calculate the height of the test lens according to the following equation:

$$h = \frac{\lambda}{2\pi} \frac{\varphi}{\left(n_L - n_S\right)} \tag{6.18}$$

where
λ is the wavelength of the light
φ is the phase given by the SCBS microscope
n_L is the refractive index of the lens
n_S is the refractive index of the medium around the test lens

As shown in Figure 6.20, the linear microlens array can be tested in different magnifications as required.

The microscope is calibrated by using the 0.1 mm/100 div calibration microstage for the actual magnification it can provide. For example, the calibration results for a 40× microscope objective is shown in Figure 6.21. The total pixel used in the *y* direction is 960 with a pixel size of 4.65 μm. The length of the calibration microstage is 130 μm. Consequently, the magnification of the system is about 34×.

Given that the refractive index is 1.457 at 633 nm, the height profile of the single microlens is shown in Figure 6.22(b). It was drawn from the height

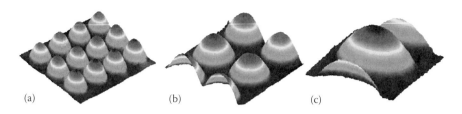

(a)   (b)   (c)

FIGURE 6.20
Three-dimensional imaging of linear microlens array in different magnifications.

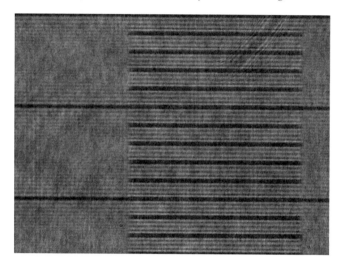

FIGURE 6.21
Magnification calibration of the 40× microscope objective.

map along with the same position as the solid line in Figure 6.22(a). Given the height profile of the lens, the ROC can be calculated by the following equation:

$$ROC = \frac{h}{2} + \frac{D^2}{8h}$$ (6.19)

where h is the height and D is the diameter of the microlens. The maximum height of the microlens, h, is read as 5.82 µm. The diameter, D, is read as 121 µm. Thus, the calculated ROC is 317 µm. It is slightly different from the ROC value of 315 mm provided by the supplier.

6.2.2.3 Characterization of Diffractive Optical Elements on Liquid Crystal Spatial Light Modulator

A diffractive optics element (DOE) is a passive optical component containing complex micro- and nanostructure patterns that can modulate or transform

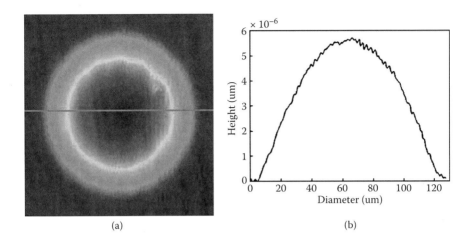

FIGURE 6.22
(See color insert.) (a) Height map of a single microlens; (b) height profile along the solid line in (a).

the light in a predetermined way. DOE works on the principle of interference and diffraction to represent desired optical elements, in contrast to refraction of light from conventional optical element. In general, DOEs fabricated on glass with wavelength-sized features involve time-consuming microfabricated semiconductor foundry resources. A computer-generated DOE displayed on the liquid crystal spatial light modulator (LCSLM) is an active and dynamic replacement of physical optical components. Diffractive optical lens and microlens array on the LCSLM with variable focal lengths and diameter make them important in many applications, including adaptive optics, optical tweezers, retinal imaging, diffraction strain sensors, etc. The computer-generated diffractive microlens array displayed on the LCSLM can replace the static physical lens array in the Shack–Hartmann wave front sensor (SHWS) and the multipoint diffraction strain sensor (MDSS) for wave front sampling. However, it is difficult to investigate the quality of these lenses by traditional metrology since the wave front profile is modulated by liquid crystal molecules on application of an electric field instead of the solid surface of the optical components. Thus, the characterization of diffractive optical microlens array on the LCSLM is necessary to quantify the effect on exiting optical wave fronts.

LCSLMs are polarization-sensitive devices and, depending on the incident polarization and wavelength, amplitude and phase of the exiting optical wave front are modulated. The diffractive microlens array displayed on the LCSLM can be characterized easily after characterizing the maximum phase modulation from the LCSLM using the DH method. The advantage of phase modulation characterization using the proposed DH over other existing methods is that the full-field quantitative phase value is calculated from the single digital hologram and thus the phase map of the full active region of the twisted nematic (TN)-LCSLM can be visualized and quantified.

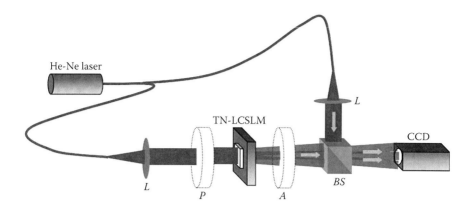

FIGURE 6.23
Digital holography experimental setup for HOLOEYE 2002 TN-LCSLM phase modulation characterization. BS = beam splitter, P = polarizer, A = analyzer, L = collimation lens, CCD = charge coupling device.

The TN-LCSLM is sandwiched between a pair of polarizers in order to separate two modes of modulation that are generally coupled. The complex wave front coming out of the polarizer–SLM–analyzer combination bears the amplitude and phase information that is evaluated quantitatively using the DH method. When an object wave interferes with a known reference wave, the recorded interference pattern is called a hologram. The amplitude and phase information of the object wave are encoded in the interference pattern. In DH, a hologram is recorded directly in digital form using CCD and numerically reconstructed for quantitative evaluation of amplitude and phase information acquired from the entire active area of TN-LCSLM.

The experimental setup to characterize phase modulation from the transmissive HOLOEYE LC2002 TN-LCSLM that uses the Sony LCX016AL liquid crystal microdisplay is shown in Figure 6.23. The TN-LCSLM active area of 21 mm × 26 mm contains 832 × 624 square pixels with pixel pitch of 32 μm and has a fill factor of 55%. HOLOEYE LC2002 has a twisted arrangement of nematic liquid crystal molecules between inner ends of two conductive glass plates. The output of a He-Ne laser (632.8 nm) is coupled into a bifurcated single mode fiber. A collimated beam coming from one fiber arm, called an object beam, is passed through the polarizer–SLM–analyzer combination and illuminates the entire active area of the TN-LCSLM. The complex object wave front, modulated according to the input conditions of the TN-LCSLM, interferes with the collimated reference beam coming from the second fiber arm using a beam splitter and the digital hologram recorded by the CCD sensor and is given by

$$H\left(n_x,n_y\right)=\left|O\left(n_x,n_y\right)\right|^2+\left|R\left(n_x,n_y\right)\right|^2+O^*\left(n_x,n_y\right)R\left(n_x,n_y\right)$$
$$+O\left(n_x n_y\right)R^*\left(n_x,n_y\right)$$

(6.20)

where
$n_x = 0,1,...N_x-1$ and $n_y = 0,1,...N_y-1$ are the pixel indices of the camera
$N_x \times N_y$ is the size of the CCD sensor in pixels
$O(n_x, n_y)$ is the object wave
$R(n_x, n_y)$ is the collimated reference wave

Since the active area of the TN-LCSLM comprises two-dimensional (2D) arrays of liquid crystal (LC) cells, the object wave, after passing through the TN-LCSLM active area, diffracts into a number of plane waves whose direction of propagation depends on the grating period and size of the LC cell. Using the Fresnel approximation, the complex object wave $O(x, y)$ at CCD plane (x, y) diffracted from SLM plane (x', y') is given by

$$O(x,y)\frac{e^{ikz}}{i\lambda z}\iint t(x',y')\exp\left[i\frac{2\pi}{\lambda z}\left\{(x-x')^2 + (y-y')^2\right\}\right]dx'dy' \quad (6.21)$$

Here, k denotes the propagation constant and z is distance between the SLM and CCD planes, λ is wavelength of object wave, and $t(x', y')$ is the TN-LCSLM transmittance. It should be noted here that the TN-LCSLM transmittance does not include any contribution from the nonactive area of the TN-LCSLM since it provides constant phase change only.

After recording the digital hologram, the convolution method of numerical reconstruction is used to extract quantitative amplitude and phase information from the recorded hologram. This method provides the same resolution of reconstructed image as that of a CCD sensor and is more effective for smaller recording distances. The numerically reconstructed real image wave $U_{real}(n'_x, n'_y)$ can be written as

$$U_{real}(n'_x, n'_y) = \Im^{-1}\left[\Im\left\{H'(n'_x, n'_y) \times R(n'_x, n'_y)\right\} \times \Im\left\{g(n'_x, n'_y)\right\}\right] \quad (6.22)$$

where
$n'_x = 0,1,...N'_x-1$ and $n'_y = 0,1,...N'_y-1$ are the pixel indices of the reconstructed image
$N'_x \times N'_y$ is the new size of the preprocessed hologram $H'(n'_x, n'_y)$
$\Im$ represents the Fourier transform operator
$R(n'_x, n'_y)$ is the numerically defined reconstructed plane wave
$g(n'_x, n'_y)$ is the impulse response function of coherent optical system

Finally, the phase value is quantitatively evaluated directly from the reconstructed real image wave as

$$\Delta\phi(n'_x, n'_y) = \arctan\left[\text{Im}\left\{U_{real}(n'_x, n'_y)\right\}\Big/\text{Re}\left\{U_{real}(n'_x, n'_y)\right\}\right] \quad (6.23)$$

As stated earlier, the amplitude and phase mode of modulation are coupled in the TN-LCSLM and can be separated by choosing particular orientation of the polarizer and analyzer. In order to characterize the TN-LCSLM for phase-mostly modulation, the active area is divided into two equally separated regions called the reference and modulation regions. The grayscale value addressed in the reference region is kept at 0 whereas, in the modulation region, the addressed grayscale value can be varied from 0 to 255 in equal steps. Now, the orientation of polarization axes of both polarizer and analyzer is adjusted with respect to the TN-LCSLM in such a way that the reconstructed intensity image does not show any difference in contrast of two regions from the digital hologram recorded when the reference region is addressed with 0 grayscale value and the modulation region is addressed with 255 grayscale value. At this position of the polarizer and analyzer orientation axes, the transmissive TN-LCSLM operates in phase-mostly mode and hence shows no intensity or amplitude modulation. In our experiment, phase-mostly modulation of the TN-LCSLM is attained with polarizer and analyzer orientation of 15° and 145°, respectively. It should be noted here that the addressed contrast and brightness value on the TN-LCSLM is 255.

Figure 6.24(a) shows the digital hologram recorded when the polarizer–LCSLM–analyzer combination is operated in the phase-mostly mode. Figure 6.24(b) shows the numerically reconstructed intensity image, which depicts that the TN-LCSLM is operated in phase-mostly mode since there is no difference in contrast between the two regions separated by a line due to diffraction. Figure 6.24(c) shows the numerically reconstructed quantitative phase image and Figure 6.24(d) indicates the phase step height of $1.26\,\pi$ measured on the two regions when the TN-LCSLM is addressed with 0 and 255 grayscale values in two equally separated regions.

The DH method to characterize the phase modulation of the TN-LCSLM is also very helpful in characterizing diffractive optics such as single digital lens or digital lens array. This digital lens array is a diffractive optical element computed using the iterative Fourier transform algorithm (IFTA). This digital diffractive lens array is addressed onto the TN-LCSLM active area and the phase modulation in the transmitted optical wave front is analyzed using the DH method in a manner similar to that done previously. First, single lenses with different focal lengths of 8, 120, and 160 mm are displayed on the active area of the TN-LCSLM and tested. The numerically reconstructed phase map for these diffractive optics lenses and the line profile through the center along the diameter are shown in Figure 6.25. The maximum phase modulation measured for three different diffractive optics lenses with different focal lengths is the same and is equal to maximum phase modulation of the TN-LCSLM evaluated earlier. However, the more detailed line profile through the center along the diameter for each lens is slightly different in shape and profile from one to another due to difference in focal lengths and the aberrations existing for practical use of the TN-LCSLM.

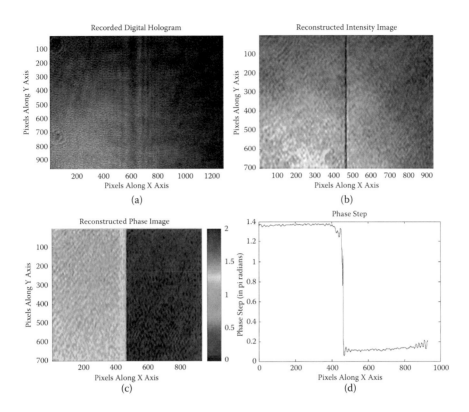

FIGURE 6.24
(See color insert.) (a) Recorded digital hologram numerically reconstructed; (b) intensity image; (c) phase image; (d) phase step height measured from digitally recorded hologram with the TN-LCSLM addressed with 0 and 255 grayscales in two equally separated regions.

Figure 6.26(a) shows the diffractive lens array—all with focal length of 80 mm. The lens array is displayed onto the active area of the TN-LCSLM and the quantitative modulated phase is extracted from the recorded digital hologram. The phase modulation value from the different lenses is the same, however; it was found that if the number of diffractive lenses is increased in an array, the phase modulation from individual digital lenses is decreased due to increase in phase modulation from the lens background. Figure 6.25(b) and (c) show the numerically reconstructed intensity and phase images from the recorded digital hologram, respectively. Figure 6.25(d) shows the phase line profile through the center along the diameter of individual lenses; this can be compared to other lenses in the lens array to ensure the quality of the array. The final results can be used as a quantitative qualification of the DOE lenses or as feedback to revise and improve the previous DOE lens design.

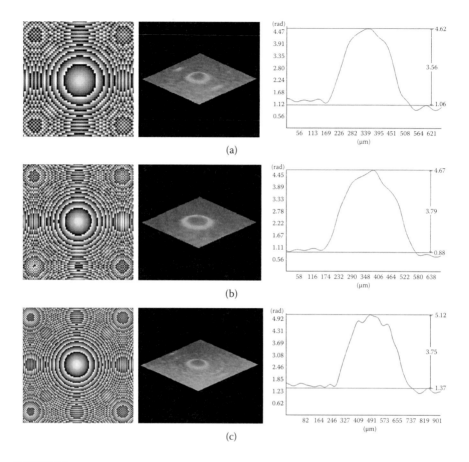

FIGURE 6.25

(See color insert.) Diffractive optics lens with focal lengths with their numerical phase reconstruction using DH method and line profile traced at the center in respective order: (a) 80 mm; (b) 120 mm; (c) 160 mm.

6.2.3 Resolution Enhancement in Digital Holography

6.2.3.1 Resolution Enhancement Methods

Compared to conventional holography, DH [31–33] has many advantages, including access to quantitative amplitude and phase information. However, the lateral resolution of DH is limited by the digital recording device. Factors contributing to lateral resolution limitation have been investigated [34–37].

Much research work on the lateral resolution enhancement has been reported. One direct method is to introduce MO into the DH system [38–49]. The drawback of this method is the reduction of the field of view. The MO introduces unwanted curvature and other aberrations that need to be compensated to get correct results.

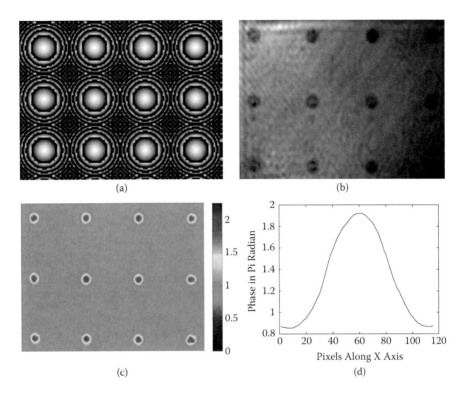

(a)

(b)

(c)

(d)

FIGURE 6.26
(a) Diffractive lens array with focal length of 80 mm; (b) numerically reconstructed intensity; (c) phase distribution from the lens array using the DH method; (d) line profile of distribution along the diameter of a typical microlens from the array.

Another method to enhance lateral resolution is to use the aperture synthesis method, which can be categorized into three approaches. One translates the CCD position and records multiple holograms at different positions to collect more object information at a larger diffraction angle. The second approach changes the illuminating light angle or translates the illuminating point source with a fixed CCD position to record multiple holograms. By changing the illuminating light direction or translating the illuminating point source, different diffraction angles of object information can be projected onto and recorded by the CCD. In the third method, the specimen is rotated with a fixed CCD position and illuminating angle. Different diffraction angles of the object are recorded onto the CCD. Larger diffraction angles correspond to higher object frequencies and larger numerical apertures. Better lateral resolution can be expected. After recording, the information recorded by different holograms in aperture synthesis methods needs to be integrated. References 1 and 50–57 illustrate the first aperture synthesis approach of translation of CCD positions.

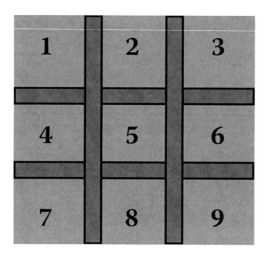

FIGURE 6.27
Nine holograms taken by shifting CCD. The original hologram is in the middle. (Yan, H. and A. Asundi. *Optics and Lasers in Engineering*, 2011. With permission.)

Here we use reference 1 as an example to illustrate this method. In the experiment, nine holograms are taken, as in Figure 6.27, with the original hologram located at the center. Before stitching, the hologram aperture is 5.952 mm in width and 4.464 mm in length. After stitching, the CCD aperture is expanded to 8.952 mm in width and 7.464 mm in length. The reconstructed intensity images before and after stitching are shown in Figure 6.28. Figure 6.28(a)–(c) are before stitching and Figure 6.28(d)–(f) are after stitching. Figure 6.28(b) and (e) are the images in the highlighted square area of Figure 6.28(a) and (d), respectively, which present the G4 and G5 groups of the USAF target. Figure 6.28(c) and (f) are the images in the highlighted square area of Figure 6.28(b) and (e), respectively, which show the G6 and G7 groups of the USAF target. Lateral resolution is enhanced from about 6.960 and 8.769 μm in x and y directions, respectively, as in Figure 6.28(c), to about 4.385 μm in the x direction and 4.922 μm in the y direction, as in Figure 6.28(f). The field of view is enlarged at the same time, as seen by comparison of Figure 6.28(a) and (b).

References 58–64 illustrate the second aperture synthesis approach by changing the illumination light direction or translating the illuminating point source. As an example, we use reference 60, which reported a super-resolving approach for off-axis digital holographic microscopy in which a microscope objective is used. In this approach, the single illumination point source shifts in sequential mode and holograms are recorded at each shift position. Holograms recorded at different illumination positions are superimposed. Each shift of the illumination beam generates a shift in the object spectrum in such a way that different spatial-frequency bands are transmitted through the objective lens as seen in Figure 6.29. The lateral resolution

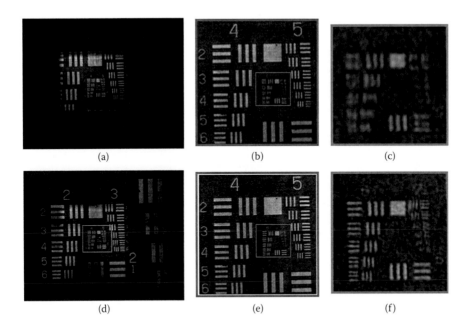

FIGURE 6.28
Reconstructed intensity images before and after stitching: (a–c) images reconstructed from a single hologram; (d–f) images reconstructed from the stitched hologram. (Yan, H. and A. Asundi. *Optics and Lasers in Engineering*, 2011. With permission.)

was enhanced by a factor of 3× in the *x* and *y* directions, and 2.4× in the oblique directions. Finally, reconstruction with lateral resolution of 1.74 µm, as in Figure 6.30, was demonstrated. The setup is a phase step digital holographic setup.

References 65 and 66 illustrate the third aperture synthesis approach by specimen rotation. Of the three methods of aperture synthesis, the first one by translation of CCD has a limitation on the object spatial frequency recorded by the hologram. When CCD is away from the optical axis, the light of higher object spatial frequency is collected with larger angle light diffraction on the CCD. The sampling interval sets a limitation on the maximum frequency to avoid spectrum overlapping. The other two methods do not suffer this limitation as the angles of the light projected on the CCD are not affected by the object spatial frequency. In the second method, object spatial frequency recorded by the CCD is controlled by the illumination light angle. Larger illumination angles diffract higher object frequency onto the CCD. In the third method, the object spatial frequency recorded by the CCD is controlled by the rotation angle of the object. Larger object rotation diffracts higher object frequencies onto the CCD. However, as the object position and phase are also changed during the rotation, additional work to compensate the position and phase change is needed.

FIGURE 6.29
Fourier transform of the addition of different recorded holograms. (Mico, V. et al. *Journal of the Optical Society of America A—Optics Image Science and Vision* 23:3162–3170, 2006. With permission.)

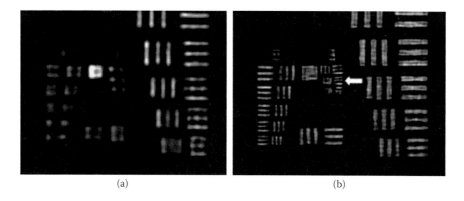

(a)   (b)

FIGURE 6.30
(a) Image obtained with 0.1 NA lens and conventional illumination; (b) super-resolved image obtained with the synthetic aperture. The G9E2 corresponding to the resolution limit using the proposed method is marked with an arrow. (Mico, V. et al. *Journal of the Optical Society of America A—Optics Image Science and Vision* 23:3162–3170, 2006. With permission.)

6.2.3.2 Differences between Hologram Stitching and Zero Padding

Some people think that hologram stitching provides a similar effect to that of zero padding [67]. Their standpoint is that if the size of the modified hologram after zero padding and stitching is the same, the NA should be the same. The fields of view with hologram stitching and zero padding are the same.

But hologram stitching does not really provide an effect similar to that of zero padding, as demonstrated in the practical experiment shown in Figure 6.31. The first column shows the reconstruction from an original recorded hologram (a) of size 1280 × 960 pixels. No padding or stitching is used here. The

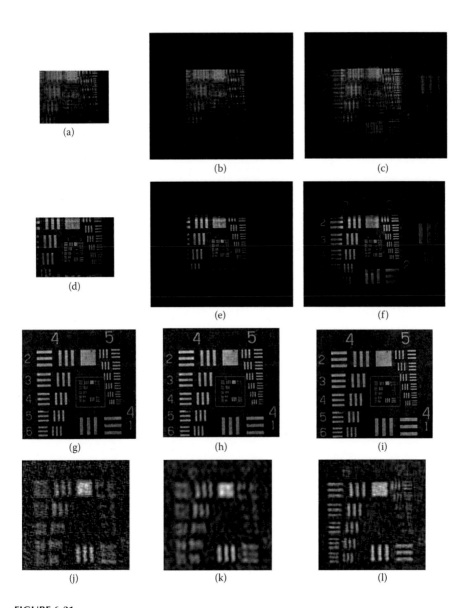

FIGURE 6.31

Reconstruction of original hologram: (a) zero-padded hologram; (b) stitched hologram; (c) using transfer function method.

second column shows the reconstruction from a zero padded hologram (b) from the original hologram (a). This hologram (b) is 2600 × 2280 pixels with zero padding. The third column shows the reconstruction from a stitched hologram (c) of 2600 × 2280 pixels. Hologram (c) is stitched by five single holograms. Each of them is 1280 × 960 pixels. The size of the modified hologram after zero padding (Figure 6.31b) and stitching (Figure 6.31c) is the same.

In reconstruction, the transfer function (or convolution) method is used. Figure 6.31(d)–(f) are the full-field reconstructed images of holograms (a), (b), and (c), respectively. Figure 6.31(g)–(i) show group 4 and group 5 in Figure 6.31(d)–(f), respectively. Figure 6.31(j)–(l) show group 6 and group 7 in Figure 6.31(g)–(i) (inside the white squares), respectively.

From the comparison of Figure 6.31(j)–(l), it can be seen that the resolution capability in (l) by hologram stitching is twice the resolution in (j) from the original hologram and the resolution in (k) from the zero padded hologram. The resolution difference in (j) and (k) is not obvious. Therefore, the lateral resolution of hologram stitching and zero padding is not the same. Hologram stitching provides better lateral resolution than just zero padding.

The aperture size that determines the resolution is not the zero padded hologram size, but rather is the aperture size before zero padding. Suppose that if the original hologram is of size $2D$, it is zero padded to size L, where L is larger than $2D$. The system lateral resolution is $\lambda z/2D$, but not $\lambda z/L$. If the zero padded hologram is used to perform reconstruction, each pixel of the reconstructed image is of the size $\lambda z/L$. But the pixel size of the reconstructed image is not the lateral resolution. They are two different concepts. In the zero padding method, $2D$ is not changed. But, by hologram stitching, the value of $2D$ increases and therefore lateral resolution can be improved.

From the physical view, zero padding does not collect additional information in the system, especially the lights of larger diffraction angles. Only the collection of lights with larger diffraction angles can provide more details of the object and therefore better resolution. Hologram stitching is one way to collect lights of larger diffraction angles. Therefore, it can provide better resolution while zero padding cannot.

Zero padding is usually used in the Fresnel reconstruction method rather than the transfer function method. In Figure 6.32, we use the Fresnel reconstruction method to reconstruct the same holograms in Figure 6.31(a)–(c). Holograms in Figure 6.32(a)–(c) are identical to Figure 6.31(a)–(c), respectively. Figure 6.32(d)–(f) are the full-field reconstructed images of holograms Figure 6.32(a)–(c), respectively. Figure 6.32(g)–(i) show group 4 and group 5 in Figure 6.32(d)–(f), respectively. Figure 6.32(j)–(l) show group 6 and group 7 in (g)–(i) (inside white squares), respectively.

From the comparison of Figure 6.32(j)–(l), it can be seen that the resolution capability in Figure 6.32(l) by hologram stitching is twice the resolution in Figure 6.32(k) by zero padding of the hologram. From the comparison of Figure 6.32(j) and (k), it seems that zero padding in Figure 6.32(k) provides better lateral resolution than without zero padding in Figure 6.32(j). But if we compare Figure 6.32(j) and (k) with Figure 6.31 (j) and (k), it can be noticed that Figure 6.31(j) and (k) and Figure 6.32(k) can provide nearly the same resolution. The resolution of Figure 6.32(j) is worse. Figure 6.31(j) and Figure 6.32(j) are reconstructed from the same hologram. The difference between them is only the reconstruction method. The reason for this is that, under certain parameters, the Fresnel reconstruction causes the loss of lateral

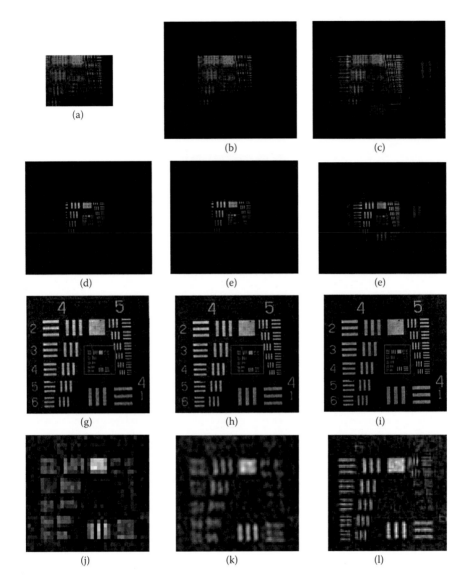

FIGURE 6.32
Reconstruction of original hologram: (a) zero-padded hologram; (b) stitched hologram; (c) using Fresnel method.

resolution; zero padding in the Fresnel reconstruction method can help to avoid this loss. But the transfer function method does not have such problems. Zero padding essentially can help to present the information recorded in the hologram fully, but it cannot add information to the image. Hence, it can only improve the lateral resolution to the value $\lambda z/2D$ where $2D$ is the hologram size before zero padding but cannot exceed it. This improvement by zero padding is reconstruction method dependent. On the other hand, in both Figure 6.31 and Figure 6.32, it can be seen that hologram stitching provides better lateral resolution than just zero padding.

6.3 Conclusion

Digital holography has been shown to have widespread applications for 3D profilometry with nanoscale sensitivity. For reflective surfaces as encountered in MEMS and the microelectronics industry, the compact digital holoscope is a portable system that can be used for in situ process measurement as well as characterizing and inspection of microsystems and devices. For transparent objects such as microlenses and other optical elements, the system can characterize these components. Indeed, some novel applications of the transmission system for ferroelectric and liquid crystal based diffractive optical element testing have been demonstrated. Finally, optical systems suffer from a lack of spatial resolution. While the axial resolution is high (~10 nm), the spatial resolution is diffraction limited. A novel approach, which enables smaller objects to be monitored, has been shown to enhance this spatial resolution.

References

1. H. Yan, and A. Asundi. Studies on aperture synthesis in digital Fresnel holography. *Optics and Lasers in Engineering* 50:(4)556–562 (2011).
2. D. Gabor. A new microscopic principle. *Nature* 161:777–778 (1948).
3. S. Kostianovski, S. G. Lipson, and E. N. Ribak. Interference microscopy and Fourier fringe analysis applied to measuring the spatial refractive-index distribution. *Applied Optics* 32:7 (1993).
4. E. Cuche, F. Bevilacqua, and C. Depeursinge. Digital holography for quantitative phase-contrast imaging. *Optics Letters* 24:291 (1999).
5. F. Zhang, I. Yamaguchi, and L. P. Yaroslavsky, Algorithm for reconstruction of digital holograms with adjustable magnification. *Optics Letters* 29:1668–1670 (2004).
6. S. D. Nicola, A. Finizio, G. Pierattini, P. Ferraro, and A. D. Alfieri. Angular spectrum method with correction of anamorphism for numerical reconstruction of digital holograms on tilted planes. *Optics Express* 13:9935–9940 (2005) (http://www.opticsinfobase.org/oe/abstract.cfm?uri=oe-13-24-9935).
7. L. Yu, and M. K. Kim. Pixel resolution control in numerical reconstruction of digital holography. *Optics Letters* 31:897–899 (2006).

8. W. Haddad, D. Cullen, J. C. Solem, J. M. Longworth, A. McPherson, K. Boyer, and C. K. Rhodes. Fourier-transform holographicmicroscope. *Applied Optics* 31:4973–4978 (1992).

9. T. Colomb, F. Montfort, J. Kühn, N. Aspert, E. Cuche, A. Marian, F. Charrière, et al. Numerical parametric lens for shifting, magnification, and complete aberration compensation in digital holographic microscopy. *Journal Optical Society America* A 23:3177 (2006).

10. T. Colomb, E. Cuche, F. Charrière, J. Kühn, N. Aspert, F. Montfort, P. Marquet, and A. C. Depeursinge. Automatic procedure for aberration compensation in digital holographic microscopy and applications to specimen shape compensation. *Applied Optics* 45:851 (2006).

11. E. Cuche, P. Marquet, and A. C. Depeursinge. Simultaneous amplitude-contrast and quantitative phase-contrast microscopy by numerical reconstruction of Fresnel off-axis holograms. *Applied Optics* 38:6994–7001 (1999).

12. A. Stadelmaier, and J. H. Massig. Compensation of lens aberrations in digital holography. *Optics Letters* 25:3 (2000).

13. D. Carl, B. Kemper, G. Wernicke, and G. V. Bally. Parameter-optimized digital holographic microscope for high-resolution living-cell analysis. *Applied Optics* 43:9 (2004).

14. P. Ferraro, S. D. Nicola, A. Finizio, G. Coppola, S. Grilli, C. Magro, and A. G. Pierattini. Compensation of the inherent wave front curvature in digital holographic coherent microscopy for quantitative phase-contrast imaging. *Applied Optics* 42:1938–1946 (2003).

15. F. Montfort, F. Charrière, T. Colomb, E. Cuche, P. Marquet, and A. C. Depeursinge. Purely numerical compensation for microscope objective phase curvature in digital holographic microscopy: Influence of digital phase mask position. *Journal Optical Society America* A 23:2944 (2006).

16. W. Qu, C. O. Choo, Y. Yingjie, and A. Asundi, Microlens characterization by digital holographic microscopy with physical spherical phase compensation. *Applied Optics* 49:6448–6454 (2010).

17. D. Malacara, ed. *Optical shop testing.* New York: Wiley (1992).

18. C. Mann, L. Yu, C.-M. Lo, and M. Kim. High-resolution quantitative phase-contrast microscopy by digital holography. *Optics Express* 13:8693–8698 (2005).

19. Z. Ya'nan, Q. Weijuan, L. De'an, L. Zhu, Z. Yu, and L. Liren. Ridge-shape phase distribution adjacent to 180° domain wall in congruent LiNbO3 crystal. *Applied Physics Letters* 89:112912 (2006).

20. W. Qu, C. O. Choo, V. R. Singh, Y. Yingjie, and A. Asundi. Quasi-physical phase compensation in digital holographic microscopy. *Journal Optical Society America* A 26:2005–2011 (2009).

21. J. A. Ferrari, and E. M. Frins. Single-element interferometer. *Optics Communications* 279:235–239 (2007).

22. Q. Weijuan, Y. Yingjie, C. O. Choo, and A. Asundi. Digital holographic microscopy with physical phase compensation. *Optics Letters* 34:1276–1278 (2009).

23. Q. Weijuan, K. Bhattacharya, C. O. Choo, Y. Yingjie, and A. Asundi. Transmission digital holographic microscopy based on a beam-splitter cube interferometer. *Applied Optics* 48:2778–2783 (2009).

24. B. Kemper, and G. V. Bally. Digital holographic microscopy for live cell applications and technical inspection. *Applied Optics* 47:10 (2008).

25. E. Cuche, P. Marquet, and C. Depeursinge. Spatial filtering for zero-order and twin-image elimination in digital off-axis holography. *Applied Optics* 39:4070 (2000).

26. Q. Xi, D. A. Liu, Y. N. Zhi, Z. Luan, and A. L. Liu. Reversible electrochromic effect accompanying domain-inversion in LiNbO3:Ru:Fe crystals. *Applied Physics Letters* 87:121103 (2005).

27. Y. N. Zhi, D. A. Liu, Y. Zhou, Z. Chai, and A. L. Liu. Electrochromism accompanying ferroelectric domain inversion in congruent RuO2:LiNbO3 crystal. *Optics Express* 13:10172 (2005).

28. S. Grilli, P. Ferraro, M. Paturzo, D. Alfieri, P. D. Natale, M. D. Angelis, S. D. Nicola, A. Finizio, and A. G. Pierattini. In-situ visualization, monitoring and analysis of electric field domain reversal process in ferroelectric crystals by digital holography. *Optics Express* 12:1832 (2004).

29. M. Jazbinsek, and M. Zgonik. Material tensor parameters of LiNbO3 relevant for electro- and elasto-optics. *Applied Physics* B 74:407–414 (2002).

30. H. Takahashi, N. Kureyama, and T. Aida. Flatbed-type bidirectional three-dimensional display system. *International Journal of Innovative Computing, Information and Control* 5:4115–4124 (2009).

31. U. Schnars. Direct phase determination in hologram interferometry with use of digitally recorded holograms. *Journal of the Optical Society of America A—Optics Image Science and Vision* 11:2011–2015 (1994).

32. U. Schnars, and W. Juptner. Direct recording of holograms by a CCD target and numerical reconstruction. *Applied Optics* 33:179–181 (1994) (<Go to ISI>://A1994MX44500005).

33. U. Schnars, and W. P. O. Juptner. Digital recording and numerical reconstruction of holograms. *Measurement Science & Technology* 13:R85–R101 (2002) (<Go to ISI>://000178298700001).

34. D. P. Kelly, B. M. Hennelly, N. Pandey, T. J. Naughton, and W. T. Rhodes. Resolution limits in practical digital holographic systems. *Optical Engineering* 48 (9)095801-1-13 (2009) (<Go to ISI>://000270882000012).

35. L. Xu, X. Y. Peng, Z. X. Guo, J. M. Miao, and A. Asundi. Imaging analysis of digital holography. *Optics Express* 13:2444–2452 (2005) (<Go to ISI>://000228180800024).

36. H. Z. Jin, H. Wan, Y. P. Zhang, Y. Li, and P. Z. Qiu. The influence of structural parameters of CCD on the reconstruction image of digital holograms. *Journal of Modern Optics* 55:2989–3000 (2008) (<Go to ISI>://000261381800008).

37. A. A. Yan Hao. Resolution analysis of a digital holography system. *Applied Optics* 50:11 (2011).

38. T. Colomb, J. K. Kuhn, F. Charriere, C. Depeursinge, P. Marquet, and N. Aspert. Total aberrations compensation in digital holographic microscopy with a reference conjugated hologram. *Optics Express* 14:4300–4306 (2006) (<Go to ISI>://000237608600011).

39. F. Charriere, J. Kuhn, T. Colomb, F. Montfort, E. Cuche, Y. Emery, K. Weible, P. Marquet, and C. Depeursinge. Characterization of microlenses by digital holographic microscopy. *Applied Optics* 45:829–835 (2006) (<Go to ISI>://000235387400003).

40. B. Rappaz, P. Marquet, E. Cuche, Y. Emery, C. Depeursinge, and P. J. Magistretti. Measurement of the integral refractive index and dynamic cell morphometry of living cells with digital holographic microscopy. *Optics Express* 13:9361–9373 (2005) (<Go to ISI>://000233334900026).

41. F. Charriere, N. Pavillon, T. Colomb, C. Depeursinge, T. J. Heger, E. A. D. Mitchell, P. Marquet, and B. Rappaz. Living specimen tomography by digital holographic microscopy: Morphometry of testate amoeba. *Optics Express* 14:7005–7013 (2006) (<Go to ISI>://000239861100004).

42. C. Liu, Y. S. Bae, W. Z. Yang, and D. Y. Kim. All-in-one multifunctional optical microscope with a single holographic measurement. *Optical Engineering* 47 (8)087001-1-7 (2008) (<Go to ISI>://000259865500025).

43. F. Charriere, A. Marian, F. Montfort, J. Kuehn, T. Colomb, E. Cuche, P. Marquet, and C. Depeursinge. Cell refractive index tomography by digital holographic microscopy. *Optics Letters* 31:178–180 (2006) (<Go to ISI>://000234665000013).

44. P. Marquet, B. Rappaz, P. J. Magistretti, E. Cuche, Y. Emery, T. Colomb, and C. Depeursinge. Digital holographic microscopy: A noninvasive contrast imaging technique allowing quantitative visualization of living cells with subwavelength axial accuracy. *Optics Letters* 30:468–470 (2005) (<Go to ISI>://000227371800006).

45. F. Montfort, T. Colomb, F. Charriere, J. Kuhn, P. Marquet, E. Cuche, S. Herminjard, and C. Depeursinge. Submicrometer optical tomography by multiple-wavelength digital holographic microscopy. *Applied Optics* 45:8209–8217 (2006) (<Go to ISI>://WOS:000241888200006).

46. F. Charriere, T. Colomb, F. Montfort, E. Cuche, P. Marquet, and C. Depeursinge. Shot-noise influence on the reconstructed phase image signal-to-noise ratio in digital holographic microscopy. *Applied Optics* 45:7667–7673 (2006) (<Go to ISI>://WOS:000241084300017).

47. F. Charriere, A. Marian, T. Colomb, P. Marquet, and C. Depeursinge. Amplitude point-spread function measurement of high-NA microscope objectives by digital holographic microscopy. *Optics Letters* 32:2456–2458 (2007) (<Go to ISI>://WOS:000249327600062).

48. I. Yamaguchi, J. Kato, S. Ohta, and J. Mizuno. Image formation in phase-shifting digital holography and applications to microscopy. *Applied Optics* 40:6177–6186 (2001) (<Go to ISI>://WOS:000172713000005).

49. A. Stern, and B. Javidi. Improved-resolution digital holography using the generalized sampling theorem for locally band-limited fields. *Journal of the Optical Society of America A—Optics Image Science and Vision* 23:1227–1235 (2006), (<Go to ISI>://000237303600029).

50. T. Kreis, M. Adams, and W. Juptner. Aperture synthesis in digital holography. In *Interferometry XI: Techniques and analysis,* ed. K. Creath and J. Schmit, 69–76, Bellingham, WA: SPIE (2002).

51. J. H. Massig. Digital off-axis holography with a synthetic aperture. *Optics Letters* 27:2179–2181 (2002) (<Go to ISI>://000179795000011).

52. L. Martinez-Leon, and B. Javidi. Improved resolution synthetic aperture holographic imaging. Art. no. 67780A. In *Three-dimensional TV, video, and display VI,* ed. B. Javidi, F. Okano, and J. Y. Son, A7780–A7780 (2007).

53. D. Claus. High resolution digital holographic synthetic aperture applied to deformation measurement and extended depth of field method. *Applied Optics* 49:3187–3198 (2010) (<Go to ISI>://000278265600027).

54. F. Gyimesi, Z. Fuzessy, V. Borbely, B. Raczkevi, G. Molnar, A. Czitrovszky, A. T. Nagy, G. Molnarka, A. Lotfi, A. Nagy, I. Harmati, and D. Szigethy. Half-magnitude extensions of resolution and field of view in digital holography by scanning and magnification. *Applied Optics* 48:6026–6034 (2009) (<Go to ISI>://000271374000044).

55. J. L. Di, J. L. Zhao, H. Z. Jiang, P. Zhang, Q. Fan, and W. W. Sun. High resolution digital holographic microscopy with a wide field of view based on a synthetic aperture technique and use of linear CCD scanning. *Applied Optics* 47:5654–5659 (2008) (<Go to ISI>://000260726000013).

56. L. Martinez-Leon, and B. Javidi. Synthetic aperture single-exposure on-axis digital holography. *Optics Express* 16:161–169 (2008) (<Go to ISI>://000252234800019).

57. T. Kreis, and K. Schluter. Resolution enhancement by aperture synthesis in digital holography, *Optical Engineering* 46 (5)055803-1-7 (2007) (<Go to ISI>://000247812900042).

58. S. A. Alexandrov, T. R. Hillman, T. Gutzler, and D. D. Sampson. Synthetic aperture Fourier holographic optical microscopy. *Physical Review Letters* 97 (16):168102 (2006), <Go to ISI>://000241405400066.

59. T. R. Hillman, T. Gutzler, S. A. Alexandrov, and D. D. Sampson. High-resolution, wide-field object reconstruction with synthetic aperture Fourier holographic optical microscopy. *Optics Express* 17:7873–7892 (2009) (<Go to ISI>://000266381900017).

60. V. Mico, Z. Zalevsky, P. Garcia-Martinez, and J. Garcia. Synthetic aperture super-resolution with multiple off-axis holograms. *Journal of the Optical Society of America* A—Optics Image Science and Vision 23:3162–3170 (2006) (<Go to ISI>://000242326400019).

61. C. J. Yuan, H. C. Zhai, and H. T. Liu. Angular multiplexing in pulsed digital holography for aperture synthesis. *Optics Letters* 33:2356–2358 (2008) (<Go to ISI>://000260970800025).

62. L. Granero, V. Mico, Z. Zalevsky, and J. Garcia. Synthetic aperture super-resolved microscopy in digital lensless Fourier holography by time and angular multiplexing of the object information. *Applied Optics* 49:845–857 (2010) (<Go to ISI>://000274444100013).

63. P. Feng, X. Wen, and R. Lu. Long-working-distance synthetic aperture Fresnel off-axis digital holography. *Optics Express* 17:5473–5480 (2009) (<Go to ISI>://000264747500064).

64. V. Mico, Z. Zalevsky, P. Garcia-Martinez, and J. Garcia. Superresolved imaging in digital holography by superposition of tilted wave fronts. *Applied Optics* 45:822–828 (2006) (<Go to ISI>://000235387400002).

65. R. Binet, J. Colineau, and J. C. Lehureau. Short-range synthetic aperture imaging at 633 nm by digital holography. *Applied Optics* 41:4775–4782 (2002) (<Go to ISI>://000177327100003).

66. Y. Zhang, X. X. Lu, Y. L. Luo, L. Y. Zhong, and C. L. She. Synthetic aperture digital holography by movement of object. In *Holography, diffractive optics, and applications II,* Pts 1 and 2, ed. Y. L. Sheng, D. S. Hsu, C. X. Yu, and B. H. Lee, 581–588 Bellingham, WA: SPIE (2005).

67. S. D. N. Pietro Ferraro, A. Finizio, G. Pierattini, and G. Coppola. Recovering image resolution in reconstructing digital off-axis holograms by Fresnel-transform method. *Applied Physics Letters* 85 (14)2709–2711 (2004).

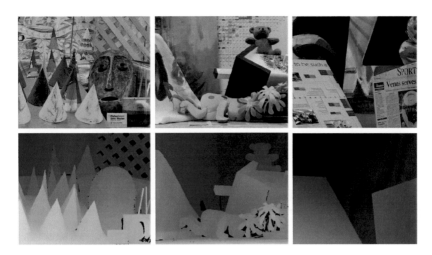

FIGURE 1.13
Stereo test images from the Middlebury database (http://vision.middlebury.edu/stereo/). Top
is original images and bottom is disparity images. From left: cone, teddy, and Venus.

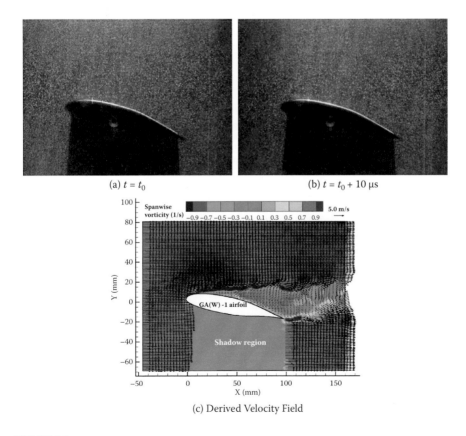

(a) $t = t_0$ (b) $t = t_0 + 10\ \mu s$

(c) Derived Velocity Field

FIGURE 4.1
A pair of PIV images and the corresponding velocity distribution. (Hu, H. and Yang, Z. 2008.
ASME Journal of Fluid Engineering 130 (5): 051101. With permission.)

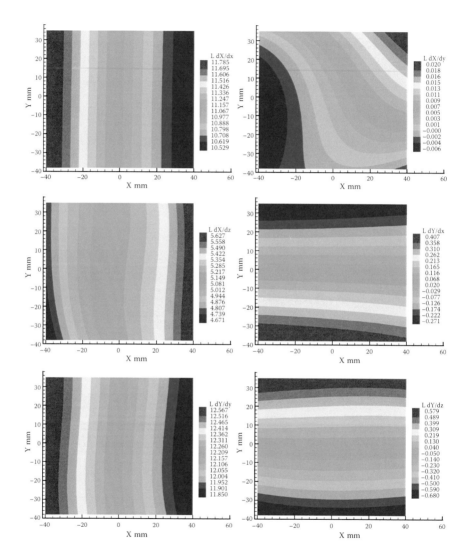

FIGURE 4.8
The gradients of the left-hand image recording camera for stereo image recording.

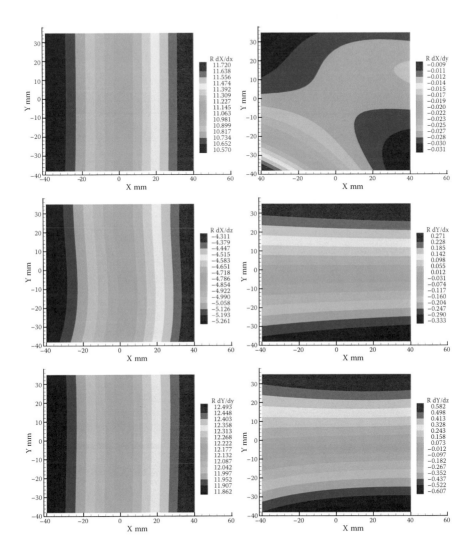

FIGURE 4.9
The gradients of the right-hand image recording camera for stereo image recording.

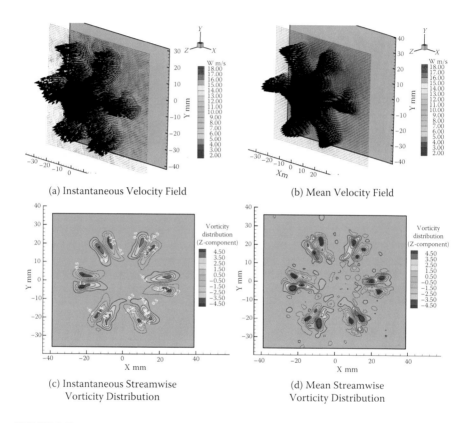

(a) Instantaneous Velocity Field

(b) Mean Velocity Field

(c) Instantaneous Streamwise
Vorticity Distribution

(d) Mean Streamwise
Vorticity Distribution

FIGURE 4.13
Stereo PIV measurement results in the $Z/D = 0.25$ ($Z/H = 0.67$) cross plane.

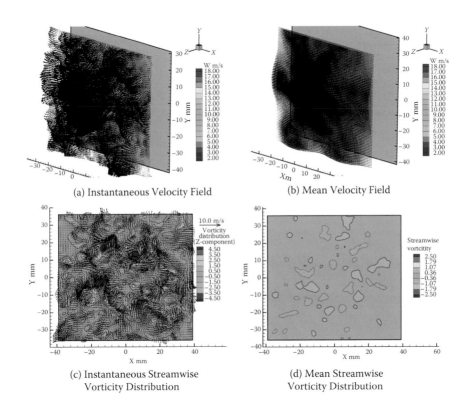

(a) Instantaneous Velocity Field

(b) Mean Velocity Field

(c) Instantaneous Streamwise
Vorticity Distribution

(d) Mean Streamwise
Vorticity Distribution

FIGURE 4.14
Stereo PIV measurement results in the Z/D = 3.0 (Z/H = 8.0) cross plane.

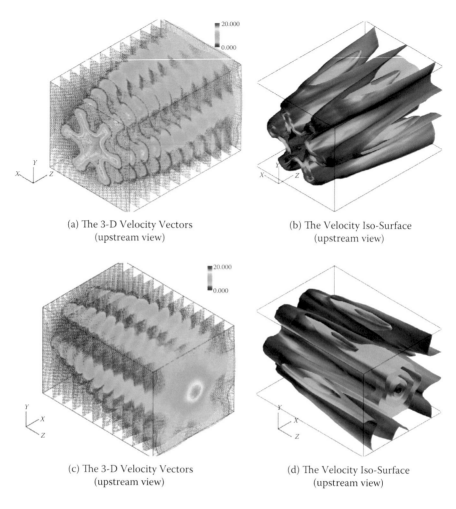

(a) The 3-D Velocity Vectors
(upstream view)

(b) The Velocity Iso-Surface
(upstream view)

(c) The 3-D Velocity Vectors
(upstream view)

(d) The Velocity Iso-Surface
(upstream view)

FIGURE 4.15
Reconstructed three-dimensional flow fields of the lobed jet mixing flow.

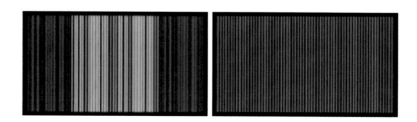

FIGURE 5.1
Pattern proposed by Pages et al. RGB pattern and luminance channel. (Pages, J. et al. *17th International Conference on Pattern Recognition, ICPR 2004* 4:284–287, 2004. With permission.)

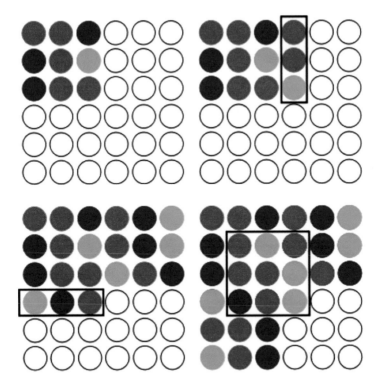

FIGURE 5.2
Code generation direction followed by Morano et al. with colored spots representation. (Morano, R. A. et al. *IEEE Transactions on Pattern Analysis and Machine Intelligence* 20 (3): 322–327, 1998. With permission.)

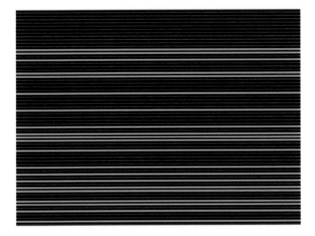

FIGURE 5.8
Pattern of the proposed method; $m = 64$ sinusoidal fringes are coded in color using a De Bruijn code generator algorithm.

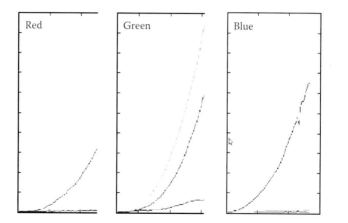

FIGURE 5.9
Received color intensities for projected increasing values of red, green, and blue, respectively.

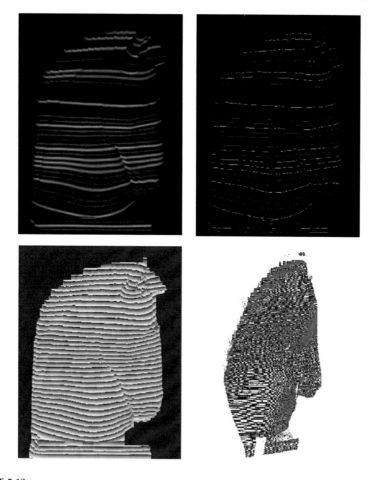

FIGURE 5.12
Proposed algorithm: input image (top left), extracted color slits (top right), combined slits and WFT fringes (bottom left), and 3D cloud of points (bottom right).

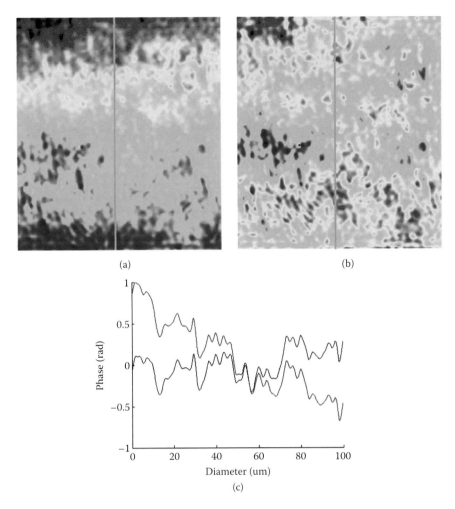

FIGURE 6.7
(a) Phase with subpixel tilt; (b) phase without subpixel tilt; (c) phase profile comparison along the dark lines in (a) and (b).

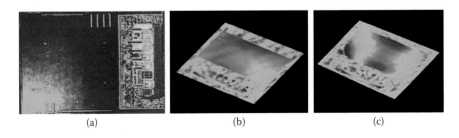

FIGURE 6.12
Profile of accelerometer: (a) bare, unmounted; (b) mounted on ceramic substrate; (c) mounted on plastic substrate.

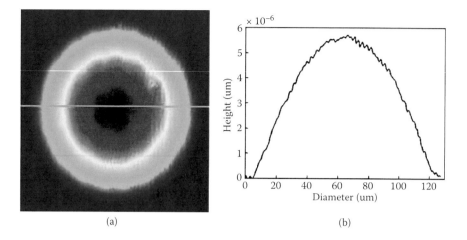

(a)   (b)

FIGURE 6.22

(a) Height map of a single microlens; (b) height profile along the solid line in (a).

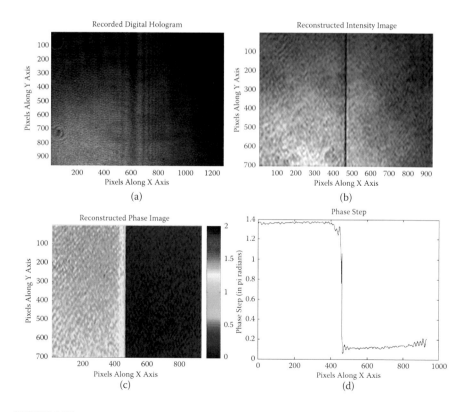

FIGURE 6.24

(a) Recorded digital hologram numerically reconstructed; (b) intensity image; (c) phase image; (d) phase step height measured from digitally recorded hologram with the TN-LCSLM addressed with 0 and 255 grayscales in two equally separated regions.

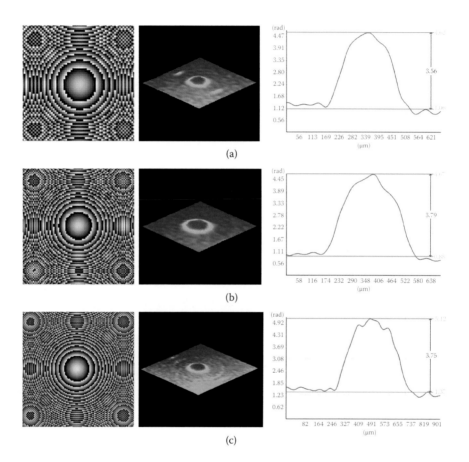

FIGURE 6.25
Diffractive optics lens with focal lengths with their numerical phase reconstruction using
DH method and line profile traced at the center in respective order: (a) 80 mm; (b) 120 mm;
(c) 160 mm.

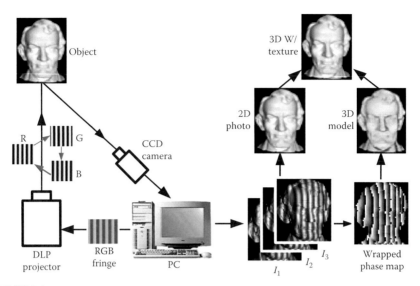

FIGURE 9.4
The layout of the real-time 3D profilometry system we developed. (Modified from Zhang, S. *Optics and Lasers in Engineering* 48:149, 2010. With permission.)

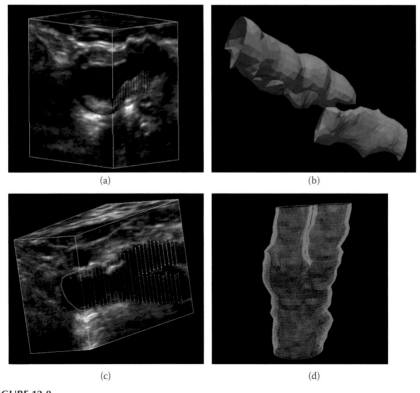

(a)

(b)

(c)

(d)

FIGURE 12.8
Manual segmentation of the CCA, ICA, and ECA lumen and outer wall boundaries from 3D US images are used to calculate the vessel wall volume (VWV). The reconstructed surfaces for the lumen and outer wall boundaries from the manual segmentations are shown.

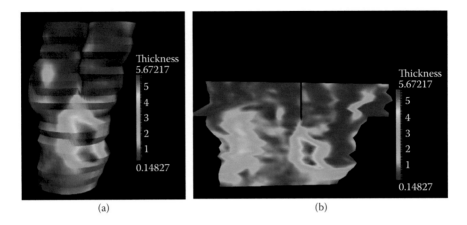

(a)   (b)

FIGURE 12.9
(a) Vessel wall thickness map for a patient with moderate stenosis. Manual segmentations of the lumen and outer wall boundaries were used to generate the thickness maps (indicated in millimeters). (b) Corresponding flattened thickness map for better visualization.

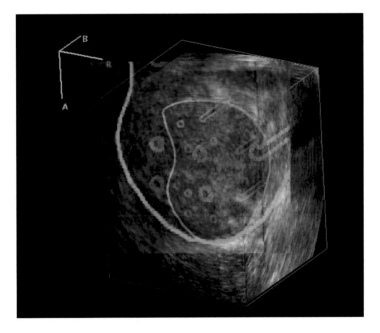

FIGURE 12.11
A 3D US image of the prostate acquired using an end-firing endocavity rotational 3D scanning approach (rotation of a TRUS transducer), which was used to guide a biopsy using the system shown in Figure 12.10. The transducer was rotated around its long axis, while 3D US images were acquired and reconstructed. The image also showed the segmented boundary of the prostate and the locations of the biopsy cores within the prostate.

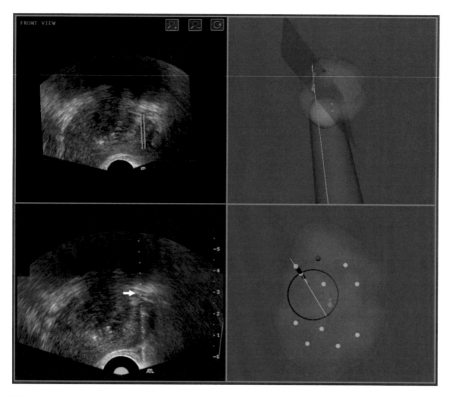

FIGURE 12.12
The 3D TRUS-guide biopsy system's interface is composed of four windows: (top left) the 3D TRUS image dynamically sliced to match the real-time TRUS transducer 3D orientation; (bottom left) the live 2D TRUS video stream; (right side) the 3D location of the biopsy core is displayed within the 3D prostate models. The targeting ring in the bottom right window shows all the possible needle paths that intersect the preplanned target by rotating the TRUS about its long axis. This allows the physician to maneuver the TRUS transducer to the target (highlighted by the red dot) in the shortest possible distance. The biopsy needle (arrow) is visible within the real-time 2D TRUS image.

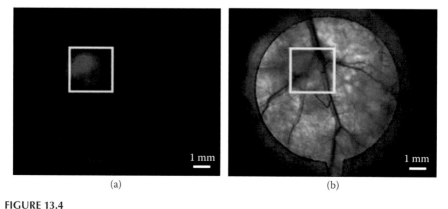

(a)   (b)

FIGURE 13.4
Wide-field (a) fluorescence and (b) corresponding white light image of a tumor-bearing (ME-180) nude mouse implanted within a mouse window chamber. The square region is imaged with Doppler and speckle variance optical coherence tomography (shown in Figure 13.5).

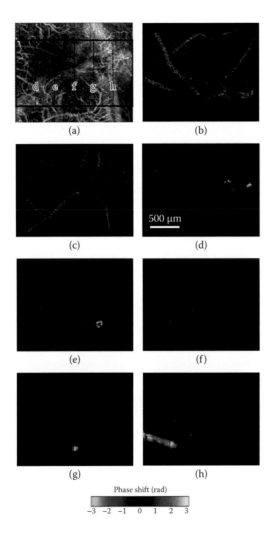

FIGURE 13.5
Functional imaging of a tumor-bearing nude mouse 1 week after ME-180 cell implantation. (a) Speckle-variance OCT image (3×3 mm²) of a tumor and surrounding blood vessels. A necrotic core is present, which manifests as a region devoid of vessels. The labels indicate the locations of DOCT image slices. (b, c) Three-dimensional reconstruction of the detected Doppler signal viewed from an angle and parallel (x–z plane) and perpendicular (x–y plane) to the imaging beam, respectively. (d–h) Corresponding DOCT image slices labeled in (a). Each image is 2.0 mm across and has a 1.7 mm depth of view.

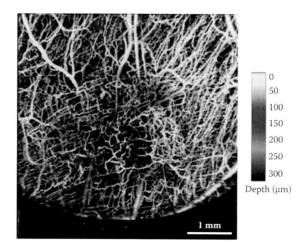

FIGURE 13.7
Second-generation speckle variance optical coherence tomography image of the mouse dorsal skin. The color indicates the relative depth of the vessels. A tumor mass (ME-180) is in the center of the image. The black border at the bottom is the edge of the plastic fastener of the tissue window.

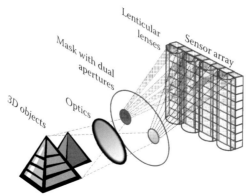

FIGURE 14.7
VisionSense stereoscopic camera: The camera has a mask with two apertures, a lenticular microlens array in front of the sensor chip. Each lenslet is 2 pixels wide. Rays from the 3D object that pass through the left aperture generate a left view on columns marked with a red color, and rays that pass through the right aperture generate a right view image on neighboring columns marked with a green color. A pair of stereoscopic images is thus generated using a single sensor chip. (Yaron, A. et al. www.visionsense.com.)

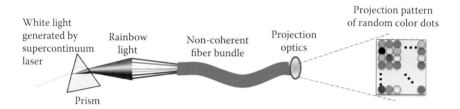

FIGURE 14.11
Spectrally encoded fiber-based structured lighting probe. (Clancy, N. T. et al. *Biomedical Optics Express* 2:3119–3128. With permission.)

7

3D Dynamic Shape Measurement Using the Grating Projection Technique

Xianyu Su, Qican Zhang, and Wenjing Chen

CONTENTS

Three-dimensional (3D) shape measuring techniques, using a combination of grating projection and a most frequently used mathematical tool—Fourier fringe analysis—have been deeply researched and are increasing in number. These kinds of techniques are based on the idea of projecting and superposing a carrier fringe pattern onto the surface of the tested object and then reconstructing its corresponding 3D shape from the deformed fringe pattern modulated by the height of the tested object and captured by a camera from another view direction. In this chapter, the basic principles and some proof-of-principle applications of 3D dynamic shape measurement using grating projection and Fourier fringe analysis will be demonstrated to review our research results of this combined approach. This chapter mainly focuses on this technology and its applications that we have developed over the past 10 years. Section 7.1 gives a brief introduction of 3D shape measurement using grating projection. Section 7.2 describes the basic principles of 3D dynamic shape measurement using grating projection and Fourier fringe analysis. Section 7.3 demonstrates some typical applications that we achieved. Section 7.4 expresses the fundamental concept of the time-average fringe method for vibration mode analysis and its experimental results. In the last section, we discuss the advantages and challenges of this technique and the current development of real-time measurement in this research field.

7.1 Introduction

Optical noncontact 3D shape measurement based on grating projection (namely, structured illumination) is concerned with extracting the geometric information from an image of the measured object. With its excellence in high speed and high accuracy, it has been widely used for industry inspection, quality control, biomedical engineering, dressmaking, and machine vision [1,2].

By means of two-dimensional (2D) grating projection, we have the possibility of recovering 3D shape information with the advantages of speed and full field. Many varieties of projected grating, such as black and white grating, grayscale grating, color-coded grating, and Gray-coded binary fringe sequences (which can be also classified as Ronchi, sinusoidal, and saw tooth grating), have been proposed and used in 3D shape measurement [3]. Actually, a more commonly used fringe is 2D Ronchi or sinusoidal grating.

There are two types of projection units. The first uses a conventional imaging system to make an image of the mask, precisely produced from a well designed pattern onto the surface of the tested object. The new type of projection unit uses a programmable spatial light modulator (SLM)—for example, a liquid crystal display (LCD) or digital micromirror device (DMD)—to control the spacing, color, and structure of the projected grating precisely.

Meanwhile, several methods of fringe analysis have been exhaustively studied for 3D shape measurement of a static object, including the moiré technique (MT) [4], phase-measuring profilometry (PMP) [5,6], Fourier transformation profilometry (FTP) [7–10], modulation measurement profilometry (MMP) [11], spatial phase detection (SPD) [12,13], etc. For 3D dynamic shape measuring, a set of grayscale random stripe patterns is projected onto a dynamic object and then the time-varying depth maps are recovered by the space–time stereo method [14]. Some researchers have succeeded in establishing a high-speed projection unit by removing the DMD's color wheel to increase the rate at which three-step phase-shifting grayscale gratings are projected to 180 Hz; they have achieved some wonderful applications using this system [15].

Among these techniques, FTP, originally conceived and demonstrated by Takeda, Ina, and Kobayashi [7], is one of the most used methods because only one or two fringes are needed, full-field analysis, high precision, etc. In FTP, a Ronchi grating or a sinusoidal grating is projected onto an object, and the depth information of the object is encoded into the deformed fringe pattern recorded by an image acquisition sensor. The surface shape can be decoded by calculating Fourier transformation, filtering in the spatial frequency domain, and calculating the inverse Fourier transformation.

Compared with MT, FTP can accomplish a fully automatic distinction between a depression and an evaluation of the object shape. It requires no fringe order assignments or fringe center determination, and it requires no interpolation between fringes because it gives height distribution at each pixel over the entire field. Compared with PMP and MMP, FTP requires only one or two images of the deformed fringe pattern, which has become one of the most popular methods in real-time data processing and dynamic data processing.

After Takeda et al., the FTP method and its applications were extensively studied [16–28]. Many researchers worked to improve one-dimensional (1D) FTP method and extend its application. W. W. Macy [16] expanded 1D FTP to 2D FTP, and Bone discussed fringe pattern analysis issues using a 2D Fourier transform [17,18]. Two-dimensional Fourier transform and 2D filtering techniques were successfully introduced into specific applications [19–22]. In order to extend the measurable slope of FTP, a sinusoidal projection technique and π-phase-shifting grating technique, by which the measurable slope of height variation was nearly three times that of the original FTP, were proposed [9]. Some modified FTP permits the tested objects to have 3D steep shapes, discontinuous height step, and/or spatially isolated surfaces, or 360° entire shape [23–26]. The phase error introduced by the application of the Fourier transform method to improve the accuracy of FTP has been discussed in detail [27,28]. In addition, 3D dynamic measurement based on FTP has widely attracted research interest [29,30]. In recent years, 3D Fourier fringe analysis was proposed by Abdul-Rahman et al. [31]. In 2010, an editorial, "Fringe Projection Techniques: Whither We Are?" reviewed the fringe projection techniques and their applications [32]; this was helpful to people interested in 3D measurement based on the fringe projection technique.

We have dedicated our research effort to improving the FTP method, extending its application, and introducing it into 3D dynamic shape measurement over the past 10 years; we have developed some achievements in 3D dynamic shape measurement based on grating projection and Fourier fringe analysis (covered in this chapter and already published in either conference proceedings or journal articles). For more detailed descriptions and discussions of this technique, the references listed at the end of this chapter, especially our three published papers in *Optics and Lasers in Engineering* [10,29,30], are recommended.

7.2 Basic Principles of 3D Dynamic Shape Measurement Based on FTP

Fourier transformation profilometry for 3D dynamic shape measurement is usually implemented as follows. A Ronchi grating or sinusoidal grating is projected onto an object's surface to modulate its height distribution. Then, a sequence of dynamic deformed fringe images can be grabbed by a camera from the other view and rapidly saved in a computer. Next, data are processed by Fourier fringe analysis with three steps.

- By using Fourier transform, we obtain their spectra, which are isolated in the Fourier plane when the sampling theorem is satisfied
- By adopting a suitable band-pass filter (e.g., a Hann window) in the spatial frequency domain, all the frequency components are eliminated except the fundamental component. And by calculating inverse Fourier transform of the fundamental component, a sequence of phase maps can be obtained.
- By applying the phase-unwrapping algorithm in 3D phase space, the height distributions of the measured dynamic object in different time can be reconstructed under a perfect phase-to-height mapping.

7.2.1 Measuring System Based on Grating Projection

The optical geometry of the measurement system for dynamic scenes is similar to traditional FTP. Two options of optical geometry are available in FTP, including crossed optical axes geometry and parallel optical axes geometry. Each has its own merits as well as disadvantages [5]. The crossed optical axes geometry, as shown in Figure 7.1, is easy to construct because both a grating and an image sensor can be placed on the optical axes of the projector and the camera, respectively. It is very popular in FTP. Here, we use it as an example to describe the principle of FTP.

In Figure 7.1, the optical axis $E_p'-E_p$ of a projector lens crosses the optical axis $E_c'-E_c$ of a camera lens at point O on a reference plane, which is normal to the optical axis $E_c'-E_c$ and serves as a fictitious reference to measure the object height $Z(x, y)$. d is the distance between the projector system and the

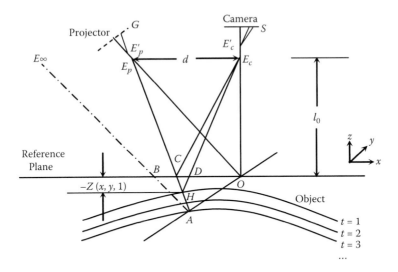

FIGURE 7.1
Crossed-optical-axes geometry of a 3D dynamic shape measurement system based on FTP.

camera system, and l_0 is the distance between the camera system and the reference plane.

By projecting a grating G onto the reference plane, its image (with period p_0) can be observed through the camera in another view and can be represented by

$$g_0(x,y) = \sum_{n=-\infty}^{+\infty} A_n r_0(x,y) \exp\{i[2n\pi f_0 x + n\phi_0(x,y)]\} \tag{7.1}$$

where
$r_0(x, y)$ is a nonuniform distribution of reflectivity on the reference plane
A_n are the weighting factors of Fourier series
f_0 ($f_0 = 1/p_0$) is the carrier frequency of the observed grating image in the x-direction
$\phi_0(x,y)$ is the original phase on the reference plane (i.e., $Z(x, y) = 0$)

The coordinate axes are chosen as shown in Figure 7.1. According to this optical geometry, an experimental setup can be established as shown in Figure 7.2.

When the measured object is stationary, the image intensity, which is obtained by the camera, is independent of the time and usually expressed as $g(x, y)$. But when a dynamic object whose height distribution varies with the time is placed into this optical field, the intensity of these fringe patterns is obviously a function of both the spatial coordinates and the time, and it can

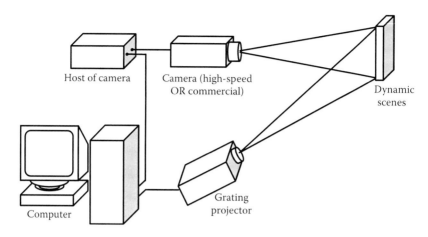

FIGURE 7.2
Schematic diagram of the experimental setup.

be marked as $g(x, y, z(t))$. The phase distribution that implicated the height variation of the measured dynamic object is also a function of the time and can be noted as $\varphi(x,y,t)$. Strictly speaking, in dynamic scenes, x and y coordinates are also changing with time and are much smaller in comparison with z coordinates. Therefore, their changes are usually ignored.

A sequence of the deformed fringe patterns can be grabbed by camera and rapidly stored in a computer. The intensity distributions of these fringe patterns in different time can be expressed as

$$g(x,y,z(t)) = \sum_{n=-\infty}^{+\infty} A_n r(x,y,t)\exp\{i[2n\pi f_0 x + n\phi(x,y,t)]\} \quad (t = 1,2,...,m) \quad (7.2)$$

where $r(x,y,t)$ and $\varphi(x,y,t)$ respectively represent a nonuniform distribution of reflectivity on the object surface and the phase modulation caused by the object height variation in the different time; m is the total number of all fringe images grabbed by the camera.

Fourier transform (filtering only the first-order term ($n = 1$) of the Fourier spectra) and inverse Fourier transform are carried out to deal with each fringe pattern grabbed by the camera at a different time. Complex signals at different times can be calculated:

$$\hat{g}(x,y,z(t)) = A_1 r(x,y,t)\exp\{i[2\pi f_0 x + \phi(x,y,t)]\} \quad (7.3)$$

The same operations are applied to the fringe pattern on the reference plane to obtain the complex signal of the reference plane:

$$\hat{g}_0(x,y) = A_1 r_0(x,y) \exp\{i[2\pi f_0 x + \phi_0(x,y)]\} \tag{7.4}$$

By knowledge of the geometrical and optical configuration of the system in Figure 7.1, the phase variation resulting from the object height distribution is

$$\Delta\phi(x,y,t) = \phi(x,y,t) - \phi_0(x,y) = 2\pi f_0 (\overline{BD} - \overline{BC}) = 2\pi f_0 \overline{CD}$$

$$= 2\pi f_0 \frac{-dZ(z,y,t)}{l_0 - Z(x,y,t)} \tag{7.5}$$

Since the phase calculated by computer gives principal values ranging from $-\pi$ to π, $\Delta\phi(x,y,t)$ is wrapped into this range and consequently has discontinuities with 2π-phase jumps. A phase-unwrapping process is required to correct these discontinuities by adding or subtracting 2π according to the phase jump ranging from π to $-\pi$ or vice versa.

Solving Equation (7.5) for $Z(x, y, t)$, the formula of height distribution can be obtained:

$$Z(x,y,t) = \frac{l_0 \Delta\Phi(x,y,t)}{\Delta\Phi(x,y,t) - 2\pi f_0 d} \tag{7.6}$$

where $\Delta\Phi(x,y,t)$ is the unwrapped phase distribution of $\Delta\phi(x,y,t)$. In practical measurements, usually $l_0 \gg Z(x,y,t)$, Equation (7.6) can be simplified as

$$Z(x,y,t) \approx -\frac{l_0}{2 f_0 d} \Delta\Phi(x,y,t) = -\frac{1}{2\pi\lambda_e} \Delta\Phi(x,y,t) \tag{7.7}$$

where λ_e is the equivalent wavelength of the measuring system. When the measuring system is not perfect, some research work has been carried out to solve the problems.

7.2.2 3D Phase Calculation and Unwrapping

The 3D wrapped phase can be calculated by two approaches. In the first one, the recorded 3D fringe data (time-sequence fringes) are regarded as a collection of many individual 2D fringes, and the wrapped phase can be calculated frame by frame in 2D spatial space according to the sampling time. Directly calculating the multiplication of $\hat{g}(x,y,z(t))$ with $\hat{g}^*(x,y,t=0)$ in each 2D space at different sampling times, the phase distribution $\Delta\phi(x,y,t)$ can be calculated by

$$\Delta\phi(x,y,t) = \phi(x,y,t) - \phi(x,y,0) = arctg \frac{\mathrm{Im}[\hat{g}(x,y,z(t))\hat{g}^*(x,y,t=0)]}{\mathrm{Re}[\hat{g}(x,y,z(t))\hat{g}^*(x,y,t=0)]} \tag{7.8}$$

where *Im* and *Re* represent the imaginary part and real part of $\hat{g}(x,y,z(t))\hat{g}*$ $(x,y,t=0)$, respectively. By this method, we can obtain a sequence of phase distributions contained in each deformed fringe that include the fluctuation information of a dynamic object. If the sampling rate is high enough, the phase difference between two adjacent sampling frames will be smaller, and it will lead to an easy phase-unwrapping process in 3D space. On the other hand, the fringe analysis could also be done in a single 3D volume rather than as a set of individual 2D frames that are processed in isolation. Three-dimensional FFT and 3D filtering in the frequency domain to obtain the 3D wrapped phase volume are equivalent in performance to 2D fringe analysis [31].

The phase calculation by any inverse trigonometric function (i.e., arctangents) provides the principal phase values ranging from $-\pi$ to π; consequently, phase values have discontinuities with 2π phase jumps. For phase data without noise, these discontinuities can be easily corrected by adding or subtracting 2π according to the phase jump ranging from $-\pi$ to π or vice versa. This is called the phase-unwrapping procedure [33–37]. Research on phase-unwrapping algorithm is also an attractive branch in FTP. Some achievements have been gained and employed in special application results [38,39].

Three-dimensional phase unwrapping must be conducted along *x*, *y*, and *t* directions. Compared to 2D phase unwrapping, it can provide more choices of the path for the unwrapping process. Some discontinuity points, which result from noise, shadow, and undersampling and cannot be unwrapped along *x* or *y* directions in their own frame, can be unwrapped successfully along the *t* direction. Therefore, compared with 2D phase unwrapping, 3D phase unwrapping is easier and more accurate.

If the wrapped phase is reliable everywhere and the phase difference between the neighboring pixels in a 3D phase space is less than π (the camera's frame rate is high enough), the unwrapping problem is trivial. The precise phase values in the whole 3D phase field can be obtained by calculating the sum of the phase difference along the *t* direction under this condition.

In an actual measurement, many factors, such as noise, fringe break, shadow, undersampling resulting from very high fringe density, and undermodulation from very sparse fringe density, make the actual unwrapping procedure complicated and path dependent. Combined with a modulation analysis technique, the 2D phase-unwrapping method based on modulation ordering has been discussed well in references 32 and 35. This 2D phase-unwrapping procedure based on modulation can be extended as a reliability function to 3D phase space. The unwrapping path is always along the direction from the pixel with a higher modulation to the pixel with lower modulation until the entire 3D wrapped phase is unwrapped. The unwrapping scheme is shown in Figure 7.3; the lines with arrows display one of the phase-unwrapping paths based on modulation ordering.

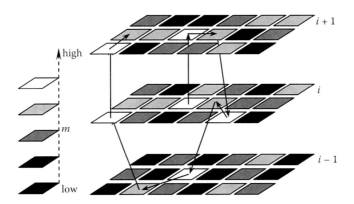

FIGURE 7.3
Sketch map of 3D phase unwrapping based on modulation ordering.

7.2.3 System Calibration

In practice, the system calibration will be completed through two steps. The first is to calibrate the used camera accurately. We adopt Zhang's calibration model [40] to find the intrinsic parameters and extrinsic parameter and the transformation between the (x, y, z) coordinates in a world coordinate system and that in the camera coordinate system. After the camera is well calibrated, the second step is to establish a phase-to-height mapping table, which is used to determine the height of each point of the object. The dual-direction nonlinear phase-to-height mapping technique is usually adopted (see detailed description in reference 41). Generally, the relation between the unwrapped phase and the height $Z(x, y, t)$ can be written as

$$\frac{1}{Z(x,y,t)} = a(x,y) + \frac{b(x,y)}{\Delta\Phi_r(x,y,t)} + \frac{c(x,y)}{\Delta\Phi_r^2(x,y,t)} \tag{7.9}$$

where, for a sampling instant t, $\Delta\Phi_r(x,y,t) = \Phi(x,y,t) - \Phi_r(x,y)$ is the phase difference between the two unwrapped phase distributions, $\Phi(x,y,t)$ for the measured object and $\Phi_r(x,y)$ for the reference plane. $Z(x,y,t)$ is the relative height from the reference plane; $a(x,y)$, $b(x,y)$, and $c(x,y)$ are the mapping parameters that can be calculated from the continuous phase distributions of four or more standard planes with known heights. The height distribution of the measured object at each sampling instant will be obtained by Equation (7.9), as long as its 3D phase distribution has been unwrapped.

7.2.4 Measuring Accuracy and Measurable Range

Since FTP is based on filtering for selecting only a single spectrum of the fundamental frequency component, the carrier frequency f_0 must separate this spectrum from all other spectra. This condition limits the maximum range measurable by FTP. The measurable slope of height variation of conditional FTP does not exceed the following limitation [8]:

$$\left|\frac{\partial Z(x,y)}{\partial x}\right|_{max} < \frac{l_0}{3d} \qquad (7.10)$$

When the measurable slope of height variation extends this limitation, the fundamental component will overlap the other components, and then the reconstruction will fail. In order to extend the measurable slope of FTP, the sine or quasi-sine projection technique and π-phase-shifting technique can be employed to make only fundamental component exist in a spatial frequency domain [9]. In this case, the lower frequency part of the fundamental component can extend to zero, and the higher part can extend to $2f_0$ without overlaps. The sinusoidal (quasi-sinusoidal) projection and π-phase-shifting technique result in a larger range of the measurement—that is,

$$\left|\frac{\partial h(x,y)}{\partial x}\right| < \frac{L_0}{d} \qquad (7.11)$$

The entire accuracy of a measurement system is determined by the system parameter d/l_0 and the grating period p_0. It can be improved by increasing the d/l_0 or decreasing the p_0 according to the maximum variation range of the measured height.

7.3 Some Applications of 3D Dynamic Shape Measurement

According to the speed of object's motion and the requisite precision of time in measurement from low speed to high speed, the measured dynamic objects can be divided into different types: the slow-movement process, the high-movement process, rapid rotation, and the instantaneous process. In our research work, we have proposed corresponding techniques and measurement systems for them.

7.3.1 3D Dynamic Shape Measurement for Slow-Movement Processes

For dynamic processes with slow movement, we have developed a low-cost measurement system in which a conventional imaging system is employed to produce an invariable sinusoidal optical field on the surface of the measured object; a video recording system is used to collect the deformed fringes.

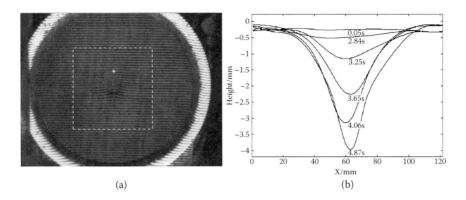

(a)   (b)

FIGURE 7.4
Dynamic measurement of vortex shape. (a) One of the deformed fringes when the stirrer is working; (b) profiles of the reconstructed vortices at different times.

This system succeeds in reconstructing the motion of vortex when the poster paint surface is stirring [42]. The deformed grating image is observed by a low-distortion TV charge-coupled device (CCD) camera (TM560, 25 frames per second [fps]) with a 12 mm focal length lens via a video-frame grabber digitizing the image in 128 × 128 pixels. One of the recorded fringes is shown in Figure 7.4(a). Figure 7.4(b) shows the profiles of the reconstructed vortices; the number under each line is the corresponding sampling instant.

7.3.2 3D Dynamic Shape Measurement for High-Movement Processes

For high-movement processes, a high-speed camera replaces the video recording system to record the deformed fringe rapidly. A measuring system is established to restore the 3D shape of a vibrating drum membrane [43]. A Chinese double-sided drum is mounted into the measuring volume and quickly hit three times. The whole vibration is recorded with 1000 fps in sampling rate speed. The vibration at the central point of the resonant side is shown in Figure 7.5(a).

The height distributions of the vibrating drumhead at their corresponding sampling instants are exactly restored. Figure 7.5(b) gives the profiles of the center row of six sampling instants in one period; the number above each line is the corresponding sampling instant.

Figure 7.5(c) and (d) show the grid charts of two sampling instants, and their sampling instants are given respectively as titles. Observing them, two modes, (1, 0) and (1, 1), can be found in one period of the principal vibration of the drumhead. This indicates that the point we hit is not the exact center of the drumhead. Furthermore, nonlinear effects (such as the uniform surface tension across the entire drumhead) exert their own preference for certain modes by transferring energy from one vibration mode to another.

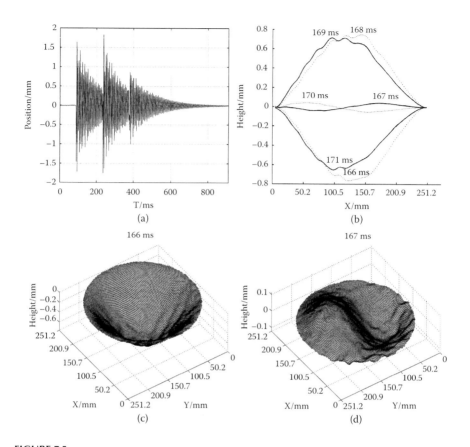

FIGURE 7.5

Three-dimensional dynamic shape measurement for a vibrating drum. (a) Vibration at the center point of the tested drumhead; (b) restored profiles of six sampling instants in one period; (c, d) grid charts of restored height distribution of the vibrating drumhead.

With the advantage of high-speed, full-field data acquisition and analysis, this system could be used in digitizing 3D shapes of those objects in rapid motion.

7.3.3 3D Dynamic Shape Measurement for Rapid Rotation and Instantaneous Process

The stroboscope [44] is an intense, high-speed light source used for the visual analysis of the objects in periodic motion and for high-speed photography [45,46]. The objects in rapid periodic motion can be studied by using the stroboscope to produce an optical illusion of stopped or slowed motion. When the flash repetition rate of the stroboscope is exactly the same as the object movement frequency or an integral multiple thereof, the moving object will appear to be stationary. This is called the *stroboscopic effect*.

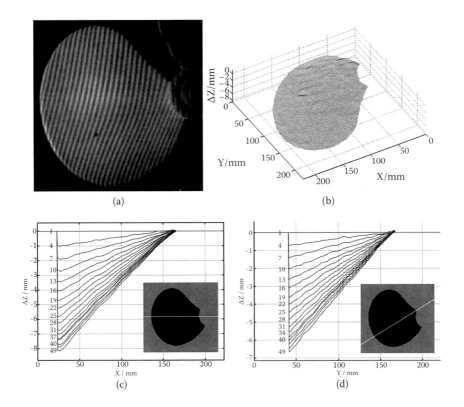

FIGURE 7.6
Three-dimensional dynamic shape measurement for a rotating blade. (a) Deformed fringes recorded at 50th revolution; (b) deformation (relative to the first revolution) at the 50th revolution; (c, d) deformations (relative to the first revolution) along the line shown in the inset.

Combining active 3D sensing with stroboscopic technology, we present another practical 3D shape measuring system for rapid rotation and instantaneous process, in which a stroboscopic sinusoidal grating is synchronously projected onto the dynamic tested object according to its motion and the deformed fringes are also synchronously collected by an ordinary commercial CCD camera or a high-speed one. The stroboscopic sinusoidal grating can "slow down" or even "freeze" the motion of the tested object and help the camera to shoot a clear, sharp, and instantaneous deformed fringe modulated by the 3D height distribution of a rapid motion object.

The 3D shape and deformation of a rotating blade with 1112 rpm (revolutions per minute) has been recovered by this method [47]. From quiescence to stable rotation, one image is synchronously captured during each revolution, as shown in Figure 7.6(a). Figure 7.6(b) shows the deformation (relative to the first revolution) of the whole reconstructed blade at the 50th revolution. Figure 7.6(c) and (d) show the deformations (relative to the first revolution also) of the reconstructed blades at difference revolutions along the line shown in each inset; the numbers marked on the left sides of the curves are their corresponding revolutions. They distinctly depict the twisting deformations of the blade with increasing speed. The farther the distance from the rotating shaft is, the bigger the deformation is.

Except for high-speed motion objects with obvious and variable repetition, such as rotation and vibration, this technique can also be used in the study of high-speed motion without obvious repetition, such as expansion, contraction, or ballistic flight and high-speed objects with known and invariable frequencies. This method can also be expanded using different detectors or sensors. For example, explosion phenomena with sound signals can be studied using a sound control unit instead of the optical position detector.

7.3.4 3D Dynamic Shape Measurement for Breaking Objects

When the measured dynamic object is deficient of sampling in a time direction or breaking into several isolated parts, the phase unwrapping is difficult. We propose a marked fringe method [48], in which a special mark is embedded into the projected sinusoidal gratings to identify the fringe order. The mark will not affect the Fourier spectra of the deformed fringe and could be extracted easily with a band-pass filter in the other direction. The phase value on the same marked strip is equivalent and known, so these phase values at the marked strips keep the relation of those separated fringes. The phase-unwrapping process of each broken part will be done from the local marked strip. In the experiment to test a breaking ceramic tile, the tile is obviously divided into four spatially isolated blocks and the fringe discontinuity is followed. The introduced marked fringe has been used to ensure that the phase-unwrapping process is error free. The whole breaking process contains 47 frames and lasts 235 ms. Six of these images are shown in Figure 7.7(a). Figure 7.7(b) shows the reconstructed 3D shape of this breaking tile at the corresponding sampling instants in Figure 7.7(a).

For the same purpose, Wei-huang Su [49] created a structured pattern in which the sinusoidal gratings are encoded with binary stripes and color grids. The binary stripes are used to identify the local fringe order, while the color grid provides additional degrees of freedom to identify the stripes. This encoding scheme provides a more reliable performance to identify the fringe orders since it is not sensitive to the observed colors. In Guo and Huang's research work [50,51], a small cross-shaped marker is embedded in the fringe pattern to facilitate the retrieval of the absolute phase map from a single fringe pattern. Another option of the marked fringe is two frequency sinusoidal gratings, in which the jumping line of the low-frequency gratings' wrapped phase could be used as the marked fringes and the same rank of high-frequency deformed grating will be tracked by this marked line.

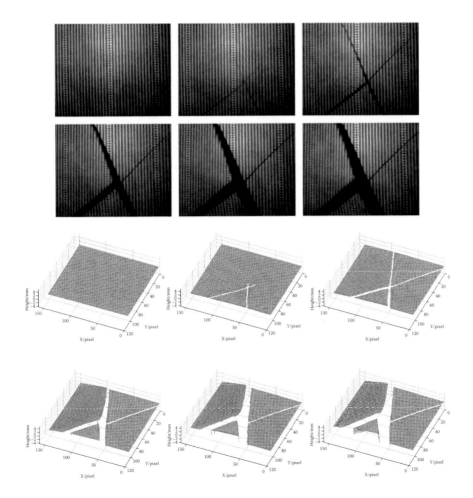

FIGURE 7.7
Three-dimensional dynamic shape measurement for a breaking tile. (a) Six frames of deformed fringe images in the breaking process; (b) the reconstructed 3D shape of the breaking tile at the corresponding sampling instants in (a).

7.4 Time-Average Fringe Method for Vibration Mode Analysis

In 3D dynamic shape measurement based on FTP, a sequence of dynamic deformed fringe images can be grabbed by a high-speed CCD camera and be processed by Fourier fringe analysis. This can efficiently measure a changing shape. When this proposed method is used for vibration analysis, it is a basic request that the time sampling rate of the detector be higher than twice that of the vibration frequency at least. It is difficult for a low-speed CCD or even a higher speed CCD to analyze the vibration with high frequency.

Similarly to time-average holographic technology for vibration analysis, the time-average fringe technology can also be used for the vibration mode analysis [52]. In this method, a sequence of the deformed and partly blurred sinusoidal fringe images on the surface of a vibrating membrane is grabbed by a low sampling rate commercial CCD camera. By Fourier transform, filtering, and inverse Fourier transform, the vibration mode is obtained from the fundamental component of the Fourier spectrum. Computer simulations and experiments have verified its validity. Under different excited vibration frequencies, the vibration modes of a vibrating surface can be qualitatively analyzed. When the excited vibration frequency changes continuously, the changing process of the vibration modes of the membrane is observed clearly.

7.4.1 Fundamental Concept

The time-average fringe method is implemented as follows. A Ronchi or sinusoidal grating is projected onto the surface of a vibrating membrane. A sequence of the deformed and partly blurred fringe images, caused by the changing shape of the tested vibrating membrane, is grabbed by a low sampling rate commercial CCD camera. Then, five steps are applied to implement the data processing.

1. Two-dimensional Fourier transform of the time-average fringe
2. Two-dimensional band-pass filtering (all frequency components of the Fourier spectrum are eliminated except the fundamental component)
3. Two-dimensional inverse Fourier transform of the filtered fundamental component
4. Calculation of the module of the complex signal extracted by 2D inverse Fourier transform
5. The same operation on the grabbed clear fringe pattern while the tested object is static

Ultimately, the vibration mode is obtained from the ratio of the modules of the two complex signals.

The optical geometry of the vibration analysis of a membrane is similar to traditional FTP, as shown in Figure 7.1. When the membrane is static, the intensity distributions of fringe patterns can still be expressed as

$$g_r(x,y) = a_r(x,y) + b_r(x,y)\cos[2\pi f_0 x + \phi_0(x,y)] \tag{7.12}$$

The vibration equation can be written as

$$Z(x,y,t) = A(x,y)\sin(2\pi ft) \tag{7.13}$$

where $A(x,y)$ is the vibration amplitudes and f is the vibration frequency. If the membrane is vibrating and each point of the membrane surface has an out-of-plane displacement, according to Equations (7.7) and (7.12), the varying phase distribution $\varphi(x,y,t)$ caused by the membrane's vibrating can be described as

$$\phi(x,y,t) = \frac{Z(x,y,t)}{\lambda_e} \cdot 2\pi \qquad (7.14)$$

The instantaneous intensity distributions of these fringe patterns in difference time can be expressed as

$$g_t(x,y,t) = a_t(x,y) + b_t(x,y)\cos[2\pi f_0 x + \phi_0(x,y,t) + \phi(x,y,t)] \qquad (7.15)$$

The Bessel functions are well known:

$$\cos(z\sin\theta) = J_0(z) + 2\sum_{n=1}^{\infty} \cos(2n\theta)J_{2n}(z) \qquad (7.16)$$

$$\sin(z\sin\theta) = 2\sum_{n=0}^{\infty} \sin[(2n+1)\theta]J_{2n+1}(z) \qquad (7.17)$$

Inserting Equations (7.13) and (7.14) into Equation (7.15) and developing the trigonometric terms into Bessel functions, the instantaneous intensity distributions is given by

$$g_t(x,y,t) = a_t(x,y) + b_t(x,y)\cos[2\pi f_0 x + \phi_0(x,y)] \cdot J_0\left[\frac{2A(x,y)\pi}{\lambda_e}\right]$$

$$+2b_t(x,y)\cos[2\pi f_0 x + \phi_0(x,y)] \cdot \sum_{n=1}^{\infty} \cos[2n \cdot (2\pi ft)]J_{2n}$$
$$\left[\frac{2A(x,y)\pi}{\lambda_e}\right] \qquad (7.18)$$

$$-2b_t(x,y)\sin[2\pi f_0 x + \phi_0(x,y)] \cdot \sum_{n=0}^{\infty} \sin[(2n+1)2\pi ft]J_{2n+1}$$
$$\left[\frac{2A(x,y)\pi}{\lambda_e}\right]$$

When these deformed fringe patterns are recorded by a low-speed commercial CCD camera with a sampling time T, which is always larger than the vibration period of the membrane, the fringe will be averaged and partly blurred during the sampling time. The averaged fringe intensity at each pixel is then given by

$$g_{averaged}(x,y) = \frac{1}{T} \int_0^T g_t(x,y,t)dt \qquad (7.19)$$

Finally, by inserting Equation (7.18) into Equation (7.19), we can obtain

$$g_{averaged}(x,y) = a_t(x,y) + b_t(x,y)\cos[2\pi f_0 x + \phi_0(x,y)] \cdot J_0\left[\frac{2A(x,y)\pi}{\lambda_e}\right] \qquad (7.20)$$

After the process of Fourier transform, filtering, and inverse Fourier transform, a complex signal will be calculated:

$$\hat{g}_{averaged}(x,y) = \frac{b_t(x,y)}{2} \cdot J_0\left[\frac{2A(x,y)\pi}{\lambda_e}\right] \exp\{i[2\pi f_0 x + \phi_0(x,y)]\} \qquad (7.21)$$

Meanwhile, the intensity distributions of fringe patterns on the static membrane (shown in Equation 7.12) can be dealt with in the same way and its corresponding complex signal can be noted as

$$\hat{g}_r(x,y) = \frac{b_r(x,y)}{2} \exp\{i[2\pi f_0 x + \phi_0(x,y)]\} \qquad (7.22)$$

Thus, the vibration mode I can be shown as

$$I = \frac{\left|\hat{g}_{averaged}(x,y)\right|}{\left|\hat{g}_r(x,y)\right|} = \left|J_0\left[\frac{2A(x,y)\pi}{\lambda_e}\right]\right| \qquad (7.23)$$

The relation between the vibration mode I and the vibration amplitudes $A(x,y)$ can be found via the zero-order Bessel function.

7.4.2 Experimental Results

A principal diagram of the experimental setup is similar to that in Figure 7.2. The experiment is performed on an aluminum membrane whose edges are tightly fixed on a loudspeaker. A function generator and a power amplifier

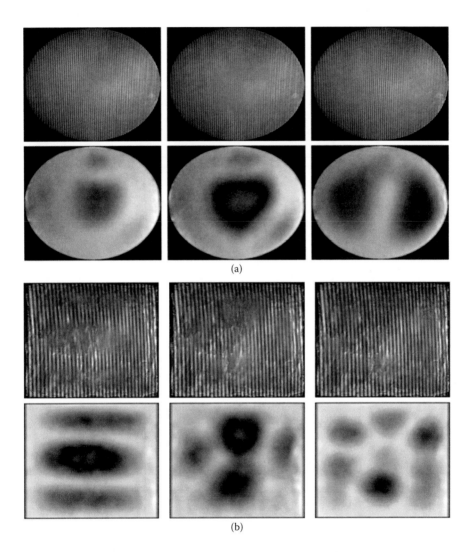

FIGURE 7.8
Deformed fringe images and their corresponding vibration mode: (a) for a circle membrane, mode (0,1) at 73.3 Hz, mode (0,2) at 82.2 Hz, mode (1,1) at 91.1 Hz; (b) for a rectangular membrane, mode (1,3) at 125.7 Hz, mode (3,1) at 163.7 Hz, mode (3,2) at 206.7 Hz.

are used to create a sine wave, which is transferred to the loudspeaker and drives the membrane to vibrate. Under different excited vibration frequencies, the vibration modes of a vibrating circle membrane and a vibrating rectangular membrane can be recovered. Figure 7.8(a) gives three deformed fringe images and their corresponding vibration modes for a circle membrane. Figure 7.8(b) gives three deformed fringe images and their corresponding vibration modes for a rectangular membrane.

The results of the theoretical analysis and experiment indicate that this method of vibration mode analysis using a time-average fringe is valid. This method has the advantage of rapid speed, high accuracy, and simple experimental setup.

7.5 Advantages and Challenges

In this chapter, we mainly reviewed our past 10+ years of research work on 3D dynamic shape measurement using grating projection and Fourier fringe analysis. The basic principles of this method were introduced, and some typical applications, such as the combination of stroboscopic effects and FTP to test the objects in rapid periodic motion, high-speed optical measurement for the vibrating membrane of the drum, dynamic measurement of a rotating vortex, and breaking tile, were also demonstrated. Furthermore, also based on grating projection and Fourier fringe analysis, the time-average fringe method was used to recover the vibration mode of a vibrating membrane. The fundamental concept and some experimental results were also described in this chapter.

There are increasing demands for obtaining 3D information; moreover, it is expected that the trend will be to expand existing techniques to deal with the dynamic process. As a noncontact shape measurement technique, Fourier transform profilometry has dealt well with the dynamic process in recent years, due to its merits of only one fringe needed, full field analysis, and high precision. This 3D dynamic shape measurement using grating projection and Fourier fringe analysis has the advantage of high-speed, full-field data acquisition and analysis. With the development of high-resolution CCD cameras and high frame rate frame grabbers, the method, as proposed here, should be a promising one in studying high-speed motion including rotation, vibration, explosion, expansion, contraction, and even shock wave.

Although this method has been well studied and applied in different fields, there are still some challenges that must be addressed in future work. Real-time 3D shape measurement, which is the key for successfully implementing 3D coordinate display and measurement, manufacturing control, and online quality inspection, is more difficult in a dynamic process. By making full use of the processing power of the graphics processing unit, 3D dynamic shape reconstruction can be performed rapidly and in real time with an ordinary personal computer [15]. A high-resolution, real-time 3D shape measurement, developed by Zhang, took advantage of digital fringe projection, phase-shifting techniques, and some developing hardware technologies and has achieved simultaneous 3D absolute shape acquisition, reconstruction, and display at an ultrafast speed [53]. A measuring system has demonstrated that it can accurately capture dynamic changing 3D scenes and it will be more promising in the future.

The requirements of the environment for dynamic measurement are tougher than 3D static shape measurement—for example, the influence of the varying environmental lights in a stroboscope and of the unexpected motion of peripheral equipment. Moreover, the measurement accuracy is also affected by the object surface color. When the color gratings are used in this proposed measuring system, the accuracy and measurable range can be improved by decreasing the overlapping spectrum. But other problems caused by the color cross and sampling rate of the color camera will affect the measuring accuracy.

Due to the varying objects, the accuracy of 3D dynamic shape measurement is usually less than that of 3D static shape measurement. Generally, the different experimental setups of each measurement will lead to different accuracy. In the applications proposed in this chapter, the accuracy of this kind of method is up to decades of micrometers. The primary task of the future work would be to improve the accuracy of 3D dynamic shape measurement to meet industrial application requirements.

Acknowledgments

First of all, we would like to thank this book's editor, Dr. Song Zhang, for his invitation. Our thanks also go to our colleagues and former graduate students in Sichuan University for their contributions to this research. We would like to thank the National Natural Science Foundation of China (nos. 60527001, 60807006, 60838002, 61177010); most of this work was carried out under NSFC's support.

References

1. Chen, F., Brown, G. M., Song, M. 2000. Overview of three-dimensional shape measurement using optical methods. *Optical Engineering* 39 (1): 10–22.
2. Jähne, B., Haußecker, H., Geißler, P. 1999. *Handbook of computer vision and applications,* volume 1: *Sensors and imaging,* chapters 17–21. San Diego: Academic Press.
3. Salvi, J., Pagès, J., J. Batlle, J. 2004. Pattern codification strategies in structured light systems. *Pattern Recognition* 37 (4): 827–849.
4. Yoshizawa, T. 1991. The recent trend of moiré metrology. *Journal Robust Mechanics* 3 (3): 80–85.

5. Srinivasan, V., Liu, H. C., Halioua, H. 1984. Automated phase-measuring profilometry of 3-D diffuse objects. *Applied Optics* 23 (18): 3105–3108.

6. Su, X. Y., Zhou, W. S., von Bally, V., et al. 1992. Automated phase-measuring profilometry using defocused projection of a Ronchi grating. *Optics Communications* 94 (6): 561–573.

7. Takeda, M., Ina, H., Kobayashi, S. 1982. Fourier-transform method of fringe-pattern analysis for computer-based topography and interferometry. *Journal Optical Society America* 72 (1): 156–160.

8. Takeda, M., Motoh, K. 1983. Fourier transform profilometry for the automatic measurement of 3-D object shapes. *Applied Optics* 22 (24): 3977–3982.

9. Li, J., Su, X. Y., Guo, L. R. 1990. An improved Fourier transform profilometry for automatic measurement of 3-D object shapes. *Optical Engineering* 29 (12): 1439–1444.

10. Su, X. Y., Chen, W. J. 2001. Fourier transform profilometry: A review. *Optics Lasers Engineering* 35 (5): 263–284.

11. Su, L. K., Su, X. Y., Li, W. S. 1999. Application of modulation measurement profilometry to objects with surface holes. *Applied Optics* 38 (7): 1153–1158.

12. Toyooka, S., Iwasa, Y. 1986. Automatic profilometry of 3-D diffuse objects by spatial phase detection. *Applied Optics* 25 (10): 3012–3018.

13. Sajan, M. R., Tay, C. J., Shang, H. M., et al. 1998. Improved spatial phase detection for profilometry using a TDI imager. *Optics Communications* 150 (1–6): 66–70.

14. Zhang, L., Curless, B., Seitz, S. 2003. Spacetime stereo: Shape recovery for dynamic senses. *Proceedings of IEEE Computer Society Conference on Computer Vision and Pattern Recognition (CVPR)*, Madison, WI, 367–374.

15. Zhang, S., Huang, P. S. 2006. High-resolution, real-time three-dimensional shape measurement. *Optical Engineering* 45 (12): 123601-1-8.

16. Macy, W. W. 1983. Two-dimensional fringe-pattern analysis. *Applied Optics* 22 (23): 3898–3901.

17. Bone, D. J., Bachor, H. A., Sandeman, R. J. 1986. Fringe-pattern analysis using a 2-D Fourier transform. *Applied Optics* 25 (10): 1653–1660.

18. Bone, D. J. 1991. Fourier fringe analysis: The two-dimensional phase unwrapping problem. *Applied Optics* 30 (25): 3627–3632.

19. Burton, D. R., Lalor, M. J. 1989. Managing some of the problems of Fourier fringe analysis. *Proceedings SPIE* 1163:149–160.

20. Burton, D. R., Lalor, M. J. 1994. Multi-channel Fourier fringe analysis as an aid to automatic phase unwrapping. *Applied Optics* 33 (14): 2939–2948.

21. Lin, J-F., Su, X-Y. 1995. Two-dimensional Fourier transform profilometry for the automatic measurement of three-dimensional object shapes. *Optical Engineering* 34 (11): 3297–3302.

22. C. Gorecki. 1992. Interferogram analysis using a Fourier transform method for automatic 3D surface measurement. *Pure Applied Optics* 1:103–110.

23. Su, X., Sajan, M. R., Asundi, A. 1997. Fourier transform profilometry for 360-degree shape using TDI camera. *Proceedings SPIE* 2921: 552–556.

24. Yi, J., Huang, S. 1997. Modified Fourier transform profilometry for the measurement of 3-D steep shapes. *Optics Lasers Engineering* 27 (5): 493–505.

25. Takeda, M., Gu, Q., Kinoshita, M., et al. 1997. Frequency-multiplex Fourier-transform profilometry: A single shot three-dimensional shape measurement of objects with large height discontinuities and/or surface isolations. *Applied Optics* 36 (22): 5347–5354.

26. Burton, D. R., Goodall, A. J., Atkinson, J. T., et al. 1995. The use of carrier frequency-shifting for the elimination of phase discontinuities in Fourier-transform profilometry. *Optics Lasers Engineering* 23 (4): 245–257.
27. Chen, W., Yang, H., Su, X. 1999. Error caused by sampling in Fourier transform profilometry. *Optical Engineering* 38 (6): 927–931.
28. Kozloshi, J., Serra, G. 1999. Analysis of the complex phase error introduced by the application of Fourier transform method. *Journal Modern Optics* 46 (6): 957–971.
29. Su, X., Chen, W., Zhang, Q., et al. 2001. Dynamic 3-D shape measurement method based on FTP. *Optics Lasers Engineering* 36: 46–64.
30. Su, X., Zhang, Q. 2010. Dynamic 3D shape measurement: A review. *Optics Lasers Engineering* 48 (2): 191–204.
31. Abdul-Rahman, H. S., Gdeisat, M. A., Burton, D. R., et al. 2008. Three-dimensional Fourier fringe analysis. *Optics Lasers Engineering* 46 (6): 446–455.
32. Siva Gorthi, S., Rastogi, P. 2010. Fringe projection techniques: Whither we are? *Optics Lasers Engineering* 48 (2): 133–140.
33. Su, X. Y., Chen, W. J. 2004. Reliability-guided phase unwrapping algorithm: A review. *Optics Lasers Engineering* 42 (3): 245–261.
34. Judge, T. R., Bryyanston-Cross, P. J. 1994. A review of phase unwrapping techniques in fringe analysis. *Optics Lasers Engineering* 21:199–239.
35. Su, X. Y. 1996. Phase unwrapping techniques for 3-D shape measurement. *Proceedings SPIE* 2866: 460–465.
36. Li, J. L., Su, X. Y., Li, J. T. 1997. Phase unwrapping algorithm-based on reliability and edge-detection. *Optical Engineering* 36 (6): 1685–1690.
37. Asundi, A. K., Zhou, W. S. 1999. Fast phase-unwrapping algorithm based on a grayscale mask and flood fill. *Applied Optics* 38 (16): 3556–3561.
38. Su, X., Xue, L. 2001 Phase unwrapping algorithm based on fringe frequency analysis in Fourier transform profilometry. *Optical Engineering* 40 (4): 637–643.
39. Takeda, M., Abe, T. 1996. Phase unwrapping by a maximum cross-amplitude spanning tree algorithm: A comparative study. *Optical Engineering* 35 (8): 2345–2351.
40. Zhang, Z. 2000. A flexible new technique for camera calibration. *IEEE Transactions Pattern Analysis Machine Intelligence* 22 (11): 1330–1334.
41. Li, W. S., Su, X. Y., Liu, Z. B. 2001. Large-scale three-dimensional object measurement: A practical coordinate mapping and imaging data-patching method. *Applied Optics* 40 (20): 3326–3333.
42. Zhang, Q. C., Su, X. Y. 2002. An optical measurement of vortex shape at a free surface. *Optics Laser Technology* 34 (2): 107–113.
43. Zhang, Q. C., Su, X. Y. 2005. High-speed optical measurement for the drumhead vibration. *Optics Express* 13 (8): 3310–3316.
44. Visionary engineer, Harold Edgerton. 2011. http://web.mit.edu/museum/exhibitions/edgertonexhibit/harolddocedgertonindex.html (accessed Oct. 26, 2011).
45. Asundi, A. K., Sajan, M. R. 1994. Low-cost digital polariscope for dynamic photoelasticity. *Optical Engineering* 33 (9): 3052–3055.
46. Asundi, A. K., Sajan, M. R. 1996. Digital drum camera for dynamic recording. *Optical Engineering* 35 (6): 1707–1713.
47. Zhang, Q. C., Su, X. Y., Cao, Y. P., et al. 2005. An optical 3-D shape and deformation measurement for rotating blades using stroboscopic structured illumination. *Optical Engineering* 44 (11): 113601-1–7.

48. Xiao, Y. S., Su, X. Y., Zhang, Q. C., et al. 2007. 3-D profilometry for the impact process with marked fringes tracking. *Opto-Electronic Engineering* 34 (8): 46–52 (in Chinese).
49. Su, W. H. 2008. Projected fringe profilometry using the area-encoded algorithm for spatially isolated and dynamic objects. *Optics Experiments* 16 (4): 2590–2596.
50. Guo, H., Huang, P. S. 2008. 3-D shape measurement by use of a modified Fourier transform method. *Proceedings SPIE* 7066: 70660E-1–8.
51. Guo, H., Huang, P. S. 2007. Absolute phase retrieval for 3D shape measurement by the Fourier transform method. *Proceedings SPIE* 6762: 676204-1–10.
52. Su, X., Zhang, Q., Wen, Y., et al. 2010. Time-average fringe method for vibration mode analysis. *Proceedings SPIE* 7522: 752257.
53. Zhang, S. 2010. Recent progress on real-time 3D shape measurement using digital fringe projection techniques. *Optics Lasers Engineering* 48 (2): 149–158.

8

Interferometry

David P. Towers and Catherine E. Towers

CONTENTS

8.1 Introduction

Interferometric techniques have found widespread use in three-dimensional (3D) profilometry. When two coherent waves of wavelength λ are brought together such that they have traversed a relative optical path difference Δx, the intensity obtained is given by [1]

$$I = I_{DC} + I_M \cos(S\Delta x)$$

where
S is introduced as a sensitivity parameter
I is the measured intensity
I_{DC} is the constant component of the intensity
I_M is the modulation depth of the interference term

In a conventional Twyman-Green interferometer [2] (see Figure 8.1a), the laser beam is split and the object beam directed to the test object with the

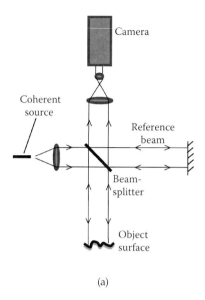

(a)

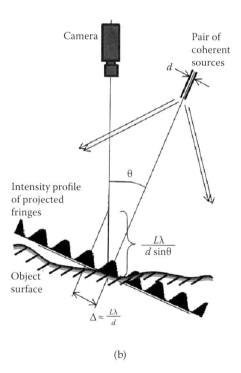

(b)

FIGURE 8.1
(a) A conventional Twyman–Green interferometer; b) a fringe projection arrangement using coherent illumination.

reflected light used to form the interferogram in combination with the reference beam. With such a configuration, the sensitivity is given by $S = 2\pi(2/\lambda)$ and an interference fringe relates to a topology change of $\lambda/2$. With a sensitivity inversely proportional to the wavelength, these techniques are suitable for inspection of, for example, microelectromechanical system (MEMS) devices, wear of materials, and optical components.

Interference fringes may also be projected onto an object in a triangulation configuration in which the illumination and viewing axes intersect at an angle θ. Assuming that the fringes originate from a pair of coherent sources, in a geometry directly analogous to Young's double-slit experiment [3], the fringe spacing Δ obtained at the object (in the fair-field approximation) is given by $\Delta \approx L\lambda/d$, where L is the distance from the sources to the object and d is the separation of the sources. By utilizing a triangulation configuration, the fringe spacing has a component along the imaging direction such that the sensitivity is given by $S = 2\pi(d\sin\theta/L\lambda)$ and the effective wavelength relates to a topology change of $L\lambda/(d\sin\theta)$. There is free choice over the values for θ and d and hence a wide range of effective wavelengths can be realized.

The field of view of conventional interferometers is normally limited by the aperture of the optical components; however, with fringe projection, this constraint does not apply. Therefore, fringe projection is more applicable for machine vision applications and will be the focus of this chapter. Coherent fringe projection was first proposed as double source contouring with holographic recording [4]. While significant flexibility arose with the advent of multimedia data projectors (see Chapter 9), coherent based fringe projection retains distinct advantages: the ability to filter out ambient light, miniaturization of the fringe projector, and the precisely known geometry of the projected patterns. This chapter will discuss the optical systems for generating projected fringe patterns, the specific calibration issues with coherent systems, and implementation with multiwavelength techniques to give absolute 3D metrology.

8.2 Coherent Fringe Projection Systems: Optical Setups and Characteristics

8.2.1 Twin-Fiber-Based Coherent Fringe Projection

One of the most direct ways to generate coherent fringes is to use a pair of single mode fibers via a single coherent source and directional coupler [5,6]. The two output fibers from the coupler are made parallel and brought close together to form the projected fringes (see Figure 8.2). By mounting one of the fibers from an appropriately oriented linear traverse, the separation of the fibers can be adjusted, thereby controlling the spacing of the projected

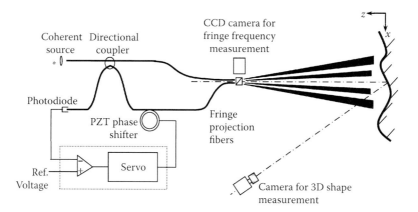

FIGURE 8.2
A twin-fiber-based fringe projector with additional components for sampling and controlling the phase of the projected fringes. (Courtesy of Towers, C. E., Towers, D. P., and Jones, J. D. C. *Optics Letters* 28 (11): 887–889, 2003. Published by Optical Society of America.)

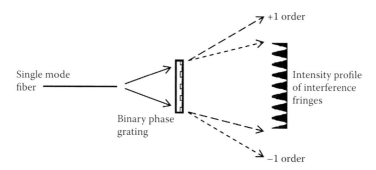

FIGURE 8.3
Interference fringe formation via a binary phase grating.

fringes. The main practical issue with fiber fringe projectors concerns the environmental sensitivity of the fibers. Vibration or thermal changes in the environment modulate the phase of the wave transmitted by the fiber and thereby cause the fringes to sweep across the object surface [7].

8.2.2 Diffraction Grating Fringe Projection

The diffraction orders from a grating can be used to form the two beams needed for fringe projection. This technique was originally introduced using a blazed grating [8], but is better implemented using a binary phase grating, which is optimized to produce only the +1 and −1 diffraction orders, thereby minimizing the energy in the other diffraction orders (see Figure 8.3) [9]. The input beam to the grating can be derived from a single mode optical fiber, thereby maintaining flexibility in positioning. The advantage of the diffraction grating approach is that the interferometer is common path in fiber with the only non-common path elements in air; hence the system is significantly less sensitive to environmental disturbances compared to the twin-fiber approach. However, the projected patterns are sensitive to the quality of the grating and the presence of overlapping diffraction orders, and they lack the precise theoretical form of those formed from a pair of single mode fibers. Phase modulation can be achieved by translating the grating and the fringe spacing can be controlled by varying the separation between the distal end of the fiber and the grating.

There are alternative bulk-optic configurations to form interference fringes that can be projected over an object in a triangulation setup. These include the use of a Michelson interferometer, a Lloyd mirror, or a Wollaston prism; however, as these methods use bulk optic components, they do not have the flexibility of the methods described before.

8.2.3 Sensitivity and Effective Wavelength

The fringe projection configurations described in the previous sections can produce a wide range of sensitivities owing to the choice of θ and d in Figure 8.1. Figure 8.4 presents a family of curves of the effective wavelength $L\lambda/(d\sin\theta)$, with $L = 1.5$ m and $\lambda = 532$ nm. It can be seen that a fringe can represent a surface depth change between >100 mm and <0.5 mm depending on the configuration.

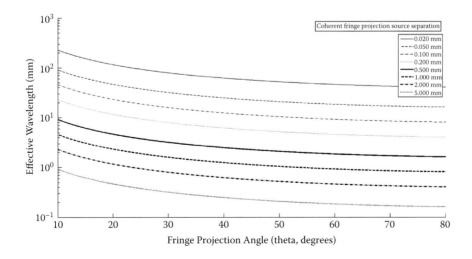

FIGURE 8.4
The effective wavelength of a fringe projection system as a function of triangulation angle and source spacing in a coherent fringe projector.

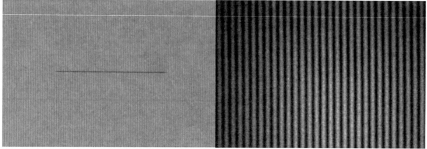

(a) Fringes from a binary phase grating interferometer, the line is 20 mm, with approximately 2.4 fringes/mm

(b) Fringes from a twin-fiber interferometer with approximately 0.2 fringes/mm

FIGURE 8.5
Examples of interferometric projected fringes. (Images courtesy of D. Dipresa.)

Examples of the patterns produced using interferometric fringe projection are given in Figure 8.5 using both twin-fiber and grating-based interferometers.

8.3 Interferometric Fringe Processing

The goal of fringe analysis in fringe projection is to determine a 3D coordinate at each pixel on the detector. An interferometric phase in the fringe pattern defines a plane from the projector. When this phase is measured at a particular pixel on the detector, the back propagation of that pixel through the imaging lens will intersect the phase plane at a unique 3D coordinate. This analysis relies on knowing the absolute phase within the projected pattern. As each fringe in the pattern has a similar sinusoidal intensity profile, knowledge of the position within the projected fringes relies on both the integer fringe number, or order, and the wrapped or fractional phase value.

8.3.1 Wrapped-Phase Calculation

The wrapped phase may be determined by carrier-fringe-based methods [10], which require a single image. This is directly compatible with either twin-fiber- or grating-based fringe projectors. The exposure time of the camera can be set according to a compromise between fringe stability (for twin-fiber projectors) and the available optical power. The disadvantage of most single image carrier fringe analysis methods is that ringing can appear in the vicinity of any edges or object discontinuities within the image, giving errors in the calculated phase.

Alternatively, phase-stepping techniques can be applied to determine the fringe phase [11]. Phase stepping requires three or more images in order to determine the fringe phase and hence the data acquisition time is increased. (The interested reader is directed to the work of Surrel for a review of the performance of various phase-stepping algorithms [12].) Phase shifting can be implemented in twin-fiber projectors by winding one of the output fibers around a cylindrical piezoelectric transducer (PZT) and applying a voltage [5]. With open loop adjustment of the phase, the actual phase shifts obtained may be different from those desired due to environmental effects. Phase stabilization can be achieved by monitoring the intensity at the fourth arm of the coupler using a photodiode (see Figure 8.2). The interference signal obtained is derived from waves that are reflected from the distal ends of the output fibers. An error signal is found by subtraction of a fixed reference voltage and a suitably tuned servo can drive the PZT in one of the output arms of the interferometer to fix the phase of the projected pattern in a closed loop mode [13].

It has also been shown that such a configuration can be used to produce exact phase shifts of 90° [13]. While this method demonstrates the potential to use twin-fiber-based projectors in more hostile environments, in practice, instabilities may remain between the projector and object that cannot be compensated by such an approach and hence single image analysis is preferred [14]. Phase-shifting techniques may also be applied to grating-based projectors by translation of the grating. The fringe projection setup and test object must remain stable during the time needed for mechanical translation of the grating and acquisition of the phase-stepped images.

8.3.2 Fringe Order Calculation

For contiguous object surfaces, the simplest approach to determine the fringe order is to utilize a reference marker within the projected field [15]. With *a priori* knowledge of surface continuity, a spatial phase-unwrapping algorithm yields the desired absolute phase distribution [16,17].

For generic artifacts or scenes composed of multiple discrete objects, the fringe order must be determined at each pixel independently. The most robust approach is to use a multiwavelength strategy [18]. Here the "wavelengths" are the effective wavelengths of the projected patterns, rather than optical wavelengths. Two strategies have evolved. The first aims to generate a beat wavelength that spans the field of view of the camera. The wrapped phase at a beat wavelength can be calculated from the difference of two wrapped-phase measurements. For example, if the original effective wavelengths are λ_0 and λ_1, at which wrapped phase measurements of $\Delta\phi_0$ and $\Delta\phi_1$ are obtained respectively, the phase $\Delta\phi_{01}$ at the beat wavelength, Λ_{01} is found from [19]

$$\Delta\phi_{01} = \Delta\phi_0 - \Delta\phi_1 = 2\pi L\left(\frac{\lambda_1 - \lambda_0}{\lambda_1\lambda_0}\right) = \frac{2\pi L}{\Lambda_{01}}$$

where L is the unknown optical path difference. From a single beat fringe there is a unique range of wrapped-phase values that correspond to each fringe order at λ_0 and thus the original fringes can be identified unambiguously.

To appreciate the formation of beat fringes, it is sometimes helpful to make an acoustic analogy. When an orchestra is warming up, each instrument will be tuned to a common frequency. When two players produce notes of slightly different pitch, a low-frequency "beat" is heard between the two. This beat frequency has a long wavelength and the aim in fringe processing is to make it span the desired measurement range of projected fringe orders. In interferometric projected fringe systems such as those depicted in Figures 8.2 and 8.3, it is not possible to obtain a single projected fringe across the field of view directly (e.g., by bringing the twin fibers close to each other). Therefore, the indirect approach using beat analysis is necessary.

The highest dynamic range in fringe projection is obtained when the best wrapped-phase resolution is combined with data over the largest number of projected fringes. However, it has been shown that the noise in the wrapped-phase data, which may be quantified by a standard deviation σ_ϕ, limits the number of projected fringes that can be reliably identified. A number of researchers in this field have presented multiwavelength methodologies for use with fringe projection and the interested reader is referred to the work of Burton and Lalor [20], Nadeborn, Andra, and Osten [21], and Towers, Towers, and Jones [22]. The theory reported in Towers et al. [22] brings together the reliability of calculating the correct fringe order with the phase noise from the interferometer to determine the maximum number of fringe orders measurable and hence defines a framework for optimization. It is shown that the number of fringes N_{f0} at wavelength λ_0 that can be ordered to 6σ reliability (i.e., 99.73%, within each beat fringe at Λ_{01}) is limited by

$$\frac{N_{f0}}{N_{f01}} \leq \frac{2\pi}{6\sqrt{2}\sigma_\phi}$$

Hence, if the phase resolution corresponds to 1/100th of a fringe, the ratio N_{f0}/N_{f01} is limited to ≤ 11.8. For a larger dynamic range, additional measurement wavelengths must be added. An optimal configuration is found when the series of beat wavelengths and the highest resolution effective wavelength, λ_0, form a geometric series with a common ratio given by N_{f0}/N_{f01}. In turn, this defines the number of projected fringes at each wavelength, N_{fi}, by [22]

$$N_{fi} = N_{f0} - \left(N_{f0}\right)^{\frac{i-1}{n-1}}$$

where n is the number of measurement wavelengths, i denotes the ith measurement wavelength, and N_{f0} is the maximum number of projected fringes at the shortest effective wavelength λ_0. For example, with the same phase noise of $1/100^{th}$ of a fringe, but with $n = 3$, the numbers of projected fringes required are $N_{f0} = 121$, $N_{f1} = 120$, and $N_{f2} = 110$. In this case, the beat is formed by N_{f0} N_{f1} and contains one beat fringe. The beat from N_{f0} and N_{f2} gives 11 beat fringes, which, together with $N_{f0} = 121$, form a geometric series. For typical megapixel resolution detectors, three wavelengths of projected fringes are sufficient and give a dynamic range for depth measurement in the region of 10,000:1 (from 100 fringes at a resolution of $1/100^{th}$ of a fringe).

An alternative multiwavelength approach is to use the uniqueness of the set of wrapped-phase values obtained at the measurement wavelengths along with the method of excess fractions [23] or the Chinese remainder theorem [24] to determine the fringe order. Recent results with the method of excess fractions have demonstrated that similar performance can be obtained as from the beat approach but with significantly greater choice of measurement wavelengths [25].

8.4 Automation and Calibration

It is essential to automate the measurement process, particularly with multiwavelength analysis, in order to obtain repeatable results and increase usability. With a twin-fiber interferometer, a PZT driven traverse with a range of up to 100 μm is suitable to adjust the effective wavelength, whereas with grating projection a larger range of motion is typically needed via a piezomotor or other motorized traverse. With three projected fringe frequencies and carrier fringe analysis, a minimum of 3 images is needed, whereas with phase step analysis 12 images are typically necessary. In either case the multiwavelength analysis algorithms require accurate knowledge of the effective wavelength in order to determine the fringe order [22].

The effective wavelength in the projected fringes can be determined by splitting off part of the projected pattern to fall onto a second area detector [13]. This is shown schematically in Figure 8.2 for a twin-fiber setup. A wrapped-phase map from this second camera can be obtained by carrier frequency or phase-shifting analysis and then spatially unwrapped. The resulting phase distribution is then fitted to the exact hyperbolic fringe function in order to recover the fiber separation parameters in all three orthogonal axes. As an example, the data in Figure 8.6 show the fiber separation in the x-axis after the fiber was moved 20 μm using a PZT actuator. The gradual movement of the fiber from the desired position can be seen and is a characteristic of the actuator; however, this behavior is largely repeatable and hence, by timing image acquisition to occur after the initial high rate of change, it is possible to obtain stable results at the desired fiber separations. It is also clear from Figure 8.6 that the fiber separation measurement is highly repeatable with an uncertainty < 50 nm.

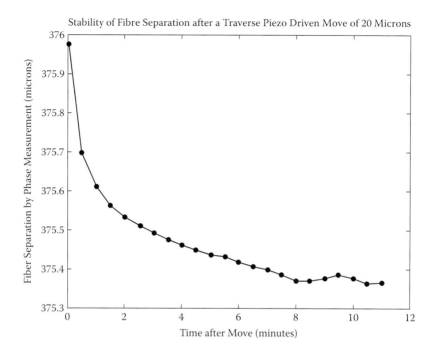

FIGURE 8.6

Fiber separation measurements from a twin-fiber interferometric projector following translation of one fiber using a PZT actuator.

Having obtained the source separation information, the fringes are defined as hyperbolic functions in 3D space. It is also possible to conduct a camera calibration for a set of internal and external parameters (see, for example, Bouguet [26]). Hence, a parametric calibration can be achieved once the triangulation angle, θ, is known. This angle is best found by using a known artifact (e.g., containing a known step height) and iteratively modifying the angle until the expected depth parameter is obtained. Alternative techniques for calibration in fringe projection are discussed in Chapter 9.

8.5 Results

Two sets of example results are shown in Figure 8.7. Figure 8.7(a) shows the calculated fringe order information on a casting insert that is measured in order to assess wear. The black fringe is the zero-order fringe with the fringe order increasing positively and negatively on either side of this. The sharply defined boundaries of each fringe show the quality of the multiwavelength processing. After calibration, the fringe data are

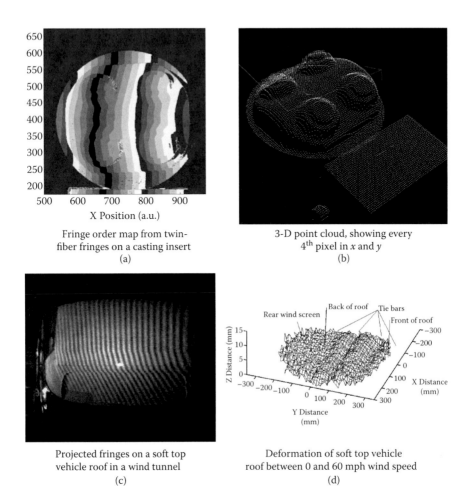

Fringe order map from twin-fiber fringes on a casting insert

(a)

3-D point cloud, showing every 4th pixel in x and y

(b)

Projected fringes on a soft top vehicle roof in a wind tunnel

(c)

Deformation of soft top vehicle roof between 0 and 60 mph wind speed

(d)

FIGURE 8.7

Example data from a twin-fiber-based interferometric fringe projection system. (a, b) Reprinted from Towers, C. E. et al., *Optics and Lasers in Engineering* 43: 788–800, copyright 2005, with permission from Elsevier. (c) Reprinted from Buckberry, C. H. et al. *Optics and Lasers in Engineering* 25:433–453, copyright 1996, with permission from Elsevier.

converted into a matrix of xyz coordinates and every fourth pixel is plotted in x and y, as plotted in Figure 8.7(b). The casting insert was mounted on a metal block. A second example is given in Figure 8.7(c), which shows interference fringes projected onto a passenger vehicle soft-top roof in a wind tunnel. The field of view is approximately 1.5 m. The 3D shape of the roof was measured with the wind off and then with a wind speed of 60 mph, and the deformation of the roof was then determined (shown in Figure 8.7d). The fringes were processed using a single image Fourier transform algorithm due to the instability of the setup. The presence of tie bars can be seen in the deformation map where the height change is close to zero, with the material bulging by approximately 5 mm between the tie bars.

The primary difficulty with using coherent light for interferometric fringe projection is the formation of speckle, which appears as noise within the image. Therefore, under the same experimental conditions, a white-light fringe projector will produce a lower noise level compared to a laser-based projector. The main mechanism to control the level of speckle noise is via the imaging lens aperture, or f-number. Increasing the aperture, corresponding to a lower f-number, produces a smaller speckle size and hence increases the number of speckles per pixel on the detector. The benefit of opening the lens aperture is illustrated in Figure 8.8, which shows the noise in the depth measurements to one standard deviation against the f-number for a narrowband 532 nm laser-based fringe projection setup with a field of view of approximately 500 mm. Opening the lens aperture has the disadvantage of reducing the depth of field. Therefore, there is a compromise in setting up an interferometric system between reducing the speckle noise, and hence the noise on the depth measurements, while achieving sufficient depth of field.

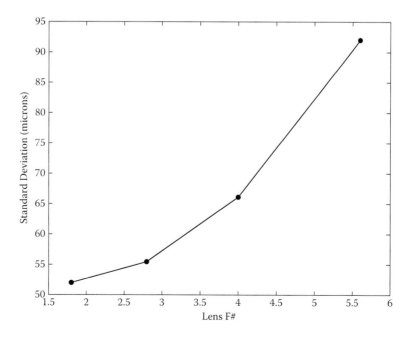

FIGURE 8.8
Resolution of shape data as a function of imaging lens aperture.

8.6 Summary

This chapter has presented the key features of interferometric fringe projection systems employing laser illumination. There are a number of optical configurations that can be used to generate the projected fringes that employ single mode fibers and hence offer considerable flexibility. While white-light-based projectors dominate the research and commercial exploitation of fringe projection, laser-based devices still offer some unique advantages: The projector can be miniaturized to the mounts required for single mode fibers and the use of narrow band filters enables operation in environments with significant levels of ambient light.

References

1. C. Vest. *Holographic interferometry.* New York: Wiley-Blackwell, 1979.
2. P. Hariharan. *Optical interferometry,* 2nd ed. San Diego: Academic Press Inc., 2003.
3. T. Young. Experimental demonstration of the general law of the interference of light. *Philosophical Transactions of the Royal Society of London* 94, 1804.
4. B. P. Hildebrand, K. A. Haines. Multiple-wavelength and multiple-source holography applied to contour generation. *Journal Optical Society of America* 57 (2): 155–157, 1967.
5. J. D. Valera, J. D. C. Jones. Phase stepping in projected-fringe fiber-based moiré interferometry. *Electronics Letters* 29 (20): 1789–1791, 1993.
6. M. J. Lalor, J. T. Atkinson, D. R. Burton, P. Barton. A fiber-optic computer-controlled fringe projection interferometer for surface measurement. *Proceedings Fringe '93, Automatic Processing of Fringe Patterns,* W. Juptner and W. Osten, eds., Akademie Verlag, Berlin, 1993.
7. D. Jackson, R. Priest, A. Dandridge, A. Tveten. Elimination of drift in a single-mode optical fiber interferometer using a piezoelectrically stretched coiled fiber. *Applied Optics* 19 (17): 2926–2929, 1980.
8. G. Schirripa-Spagnolo, D. Ambrosini, Surface contouring by diffractive optical element-based fringe projection. *Measurement Science Technology* 12: N6–N8, 2001.
9. M. Reeves, A. J. Moore, D. P. Hand, J. D. C. Jones. Dynamic shape measurement system for laser materials processing. *Optical Engineering* 42 (10): 2923–2929, 2003.
10. M. Takeda, K. Mutoh. Fourier transform profilometry for the automatic measurement of 3-D object shapes. *Applied Optics* 22: 3977–3982, 1983.
11. J. Schmit, K. Creath. Extended averaging technique for derivation of error-compensating algorithms in phase-shifting interferometry. *Applied Optics* 34 (19): 3610–3619, 1995.
12. Y. Surrel. Additive noise effect in digital phase detection. *Applied Optics* 36 (1): 271–276, 1997.

13. A. J. Moore, R. McBride, J. S. Barton, J. D. C. Jones. Closed loop phase-stepping in a calibrated fibre optic fringe projector for shape measurement. *Applied Optics* 41 (16): 3348–3354, 2002.

14. D. P. Towers, C. H. Buckberry, B. C Stockley, M. P. Jones. Measurement of complex vibrational modes and surface form—A combined system. *Measurement Science Technology* 6: 1242–1249, 1995.

15. S. Zhang, S.-T. Yau. High-resolution, real-time 3D absolute coordinate measurement based on a phase-shifting method. *Optics Express* 14 (7): 2644–2649, 2006.

16. D. C. Ghiglia M. D. Pritt. *Two-dimensional phase unwrapping.* New York: John Wiley & Sons, 1998.

17. D. P. Towers, T. R. Judge, P. J. Bryanston-Cross. Automatic interferogram analysis techniques applied to quasi heterodyne holography and ESPI. *Optics and Lasers in Engineering,* special issue on fringe pattern analysis 14: 239–281, 1991.

18. C. E. Towers, D. P. Towers, J. D. C. Jones. Optimum frequency selection in multifrequency interferometry. *Optics Letters* 28 (11): 887–889, 2003.

19. Y. Y. Cheng, J. C. Wyant. Two-wavelength phase shifting interferometry. *Applied Optics* 23 (24): 4539–4543, 1984.

20. D. R. Burton, M. J. Lalor. Multichannel Fourier fringe analysis as an aid to automatic phase unwrapping. *Applied Optics* 33: 2939–2948, 1994.

21. W. Nadeborn, P. Andra, W. Osten, A robust procedure for absolute phase measurement. *Optics Lasers Engineering* 24: 245–260, 1996.

22. C. E. Towers, D. P. Towers, J. D. C. Jones. Absolute fringe order calculation using optimized multi-frequency selection in full field profilometry. *Optics Lasers Engineering* 43: 788–800, 2005.

23. M. Takeda, Q. Gu, M. Kinoshita, H. Takai, Y. Takahashi. Frequency-multiplex Fourier-transform profilometry: A single-shot three-dimensional shape measurement of objects with large height discontinuities and/or surface isolations. *Applied Optics* 36: 5347–5354, 1997.

24. C. E. Towers, D. P. Towers, J. D. C. Jones. Time efficient Chinese remainder theorem algorithm for full-field fringe phase analysis in multi-wavelength interferometry. *Optics Express* 12 (6): 1136–1143, 2004.

25. Z. H. Zhang, D. P. Towers, C. E. Towers. Snapshot color fringe projection for absolute three-dimensional metrology of video sequences. *Applied Optics* 49 (31): 5947–5953, 2010.

26. J.-Y. Bouguet. Camera calibration toolbox for MATLAB. http://www.vision.caltech.edu/bouguetj/calib_doc/

9

Superfast 3D Profilometry with Digital Fringe Projection and Phase-Shifting Techniques

Laura Ekstrand, Yajun Wang, Nikolaus Karpinsky, and Song Zhang

CONTENTS

Recent years have seen the rise of digital fringe projection (DFP) techniques and their subsequent employment in diverse areas, including manufacturing, medicine, and homeland security. This chapter focuses on some of the recent advancements in high-speed three-dimensional (3D) optical profilometry with DFP and phase-shifting techniques. Over the past few years, we have developed a high-resolution real-time 3D profilometry system that has achieved simultaneous 3D shape acquisition, reconstruction, and display at 40 Hz with more than 250,000 points per frame. More recently, we have developed novel binary defocusing techniques that give DFP systems the potential to achieve tens of kilohertz 3D profilometry. In this chapter, we will explain the principles of these techniques, discuss their merits and limitations, and present experimental results achieved with these techniques.

9.1 Introduction

Three-dimensional optical profilometry techniques for static or quasi-static events have been extensively studied over the past few decades and have seen great success in video game design, animation, movies, music videos, virtual reality, telesurgery, and many engineering disciplines [1]. Though numerous 3D profilometry techniques exist, they can be classified into two categories: surface contact methods and surface noncontact methods. Both the coordinate measurement machine (CMM) and the atomic force microscope (AFM) require contact with the measuring surface to obtain 3D profiles at high accuracy. This requirement places severe restrictions on the speed of contact methods. They cannot reach kilohertz measurement speed with thousands of points per scan.

Surface noncontact techniques typically utilize optical triangulation methods (e.g., stereo vision [2], space–time stereo [3], structured light [4]), although some can retrieve depth from the same view (e.g., shape from focus/defocus [5,6], time of flight [7]). Triangulation-based systems can recover depth by matching and geometrically relating distinct regions of a scene viewed from different angles. This is usually realized by analyzing two images with a digital image correlation (DIC) algorithm [8]. However, because the DIC approach relies upon the unique features of corresponding region pairs to perform matching, the measurement accuracy is low for surfaces without strong local texture variations.

Speckle technology [9] can resolve this problem by actively projecting or painting the scene with random patterns [10] and then applying the DIC algorithm to recover the depth. Such speckle systems can achieve high temporal resolution (kilohertz or better). However, it is difficult for stereo DIC to reach camera-pixel spatial resolution because the DIC algorithm must match regions larger than one camera pixel. Also, the surface treatment (e.g., paint speckles) often causes surface damage.

Structured-light technology increases the measurement capability by actively projecting known patterns onto the objects [4]. Therefore, it can be used to measure surfaces without strong local texture variations. Though discrete-coded structured-light techniques have been able to achieve tens of hertz [11–13], their spatial resolution is limited to being larger than a single projector pixel. Fringe analysis is a special group of structured-light techniques that uses sinusoidal structured patterns (also known as fringe patterns). Because these patterns have intensities that vary continuously from point to point in a known manner, they boost the structured-light techniques from projector-pixel resolution to camera-pixel resolution.

In the recent past, fringe analysis techniques were instrumental in achieving high-speed and high-resolution 3D profilometry. For instance, a single fringe pattern could be used to recover dynamic 3D shape measurements [14] using the Fourier transform method [15] or the windowed Fourier transform method [16]. However, the Fourier method is limited to measuring surfaces that are "smooth" in both geometry and texture. Other fringe analysis methods [17] use coherent light (laser) or white light interference to generate sinusoidal patterns and have been extensively used in high-precision 3D profilometry. Though laser and interference systems have high accuracy, they typically require mechanical adjustments, making their measurement speed very slow.

The DFP technique uses digital video projectors instead of interference to generate sinusoidal fringe patterns. This technique has the merits of lower cost, higher speed, and simplicity of development, and it has been a very active research area within the past decade [14,18–21]. We developed systems to acquire 3D video at 40 Hz [20] and then later at 60 Hz [22,23]. More recently, various researchers have developed real-time 3D profilometry systems of their own through several different methods [21,24,25]. However, because the speed of switching multiple 8-bit grayscale patterns is limited by the projector's refresh rate, the maximum achievable 3D shape measurement rate is typically 120 Hz [24,26].

To overcome the speed limitation of the conventional DFP technique, we have invented a new method for 3D profilometry [27] that utilizes the defocusing effect of the projector lens to convert black and white stripes to pseudosinusoids. This defocusing technique significantly increases measurement speeds because the projector loads 1-bit binary structured images instead of 8-bit sinusoidal structured images for each 3D frame. This technique shows great potential for capturing 3D profilometry at unprecedented rates and has already enabled us to develop a 667 Hz 3D profilometry system [28] with the digital light processing (DLP) Discovery platform.

Nevertheless, the binary defocusing technique has the following limitations: (1) The DLP Discovery is very expensive, (2) the measurement accuracy is lower, and (3) the measurement depth range is smaller. We are currently striving to overcome these limitations while preserving the technique's merits.

This chapter will introduce the fundamental principles of DFP and provide examples of conventional, real-time, and superfast DFP systems.

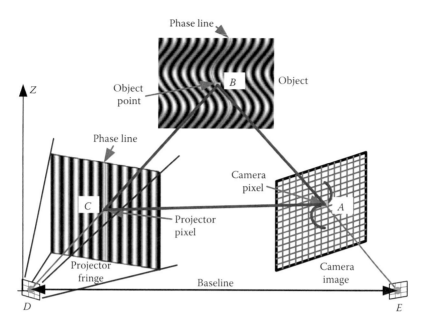

FIGURE 9.1
Schematic diagram of a 3D profilometry system using a DFP technique. (Modified from Zhang, S. *Optics and Lasers in Engineering* 48:149, 2010. With permission.)

9.2 Fundamentals

9.2.1 Digital Fringe Projection Technique

The DFP technique is a special kind of triangulation-based structured-light method where the structured patterns vary in intensity sinusoidally. This structured-light method is similar to a stereo-based method, except that it uses a projector to replace one of the cameras [4]. Figure 9.1 shows the schematic diagram of a 3D profilometry system based on a DFP technique. The image acquisition unit (a), projection unit (c), and the three-dimensional object (b) form a triangulation base. The projector shines vertical (varying horizontally) straight fringe stripes (phase lines) onto the object. The object surface distorts the fringe images from straight phase lines to curved ones. A camera captures the distorted fringe images from another angle. In such a system, the correspondence is established by analyzing the distortion of the structured patterns (fringe patterns) through fringe analysis techniques.

9.2.2 Phase-Shifting Techniques

Phase-shifting algorithms are widely used in optical metrology because of their numerous merits [17]: (1) *point-by-point measurement*, allowing camera-pixel-level spatial resolution; (2) *lower sensitivity to surface reflectivity variations*, facilitating the measurement of very complex objects with strong texture variations; and (3) *lower sensitivity to ambient light*, reducing the requirements on measurement conditions. Numerous phase-shifting algorithms have been developed including three step, four step, double three step, and five step. In general, the fringe patterns for an N-step phase-shifting algorithm with equal phase shifts can be described as

$$I_n(x,y) = I'(x,y) + I''(x,y)\cos(\phi + 2\pi n/N) \tag{9.1}$$

where $I'(x,y)$ is the average intensity, $I''(x,y)$ the intensity modulation, $\phi(x,y)$ the phase to be solved for, and $n = 1, 2, \ldots, N$. Solving these equations leads to

$$\phi(x,y) = \tan^{-1}\frac{\sum_{n=1}^{N} I_n(x,y)\sin(2\pi n/N)}{\sum_{n=1}^{N} I_n(x,y)\cos(2\pi n/N)} \tag{9.2}$$

The phase obtained in Equation (9.2) ranges from $-\pi$ to $+\pi$ with 2π discontinuities. A spatial phase-unwrapping algorithm [29] can be used to obtain the continuous phase, which can be converted to 3D shape data if the system is calibrated [30].

9.2.3 Phase-to-Height Conversion

The simplest system calibration method compares the unwrapped phase map for the 3D object with that for a reference plane taken with the same system. Figure 9.2 illustrates the derivation of this method. The digital micromirror device (DMD) pixel N in the DLP projector illuminates point A on the reference plane and point D on the object. Therefore, the phase value Φ_D at point D is the same as that at point A on the reference plane ($\Phi'_A = \Phi_D$). From the point of view of the CCD (M), the phase value that was at point A on the reference plane now appears to come from point C because of the object ($\Phi_C \leftarrow \Phi'_A$). The phase difference for this particular camera pixel is then $\Delta\Phi_{DC} = \Phi_D - \Phi'_C = \Phi'_A - \Phi'_C = \Delta\Phi'_{AC}$.

Assuming that the line of the device $\overline{MN}$ is parallel to the reference plane, by similar triangles we obtain

$$\frac{d}{\overline{CA}} = \frac{s - \overline{BD}}{\overline{BD}} = \frac{s}{\overline{BD}} - 1 \approx \frac{s}{\overline{BD}} \tag{9.3}$$

Here, we assume that, for real measurement, $s \gg \overline{BD}$. Therefore, the depth becomes

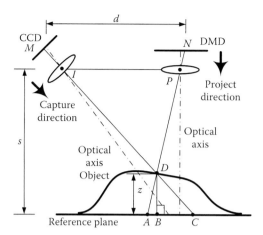

FIGURE 9.2
Calibration with a reference plane to retrieve depth for a 3D object. (Modified from Xu, Y. et al. *Applied Optics* 50:2572, 2011. With permission.)

$$z(x,y) = \overline{BD} \approx \frac{s}{d}\overline{CA} \tag{9.4}$$

This distance $\overline{CA}$ is proportional to $\Delta\Phi^r_{AC}$. Therefore, for the whole scene,

$$z(x,y) = c_0\left[\Phi(x,y) - \Phi^r(x,y)\right] \tag{9.5}$$

Here, c_0 is a constant determined by imaging a cube with known dimensions, Φ is the unwrapped phase map for the 3D scene, and Φ^r is the unwrapped phase for the reference plane. This calibration method has relatively low accuracy since it is an approximation, and it cannot provide (x,y) coordinates [25].

9.2.4 Phase-to-Coordinates Conversion

More complex calibration techniques have been developed to achieve better accuracy. Typically, the camera and the projector are calibrated through optimization while imaging a known scene such as a flat checkerboard [31]. For calibration purposes, we can model the DFP system as a stereo vision system, modeling the projector's DMD as an imaging sensor. After calibration, the linear model of the camera and the projector can be represented as

$$P = \begin{bmatrix} f_u & \alpha & u_0 \\ 0 & f_v & v_0 \\ 0 & 0 & 1 \end{bmatrix}\begin{bmatrix} r_{00} & r_{01} & r_{02} & t_x \\ r_{10} & r_{11} & r_{12} & t_y \\ r_{20} & r_{21} & r_{22} & t_z \end{bmatrix} = A[R,t] \tag{9.6}$$

where

$$A = \begin{bmatrix} f_u & \alpha & u_0 \\ 0 & f_v & v_0 \\ 0 & 0 & 1 \end{bmatrix}; \quad R = \begin{bmatrix} r_{00} & r_{01} & r_{02} \\ r_{10} & r_{11} & r_{12} \\ r_{20} & r_{21} & r_{22} \end{bmatrix}; \quad t = \begin{bmatrix} t_x \\ t_y \\ t_z \end{bmatrix} \qquad (9.7)$$

Here, (u_0, v_0) is the coordinate of the principal point, f_u and f_v are the focal lengths along the u and v axes of the image plane, and α is the parameter that describes the skewness of the two image axes. Since these parameters are intrinsic to the lenses and the imaging sensors of the camera and the projector, they form the intrinsic parameter matrix A. The matrix formed by R and t represents the rotation and translation, respectively, from the lens coordinates (x^c, y^c, z^c) to the world coordinates (x^w, y^w, z^w). It should be noted that this linear model does not account for lens distortions; those must be accounted for separately [31].

In the DFP system, the absolute phase map is usually used to establish the relationship between the camera sensor and the projector sensor as a one-to-many mapping (i.e., one point on the camera sensor corresponds to one line on the projector sensor with the same absolute phase value). This constraint was found sufficient to solve for (x^w, y^w, z^w) coordinates from the phase value point by point if the camera and the projector were calibrated in the same world coordinate system [30].

9.3 Real-Time 3D Profilometry with DLP Projectors

As was introduced in Section 9.1, we have developed real-time 3D profilometry systems based upon the DFP technique and DLP projectors. These systems take advantage of the unique projection mechanism of DLP technology, which will be addressed next.

9.3.1 DLP Technology

The core of the DLP projector is the DMD. Each micromirror can rotate between $+\theta_L$ (ON) and $-\theta_L$ (OFF). The grayscale value of each pixel is realized by controlling the ON time ratio: 0% ON time represents 0, 50% ON time means 128, and 100% ON time is 255. Therefore, a DLP projector produces a grayscale value by time modulation [32].

We have carried out a simple experiment to verify the time modulation behavior of one type of DLP projector used in our systems, the PLUS U5-632h. In this experiment, we connected a photodiode to a resistor to

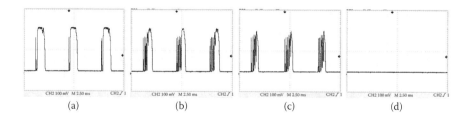

(a)   (b)   (c)   (d)

FIGURE 9.3
Examples of DLP time modulation for images with different uniform red values: (a) 255; (b) 128; (c) 64; (d) 0.

sense the output light of the projector. An oscilloscope was used to monitor the voltage of the photodiode system as the projector projected uniform red images at values of 255, 128, 64, and 0. Figure 9.3 shows the resulting oscilloscope output. It should be noted that the large gaps in the voltage data are due to the absence of the green and blue color channels. In Figure 9.3(a) and (d), the light level from the projector appears relatively constant at a high level and a low level, respectively. At a value of 128, as shown in Figure 9.3(b), the light level alternates between low and high, remaining at each level for approximately equal amounts of time. At a value of 64, the light level again alternates between the two levels, but it only jumps to the high level for about one-quarter of the time.

This experiment demonstrates that for images with values between 0 and 255, such as sinusoidal fringe patterns, the entire projection period of the image needs to be captured in order to recover the correct pattern. Therefore, to capture a conventional sinusoidal fringe pattern encoded in the red channel of a 120 Hz color DLP projector correctly, the camera exposure time must be a multiple of 1/360 second.

9.3.2 Principle

As addressed briefly in Section 9.2.2, only three fringe images are required to recover one 3D frame, and these can therefore be encoded in the three primary color channels (red, green, and blue, or RGB) and projected at once. However, the measurement quality degrades if the measured surface has strong color variations. In addition, the color coupling between RG and GB also affects the measurement quality if no filtering is used [33].

Replacing the color fringe patterns with three monochromatic fringe patterns projected in rapid succession eliminates these color-related issues. The unique projection mechanism of a single-chip DLP projection system makes this rapid switching feasible, thus facilitating real-time 3D data capture. Figure 9.4 shows the layout of such a system. In this system, the three phase-shifted images are still encoded in the projector's RGB channels for projection in rapid succession. However, the color filters of the DLP projector are first removed so that the projected fringes are monochromatic. A high-speed

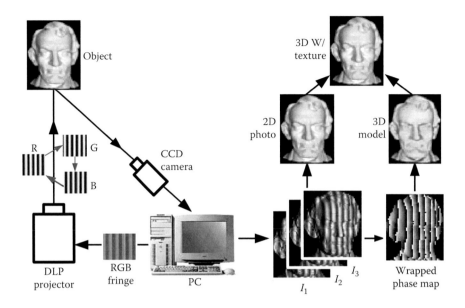

FIGURE 9.4
(See color insert.) The layout of the real-time 3D profilometry system we developed. (Modified from Zhang, S. *Optics and Lasers in Engineering* 48:149, 2010. With permission.)

CCD camera synchronized with the projector is used to capture the three channel images one by one. Applying the three-step phase-shifting algorithm to the three fringe images yields the computed phase and therefore the associated 3D shape measurements. With this type of system, we achieved 3D data acquisition at 40 frames per second (fps) [20] and later at 60 fps [34].

Furthermore, real-time 3D profilometry requires phase wrapping and phase unwrapping in real time. We have developed a fast three-step phase-shifting algorithm that improves the phase-wrapping speed by about 3.4 times [35]. This algorithm essentially approximates the arctangent function with an intensity ratio calculation in the same manner as that of the trapezoidal phase-shifting algorithm [36]. The approximation error is then compensated with a small lookup table.

Phase unwrapping removes the 2π discontinuities from the phase to yield the smooth phase map. Though numerous robust phase-unwrapping algorithms have been developed that include the branch-cut algorithms [37,38], the discontinuity minimization algorithm [39], the L^p-norm algorithm [40], and the least-squares algorithms [41], a conventional robust phase-unwrapping algorithm is usually too slow for real-time processing: It takes anywhere from a few seconds, to a few minutes, to even a few hours to process a standard 640 × 480 phase map [29]. Moreover, these algorithms are usually very sensitive to noise in the phase map.

We have developed a rapid yet robust phase-unwrapping algorithm by combining the advantages of the rapid scan-line algorithm with those of a robust quality-guided algorithm [42]. In our algorithm, a quality map is first computed from the phase gradient; higher quality points experience less noise and therefore yield lower phase gradients. Thresholds are then used to segregate points into multiple quality levels. Finally, the scan-line algorithm is used to unwrap the points in each quality level, beginning with the highest quality level. In this way, the unwrapping is both fast and accurate.

Finally, real-time 3D profilometry requires real-time conversion of the phase map to 3D coordinates and real-time visualization of the results. We found that it was very challenging to accomplish these tasks with central processing units (CPUs). Since the phase-to-coordinate conversion consists of simple point-by-point matrix operations, it can be efficiently calculated in parallel on a graphics processing unit (GPU). Data transfer rates between the CPU and the GPU do not pose a significant problem since the input data set to the GPU consists of a single phase value at each point instead of computed 3D coordinates. In addition, because the coordinates are computed on the GPU, they can be rendered immediately without accessing CPU data, again preventing a data transfer bottleneck between the CPU and the GPU. The adoption of this GPU technique permits real-time 3D coordinate calculation and visualization [43].

9.3.3 Experimental Results

We have developed a real-time 3D profilometry system that uses a high-speed CCD camera (Pulnix TM-6740 CL) with a 16 mm focal length lens (Fujinon HF16HA-1B). The sensor pixel size is 7.4 μm × 7.4 μm. The maximum data speed for this camera is 200 frames per second (fps) with an image resolution of 640 × 480. The projector (PLUS U5-632h) has an image resolution of 1024 × 768 and a focal length of f = 18.4–22.1 mm. This projector refreshes at 120 Hz, and thus this system can theoretically reach 120 Hz 3D profilometry speed. However, due to the speed limit of the camera and the synchronization requirements between the camera and the projector, we can only capture fringe images at 180 fps. Since three fringe images are needed to reconstruct one 3D shape, the 3D profilometry speed is actually 60 Hz.

With such a high measurement speed, this system can capture high-quality natural facial expressions. Figure 9.5 presents 3D measurements of the forming of a human facial expression. Photographs of the subject's face are included for comparison. As the figure demonstrates, the system captures the facial details quite well. It should be noted that the data were processed with a 5 × 5 Gaussian smoothing filter to reduce the most significant random noise. The system was calibrated using the technique discussed in Section 9.2.4, and the 3D absolute coordinates were obtained by encoding a small cross marker into the projected fringe pattern.

As introduced in Section 9.3.2, this system not only acquires 3D shape measurements in real time, but also simultaneously processes and displays them

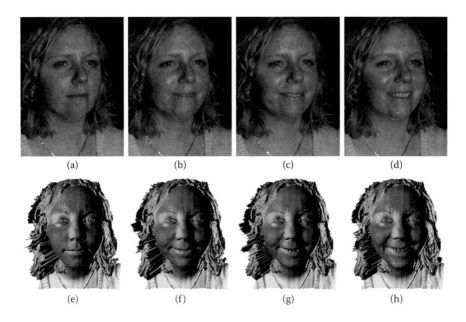

FIGURE 9.5
Real-time 3D shape measurement results using the developed system. (a)–(d) Photographs of the human subject forming a facial expression; (e)–(h) the corresponding 3D facial expressions captured at 60 fps with a spatial resolution of 640 × 480.

at high speed. Figure 9.6 illustrates the real-time measurement of a human subject. The captured 3D geometry is rendered on a computer screen to the subject's left. For this measurement, the simultaneous 3D data acquisition, reconstruction, and display speed achieved was 30 fps.

9.3.4 Discussion

As demonstrated earlier, real-time 3D profilometry can be achieved using the traditional DFP technique by taking advantage of the unique projection mechanism of single-chip DLP. However, such a technique usually requires high-quality fringe images containing at least 8 bits of precision (256 gray-scale values). The following issues arise when using 8-bit sinusoidal patterns with this real-time DFP technique:

- *Projector nonlinearity.* Since the projector is a nonlinear device, generating perfect sinusoidal fringe patterns is very difficult, even though numerous algorithms have been proposed to calibrate and correct such a problem [44–46]. In our research, we have also noticed that the computer's graphics card influences the system nonlinearity and that this influence changes over time, further complicating the problem.

FIGURE 9.6
Simultaneous 3D data acquisition, reconstruction, and display. The system achieved real-time 3D profilometry at a frame rate of 30 fps with an image resolution of 640 × 480 per frame.

- *Synchronization requirement.* As introduced in Section 9.3.1, a DLP projector is a digital device that generates images by time integration [32]. In order for the camera to image the projected pattern correctly, it must be precisely synchronized with the projector and capture the entire projection period of the image. Any mismatch will result in significant measurement error [47].

- *Minimum exposure time limitation.* The precise synchronization requirement limits the camera's exposure time to at least the duration of the projection period of the fringe image. If a fringe image is encoded in one color channel of the typical color DLP projector (which has a refresh rate of 120 Hz), the minimum exposure time is 2.78 ms. Since a much shorter exposure time is usually required for fast motion capture, this limits the potential of such a real-time DFP technique for fast motion capture applications.

If, on the other hand, 1-bit images could be used to generate sinusoidal fringes, these issues could be avoided or significantly reduced. As mentioned briefly in Section 9.3.1, for a 1-bit image, the micromirrors of the DMD remain stationary for the whole image projection period, eliminating most of the synchronization requirements. This led to the development of the binary defocusing technique for 3D profilometry, which will be addressed next.

9.4 Superfast 3D Profilometry with DLP Discovery Platform

9.4.1 Principle

9.4.1.1 Sinusoidal Fringe Pattern Generation by Defocusing

The aforementioned issues with conventional DFP render it undesirable for use in ultrafast 3D profilometry system development. Our recent research shows that a binary structured pattern becomes pseudosinusoidal when properly defocused [27]. This result is similar to that obtained by Su et al. using a Ronchi grating [48].

Figure 9.7 shows some typical results when the projector is defocused to different degrees while the camera is in focus. The pattern employed here is a basic set of black-and-white stripes referred to as a squared binary pattern (SBM). This figure shows that as the projector becomes increasingly defocused, the binary structured pattern becomes increasingly distorted. Figure 9.7(a) shows the result when the projector is in focus: clear binary structures on the image. As the degree of defocusing increases, the binary structures become less and less clear, and the sinusoidal ones become more and more obvious. However, if the projector is defocused too much, the sinusoidal structures start diminishing, as indicated in Figure 9.7(e). This experiment indicates that a pseudosinusoidal fringe pattern can indeed be generated by properly defocusing a binary structured pattern.

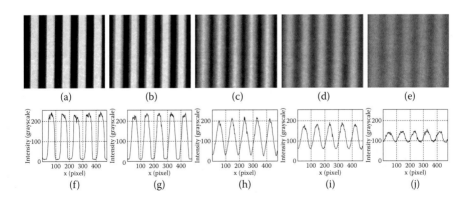

FIGURE 9.7
Example of sinusoidal fringe generation by defocusing a binary structured pattern: (a) the result when the projector is in focus; (b)–(e) the result when the projector is increasingly defocused. (f)–(j) the 240th row cross section of the corresponding image. (Modified from Ekstrand, L. and Zhang, S. *Optical Engineering* 50:123603, 2011. With permission.)

9.4.1.2 Optimal Pulse Width Modulation Technique

The seemingly sinusoidal fringe pattern generated from the SBM pattern retains some of its binary structure in the form of harmonics beyond the fundamental sinusoidal frequency [49]. Some of these harmonics induce errors in the results that limit the depth range for accurate 3D measurement [50]. An alternative approach to the defocusing technique uses fringe patterns generated by optimal pulse width modulation (OPWM), which produces binary fringe patterns by selectively removing regions of the square wave of SBM that would generate the undesired harmonics after defocusing [49]. This technique is used in the field of electrical engineering to generate sinusoidal waveforms [51]. Figure 9.8 shows an SBM pattern in comparison to an OPWM pattern, illustrating the shifting of columns that takes place during the removal of selected harmonics. Indeed, Figure 9.8(b) appears more sinusoidal even prior to defocusing. Therefore, defocused OPWM patterns can serve as an alternative to 8-bit sinusoidal patterns with all of the advantages of binary structured patterns.

9.4.2 Experimental Results

The SBM binary pattern defocusing technique was verified with the measurement of a complex sculpture. Figure 9.9 presents the results of this verification. Figure 9.9(a)–(c) shows the three fringe images employed, which are phase shifted by $2\pi/3$. To produce this phase shift, the pattern was translated in the computer by one-third of its period. The three-step phase-shifting algorithm was then used to obtain the wrapped phase map shown in Figure 9.9(d). After unwrapping via a phase-unwrapping algorithm, the unwrapped phase map was converted to depth values using the simple phase-to-height conversion algorithm given in Section 9.2.3. Figure 9.9(e) presents the recovered 3D results, which are clearly of high quality.

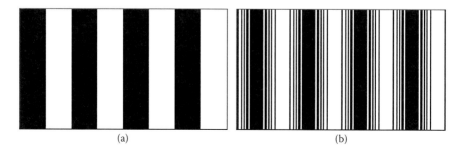

(a)   (b)

FIGURE 9.8

Comparison of computer-generated binary patterns: (a) one of the SBM patterns; (b) one of the OPWM patterns. (From Ekstrand, L. and Zhang, S. *Optical Engineering* 50:123603, 2011. With permission.)

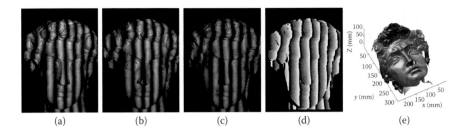

FIGURE 9.9
Three-dimensional shape measurement example using the binary defocusing technique: (a)–(c) three phase-shifted fringe patterns; (d) wrapped phase map; (e) recovered 3D shape.

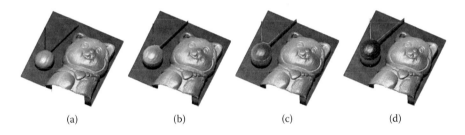

FIGURE 9.10
Three-dimensional shape measurement with the OPWM defocusing technique and a multi-frequency phase-shifting algorithm: (a)–(d) 3D frames of a swinging pendulum system next to a stationary 3D sculpture. (Modified from Wang, Y. and Zhang, S. *Optics Express* 19:5149, 2011. With permission.)

In addition to the aforementioned advantages of using binary structured patterns, the defocusing technique also leads to large breakthroughs in 3D profilometry speed. The most recently developed DLP Discovery technology is capable of switching 1-bit images at tens of kilohertz. By applying the new fringe generation technique to this DLP Discovery projection platform, we achieved an unprecedented rate for 3D shape measurement: 667 Hz [28]. Furthermore, using the OPWM technique, we implemented a multifrequency phase-shifting algorithm with fringe image capture at 5000 fps to develop a 3D profilometry system capable of measurements at 556 Hz [52]. In this system, we used a DLP Discovery projection system that included a DLP Discovery board (D4000), an ALP high-speed software package, and an S3X optical module; a high-speed CMOS camera (Phantom v9.1) that captures the fringe images at 5000 fps with an image resolution of 576 × 576; and a synchronization circuit that takes the projection timing signal and generates a timing signal to trigger the camera.

Figure 9.10 presents four 3D frames of a swinging pendulum system simultaneously measured next to a complex 3D sculpture. The measurement quality is clearly very high. These frames were captured with the 556 Hz multifrequency system; it is important to observe that this system is capable of measuring arbitrary step height objects since it utilizes temporal instead of spatial phase unwrapping.

9.4.3 Discussion

Due to the numerous advantages of defocusing binary structured patterns to generate sinusoidal fringe patterns, binary defocusing has the potential to replace the conventional fringe generation technique for 3D profilometry. However, the defocusing technique is not trouble free. A number of challenges need to be overcome before such a technology can be widely deployed:

- *Challenge 1: The hardware system cost is quite high.* Compared to commercially available DLP projectors, the DLP Discovery is approximately 10 to 20 times more expensive. The question is how to attain similar or higher speed 3D profilometry without using such an expensive platform.

- *Challenge 2: The 3D profilometry speed is limited to 10 kHz.* The binary image refresh rate of the DLP Discovery is 32 kHz, which places a limit on its 3D profilometry speed of approximately 10 kHz. The question is whether there is a way to break through this speed bottleneck.

- *Challenge 3: The calibration is difficult.* Since the projector is out of focus, all existing high-accuracy structured-light system calibration techniques that require the projector to be in focus cannot be applied. The question is how to calibrate the system accurately with an out-of-focus projector.

- *Challenge 4: The measurement depth range is much smaller.* The defocused binary structured patterns only become high-quality sinusoidal patterns within a small depth range (less than 10% of the high-contrast projection region). Therefore, any measurements made outside that range will have errors. The question is how to increase the measurement depth range without sacrificing the sensing speed.

- *Challenge 5: The high-order harmonics cause residual errors.* The defocusing technique essentially acts like a low-pass filter, suppressing the high-frequency harmonics present in the binary pattern that cause measurement errors. However, it is difficult to eliminate these harmonics completely. The question is how to remove their residual effects.

Our recent research efforts in developing OPWM [49] seem to hold the most promise for overcoming these challenges. However, the effectiveness of this OPWM technique is limited by the discrete nature of DFP; the digital fringe images cannot be continuously divided to obtain more accurate sinusoids upon defocusing. Therefore, it remains difficult to achieve the same depth range as the conventional DFP method.

9.5 Summary

This chapter has presented some of the recent advances in high-speed 3D optical profilometry with digital fringe projection and phase-shifting techniques. Two high-speed 3D profilometry systems were presented: a real-time 3D profilometry system that achieved simultaneous 3D shape acquisition, reconstruction, and display at 40 Hz with more than 250,000 points per frame; and a superfast 3D profilometry system that achieved 556 Hz measurement speed with the novel binary defocusing techniques. We have explained the principles of these techniques, presented some relevant experimental results, and addressed some challenging issues that we still face.

References

1. G. Geng. Structured-light 3D surface imaging: A tutorial. *Advances in Optics and Photonics* 3 (2):128–160 (2011).
2. U. Dhond and J. Aggarwal. Structure from stereo—A review. *IEEE Transactions Systems, Man, and Cybernetics* 19:1489–1510 (1989).
3. L. Zhang, B. Curless, and S. Seitz. Spacetime stereo: Shape recovery for dynamic scenes. *Proceedings Computer Vision and Pattern Recognition* 2:367–374 (2003).
4. J. Salvi, S. Fernandez, T. Pribanic, and X. Llado. A state of the art in structured light patterns for surface profilometry. *Pattern Recognition* 43:2666–2680 (2010).
5. S. Nayar and Y. Nakagawa. Shape from focus. *IEEE Transactions Pattern Analysis and Machine Intelligence* 16:824–831 (1994).
6. M. Subbarao and G. Surya. Depth from defocus: A spatial domain approach. *International Journal of Computer Vision* 13:271–294 (1994).
7. W. C. Wiley and I. H. McLaren. Time-of-flight mass spectrometer with improved resolution. *Review of Scientific Instruments* 26(12) (1955).
8. F. Ackermann. Digital image correlation: Performance and potential application in photogrammetry. *Photogrammetric Record* 11:429–439 (1984).
9. J. P. Siebert and S. J. Marshall. Human body 3D imaging by speckle texture projection photogrammetry. *Sensor Review* 20(3):218–226 (2000).
10. A. Anand, V. K. Chhaniwal, P. Almoro, G. Pedrini, and W. Osten. Shape and deformation measurements of 3D objects using volume speckle field and phase retrieval. *Optics Letters* 34 (10):1522–1524 (2009).
11. S. Rusinkiewicz, O. Hall-Holt, and M. Levoy. Real-time 3D model acquisition. *ACM Transactions Graphics* 21 (3):438–446 (2002).
12. L. Zhang, N. Snavely, B. Curless, and S. M. Seitz. Space-time faces: High-resolution capture for modeling and animation. *ACM Annual Conference on Computer Graphics* 23(3):548–558 (2004).
13. J. Davis, R. Ramamoorthi, and S. Rusinkiewicz. Space-time stereo: A unifying framework for depth from triangulation. *IEEE Transactions Pattern Analysis Machine Intelligence* 27 (2):1–7 (2005).

14. X. Su and Q. Zhang. Dynamic 3-D shape measurement method: A review. *Optics Lasers Engineering* 48:191–204 (2010).

15. M. Takeda and K. Mutoh. Fourier transform profilometry for the automatic measurement of 3-D object shape. *Applied Optics* 22:3977–3982 (1983).

16. K. Qian. Windowed Fourier transform for fringe pattern analysis. *Applied Optics* 43 (13):2695–2702 (2004).

17. D. Malacara. ed. *Optical shop testing*, 3rd ed. New York: John Wiley & Sons (2007).

18. P. S. Huang, C. Zhang, and F.-P. Chiang. High-speed 3-D shape measurement based on digital fringe projection. *Optics Engineering* 42 (1):163–168 (2002).

19. C. Zhang, P. S. Huang, and F.-P. Chiang. Microscopic phase-shifting profilometry based on digital micromirror device technology. *Applied Optics* 41:5896–5904 (2002).

20. S. Zhang and P. S. Huang. High-resolution real-time three-dimensional shape measurement. *Optics Engineering* 45 (12):123, 601 (2006).

21. K. Liu, Y. Wang, D. L. Lau, Q. Hao, and L. G. Hassebrook. Dual-frequency pattern scheme for high-speed 3-D shape measurement. *Optics Express* 18 (5):5229–5244 (2010).

22. S. Zhang and S.-T. Yau. High-resolution, real-time 3-D absolute coordinate measurement based on a phase-shifting method. *Optics Express* 14 (7):2644–2649 (2006).

23. S. Zhang, D. Royer, and S.-T. Yau. GPU-assisted high-resolution, real-time 3-D shape measurement. *Optics Express* 14 (20):9120–9129 (2006).

24. Y. Li, C. Zhao, Y. Qian, H. Wang, and H. Jin. High-speed and dense three-dimensional surface acquisition using defocused binary patterns for spatially isolated objects. *Optics Express* 18 (21):21, 628–631, 635 (2010).

25. S. Zhang. Recent progress on real-time 3-D shape measurement using digital fringe projection techniques. *Optics Lasers Engineering* 48 (2):149–158 (2010).

26. Y. Gong and S. Zhang. High-speed, high-resolution three-dimensional shape measurement using projector defocusing. *Optics Engineering* 50 (2):023, 603 (2011).

27. S. Lei and S. Zhang. Flexible 3-D shape measurement using projector defocusing. *Optics Letters* 34 (20):3080–3082 (2009).

28. S. Zhang, D. van der Weide, and J. Oliver. Superfast phase-shifting method for 3-D shape measurement. *Optics Express* 18 (9):9684–9689 (2010).

29. D. C. Ghiglia and M. D. Pritt. *Two-dimensional phase unwrapping: Theory, algorithms, and software.* New York: John Wiley & Sons, Inc. (1998).

30. S. Zhang and P. S. Huang. Novel method for structured light system calibration. *Optics Engineering* 45 (8):083, 601 (2006).

31. Z. Zhang. A flexible new technique for camera calibration. *IEEE Transactions Pattern Analysis Machine Intelligence* 22 (11):1330–1334 (2000).

32. L. J. Hornbeck. Digital light processing for high-brightness, high-resolution applications. *Proceedings SPIE* 3013:27–40 (1997).

33. J. Pan, P. S. Huang, and F.-P. Chiang. Color-coded binary fringe projection technique for 3-D shape measurement. *Optics Engineering* 44 (2):023, 606 (2005).

34. S. Zhang and S.-T. Yau. High-speed three-dimensional shape measurement system using a modified two-plus-one phase-shifting algorithm. *Optics Engineering* 46 (11):113, 603 (2007).

35. P. S. Huang and S. Zhang. Fast three-step phase-shifting algorithm. *Applied Optics* 45 (21):5086–5091 (2006).

36. P. S. Huang, S. Zhang, and F.-P. Chiang. Trapezoidal phase-shifting method for three-dimensional shape measurement. *Optics Engineering* 44 (12):123, 601 (2005).

37. J. M. Huntley. Noise-immune phase unwrapping algorithm. *Applied Optics* 28:3268–3270 (1989).

38. M. F. Salfity, P. D. Ruiz, J. M. Huntley, M. J. Graves, R. Cusack, and D. A. Beauregard. Branch cut surface placement for unwrapping of undersampled three-dimensional phase data: Application to magnetic resonance imaging arterial flow mapping. *Applied Optics* 45:2711–2722 (2006).

39. T. J. Flynn. Two-dimensional phase unwrapping with minimum weighted discontinuity. *Journal Optical Society America* A 14:2692–2701 (1997).

40. D. C. Ghiglia and L. A. Romero. Minimum L^p-norm two-dimensional phase unwrapping. *Journal Optical Society America* A 13:1–15 (1996).

41. J.-J. Chyou, S.-J. Chen, and Y.-K. Chen. Two-dimensional phase unwrapping with a multichannel least-mean-square algorithm. *Applied Optics* 43:5655–5661 (2004).

42. S. Zhang, X. Li, and S.-T. Yau. Multilevel quality-guided phase unwrapping algorithm for real-time three-dimensional shape reconstruction. *Applied Optics* 46 (1):50–57 (2007).

43. S. Zhang, D. Royer, and S.-T. Yau. GPU-assisted high-resolution, real-time 3-D shape measurement. *Optics Express* 14 (20):9120–9129 (2006).

44. S. Zhang and P. S. Huang. Phase error compensation for a three-dimensional shape measurement system based on the phase shifting method. *Optics Engineering* 46 (6):063, 601 (2007).

45. S. Zhang and S.-T. Yau. Generic nonsinusoidal phase error correction for three-dimensional shape measurement using a digital video projector. *Applied Optics* 46 (1):36–43 (2007).

46. B. Pan, Q. Kemao, L. Huang, and A. Asundi. Phase error analysis and compensation for nonsinusoidal waveforms in phase-shifting digital fringe projection profilometry. *Optics Letters* 34 (4):2906–2914 (2009).

47. S. Lei and S. Zhang. Digital sinusoidal fringe generation: Defocusing binary patterns vs. focusing sinusoidal patterns. *Optics Lasers Engineering* 48:561–569 (2010).

48. X.-Y. Su, W.-S. Zhou, G. von Bally, and D. Vukicevic. Automated phase-measuring profilometry using defocused projection of a Ronchi grating. *Optics Communications* 94 (13):561–573 (1992).

49. Y. Wang and S. Zhang. Optimum pulse width modulation for sinusoidal fringe generation with projector defocusing. *Optics Letters* 35 (24):4121–4123 (2010).

50. Y. Xu, L. Ekstrand, J. Dai, and S. Zhang. Phase error compensation for three-dimensional shape measurement with projector defocusing. *Applied Optics* 50 (17):2572–2581 (2011).

51. V. G. Agelidis, A. Balouktsis, and I. Balouktsis. On applying a minimization technique to the harmonic elimination PWM control: The Bipolar Waveform. *IEEE Power Electronics Letters* 2:41–44 (2004).

52. Y. Wang and S. Zhang. Superfast multifrequency phase-shifting technique with optimal pulse width modulation. *Optics Express* 19 (6):5143–5148 (2011).

10

Time-of-Flight Techniques

Shoji Kawahito

CONTENTS

10.1 Introduction

Time-of-flight (TOF) range imaging is a three-dimensional (3D) imaging method with moderate accuracy and simple configuration. In the past 20 years, great progress has been made in TOF range imaging. There are two methods for measuring the TOF: direct and indirect. In the direct TOF measurement, the TOF is measured directly in a time domain, and a semiconductor device—the so-called single photon avalanche diode—has enabled all electronic range cameras without any mechanically moving components. Indirect TOF range imaging has been realized by the invention of a lock-in pixel with charge-domain demodulation of the modulated light signal. In the indirect TOF measurement, the TOF is converted to a physical quantity and then the range is calculated by the amount of physical quantity modulation due the TOF and a time reference. The lock-in pixel used for the TOF range imagers is based on the concept of rapid charge transfer in a fully depleted semiconductor, which originated in charge-coupled device (CCD) image sensors and now is realized mostly in CMOS (complementary metal oxide semiconductor) image sensors.

Using the round-trip TOF of light T, the range to an object L is expressed as

$$\text{Mass of 1 aluminum atom} = \frac{27\,\text{g/mol}}{N_A} = 4.5 \times 10^{-23}\,\text{g/atom} \qquad (10.1)$$

where c (= 3×10^8 m/s) is the speed of light.

A high time resolution is required for precise range measurement because of the speed of light. For instance, to resolve the depth of 1.5 mm, the TOF to be resolved is 10 ps. Progress in semiconductor integrated circuits technology is enabling a high time resolution of less than 10 ps. Practical applications of the TOF range imager require high tolerance to background light as well as the range accuracy or resolution. The TOF imaging system using a mechanical scanner has relatively high tolerance to background light because of the high light energy concentration to the point of measurement. On the other hand, spatial resolution and speed of imaging are limited by the performance of the scanner. All-electronic 3D imaging systems using an area light projector and an area TOF sensor have practical advantages of cost effectiveness, high reliability due to nonmechanical implementation, higher frame rate, and higher 2D spatial resolution.

This chapter describes possible system configurations for TOF range imaging, but mainly focuses on all-electronic range cameras using direct and indirect TOF measurements. Devices, circuits, and systems for TOF range imaging are described and range resolution and tolerance to background light in indirect TOF measurements are discussed.

10.2 System Configurations for TOF Range Imaging

TOF range imaging methods are categorized into three types, depending on the types of mechanical scanners and photo detectors, as shown in Table 10.1. The *SPD* uses a two-dimensional (2D) mechanical scanner with a single-point detector The *LD* uses a one-dimensional (1D) mechanical scanner with a 1D array detector or a linear imager. An all-electronic (*AE*) type or a mechanical scannerless range imager can be implemented with a 2D array detector or an area imager. The *SPD* and *AE* types are treated hereafter.

TABLE 10.1

System Configurations for TOF Range Imaging

Type	Mechanical Scanner	Detector
SPD	2D	Single point
LD	1D	1D array (linear imager)
AE	Scannerless	2D array (area imager)

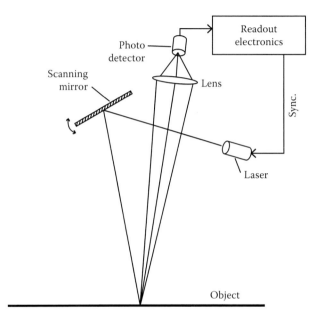

FIGURE 10.1
TOF system using a single-point detector and a laser 2D scanner.

10.2.1 Single-Point Detector with Laser Scanner

Figure 10.1 shows a configuration of the single-point TOF range imaging system. A laser beam is scanned by a 2D scanning mirror and the back-reflected light is received by a photo detector. The 2D spatial position is determined by the amount of the diffraction of the mirror and the depth is measured by the time of flight. The laser pulse generation is synchronized with the readout electronics for the received light to measure the TOF.

The advantage of the single-point detection method is a high tolerance to the background light because the laser power is concentrated at the point of measurement. The range resolution is closely related to the signal-to-noise ratio (SNR). One of the noises is due to background light. In the system shown in Figure 10.1, the photo detector area must cover the field of view of imaging and receives back-reflected background light from all the scanning

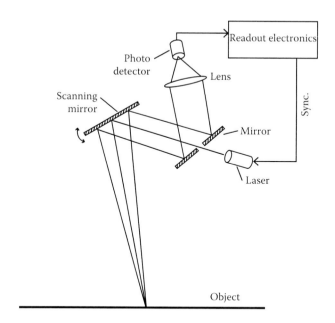

FIGURE 10.2
TOF system using coaxial optical system.

areas. Therefore, even though the laser power is concentrated at the point of measurement, this system is not always ideal for maximizing the SNR.

An optical system shown in Figure 10.2 offers the best solution for minimizing the influence of background light [1]. In this system, the laser beam emitted and light received use a coaxial optical system. This enables the field of view of the photo detector to be concentrated on the measuring point and only a small portion of background light to be influenced to the signal, resulting in a high SNR in the range calculation.

Electronics for the single-point detector to measure the TOF often use a direct measurement of the delay of the back-reflected light pulse in the time domain. The detailed circuit implementation will be discussed in the next section.

10.2.2 Scannerless System with Area Detector and Area Light Projector

Using an area image sensor or a 2D array photo detector, a scannerless TOF range imaging system can be implemented as shown in Figure 10.3. The entire field of view is illuminated by an area light projector and the back-reflected light image is taken by a TOF range camera that has a time-resolving capability in each pixel. One of the methods for time-resolved imaging is to use a gated image intensifier (GII) in front of a CCD camera [2]. A switching of a few nanosecond time windows with a modulation period of 20 to 100 ns is possible using the GII.

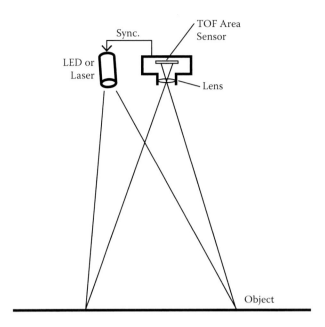

FIGURE 10.3
TOF system with area detector and area light projector.

Another time-resolved imaging method is to use time-resolved CCD or CMOS imagers in which each pixel has a dedicated design for synchronized operation and high-speed photo response to a high-frequency modulated light. Such pixel technology is often called a lock-in pixel [9]. The time-resolved photo response in each pixel is used for measuring the TOF directly or indirectly. The "direct TOF" means that the TOF is measured directly in time domain. The "indirect TOF" means that the TOF is once measured as a result of modulation of physical quantity in the receiver, such as a modulation of photo-generated charge, current, or voltage; then, the TOF is calculated by the conversion of the physical quantity to time-domain signal change using a time reference. The devices, circuits, and systems for the direct and indirect TOF measurements will be described in the following two sections.

10.3 Direct TOF Range Imaging

10.3.1 2D Array Detector with SPAD

A CMOS LSI (large scale integration) technology allows us to implement a 1D or 2D array light receiver with control circuits and makes a direct TOF system without any mechanical scanner but with an area projector possible.

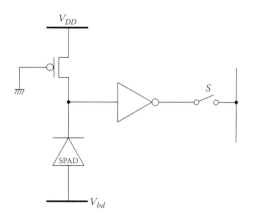

FIGURE 10.4
Single photon avalanche diode.

The scannerless direct TOF system requires photo-charge detection mechanism with very high sensitivity and high-speed response. A single photon avalanche diode (SPAD) working in Geiger mode fulfills this purpose [3–5].

The schematic of the single photon receiver pixel using the SPAD and the operation are shown in Figures 10.4 and 10.5, respectively [6]. When a p–n junction or a SPAD is suddenly reverse biased above its breakdown voltage, there will be no breakdown current until a carrier enters the depletion region. Suppose that a photon is absorbed at this moment in the depletion region. The carriers resulting from the absorption of a photon generate an extremely large number of electron-hole pairs by impact ionization, causing an avalanche multiplication due to a single photon.

A p-channel MOS (pMOS) transistor whose gate-to-source voltage is biased to V_{DD} is connected to the anode terminal of the SPAD. When the avalanche current is larger than the bias current of the pMOS transistor, the cathode terminal voltage of the SPAD goes down and hence the reverse bias is lowered so that the avalanche current is quenched. Once the avalanche current is quenched, the pMOS again recharges the depletion capacitance of the SPAD to V_{DD}. This mode of operation is commonly known as Geiger mode. A CMOS inverter connected at the anode of the SPAD is used as a comparator to generate a logical pulse right after a photon is received. The output of the inverter is connected to a common signal line by a switch S when the pixel is selected for readout.

Figure 10.6 shows the cross-sectional view of the SPAD, which is compatible to CMOS technology. To prevent a premature breakdown due to the high electric field at the periphery of the p+ junction, the p+ layer is surrounded by a guard ring using a relatively lightly doped p-well layer. A planar avalanche multiplication region is created below the p+ junction with a deep n-well region.

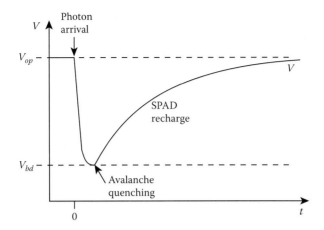

FIGURE 10.5
Operation of Geiger-mode SPAD.

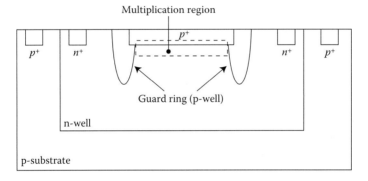

FIGURE 10.6
Cross-sectional view of the SPAD.

10.3.2 Time-to-Digital Converter

A critical component of the direct TOF range imager using the SPAD is the time-to-digital converter (TDC). A direct TOF measurement system using a TDC and operation timing is shown in Figures 10.7 and 10.8, respectively. A TDC can be implemented with a digital counter, which measures the time interval between the transmitted light pulse and the received light pulse. To do this, a gate pulse is generated by the measurement of the time interval (TOF) and the number of clock cycles is counted using the gated high-frequency clock given at a counter. To generate the gating pulse, a flip-flop with set and reset inputs can be used, as shown in Figure 10.7. The flip-flop is set by a trigger pulse for light pulse transmission and is reset by the rising edge of the received light pulse, which is generated by the output of the SPAD.

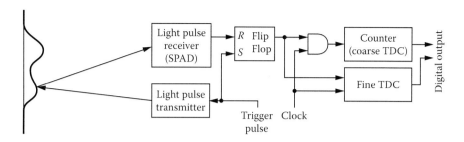

FIGURE 10.7
Direct TOF measurement system using a TDC.

The time resolution of the TDC using a counter is equal to the inverse of the clock frequency, and therefore the range resolution is determined by the clock frequency. For instance, the clock frequency of 1 GHz gives the time resolution of 1 ns and the range resolution of 15 cm. If a range resolution of 1 mm is required, the clock frequency of 150 GHz must be used for the gated counter. An implementation of the gated counter working at 150 GHz is not easy, even if state-of-the-art CMOS technology is used.

For higher time resolution without increasing clock frequency, a subranging technique is often used for the TDC. As shown in Figure 10.7, the counter is used for coarse TDC and a fine TDC measures the subrange of one clock cycle T_c of the coarse TDC.

A typical implementation of the fine TDC is shown in Figure 10.9. A start pulse is given at the input of a delay line consisting of an inverter chain and is activated at the rigging edge of the clock. The outputs of the M-stage inverter chain are latched in a flash register using a stop pulse that is activated at the falling edge of the gating pulse. The register output is expressed as a thermometer code corresponding to the subrange delay T_{SR}, which is expressed as the difference of the TOF to be measured T from its coarse measure given by $N_1 \times T_c$, where N_1 is the number of counts of the counter output. By adding up the ones of the thermometer code, a binary code M_1 is generated. Using the unit inverter delay ΔT, the binary code is expressed as

$$M_1 = \frac{T_{SR}}{\Delta T} = \frac{T - N_1 \times T_c}{\Delta T} \tag{10.2}$$

If the number of inverter stages is $M = 2^m$, where m is an integer, and an n-bit counter is used, the TDC using the subranging technique has a resolution of $n + m$ bits and the minimum time step of ΔT. The unit delay of an inverter chain using state-of-the-art CMOS technology can be smaller than 10 ps. Therefore, a range resolution of a few millimeters can be realized in the SPAD-based range imager. In actual range measurements, the TDC output is averaged over many samples and the equivalent resolution is improved to smaller than ΔT.

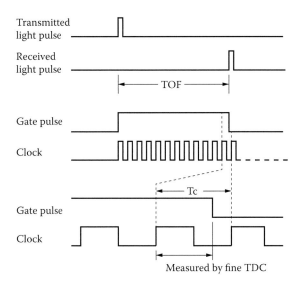

FIGURE 10.8
Operation of the TDC using a subranging technique.

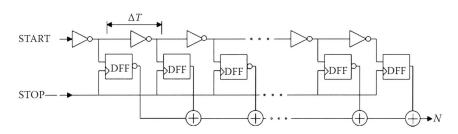

FIGURE 10.9
Time-to-digital converter using a delay line and flash register.

10.3.3 CMOS Implementations

The first TOF range image sensor with SPADs was implemented with a 0.8 μm CMOS process and had 32 × 32 pixels [6]. The pixel occupies the area of 58 × 58 μm² and the active area of the SPAD is 38 × 38 μm². The TDC is not on the sensor chip, but rather is implemented in a separated CMOS chip. Using a light source with a pulse width of 150 ps, the range resolution of 1.8 mm is attained for a distance range of 3 m.

A TOF range image sensor with a large SPAD pixel array of 128 × 128 pixels has been implemented with a 0.35 μm CMOS process and on-chip 10-bit 32-channel TDC [7]. The 10-bit TDC uses a three-level subranging technique with a counter (2 bits), DLL(delayed locked 100p)-based TDC (4 bits), and delay-line-based TDC (4 bits). The time resolution of 97 ps has been achieved with the on-chip TDC. The pixel occupies the area of 25×25 μm^2 and the active diameter of the circular SPAD is 7 μm. A range resolution (standard deviation) of 5.2 mm for the distance range of 3.75 m is attained. Though the range resolution of the TDC with 97 ps time resolution is 1.5 cm, the averaging mechanism of the system improves the range resolution to 5 mm.

10.4 Indirect TOF Range Imaging

10.4.1 Lock-in Pixel for Time-Dependent Charge Detection

In indirect TOF measurement using silicon technology, a lock-in pixel using charge domain processing is used. The first attempt of time-resolved imaging with charge domain processing was reported in Povel, Aebersold, and Stenflo [8] for an application to the 2D polarimeter, in which the modulated light is generated by a piezoelastic modulator and the modulation frequency is 50 kHz. For the application to 3D imaging with millimeter depth resolution, a modulation frequency of higher than 10 MHz is required. Dedicated pixel structures for high-frequency lock-in operation using a CCD have elevated the charge modulation frequency of lock-in pixels to be several 10 MHz [9].

Figure 10.10 shows a conceptual schematic of the charge-domain processing in a lock-in pixel for indirect TOF range imaging. The photo detector has two transfer gates for transferring photo-induced charge into two charge accumulators. When the light pulse is received at the phase as depicted by a solid line, the photo-induced charge is equally divided into two storages. If the light pulse is delayed, as depicted by a dashed line, the charge transferred to the right-side storage is increased while that of the left-side storage is decreased. The difference of the two charges depends on the delay of the light pulse and therefore the range can be measured by the amount of charge. This operation is repeated many times so that sufficient amounts of charge are accumulated in the two storages.

Figure 10.11 shows a cross-sectional view of the two-tap lock-in (range finder) CCD [10]. An incoming light pulse is received in the aperture area of the pixel and is absorbed in bulk silicon through a photo gate (PG). In TOF measurements, near infrared light is commonly used. The PG made of thin polycrystalline silicon is transparent to infrared light.

Two transfer gates $TX1$ and $TX2$ are turned on alternately, as shown in Figure 10.12. If there is no delay time in the received light, a photo-generated charge is equally divided into two storages under the gates $ST1$ and $ST2$. The delay of the received light pulse decreases the charge Q_1 and increases the charge Q_2. With photo current I_p, the light pulse width T_w, and the delay time T_d, these charges are expressed as

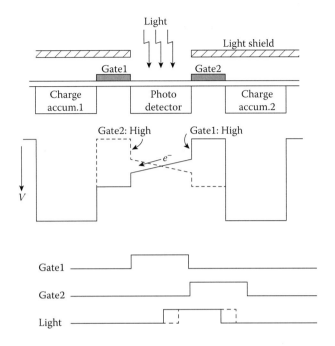

FIGURE 10.10
Operation of a lock-in pixel.

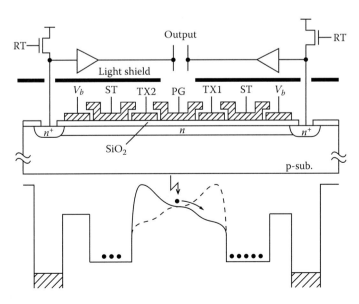

FIGURE 10.11
TOF imager using CCD process.

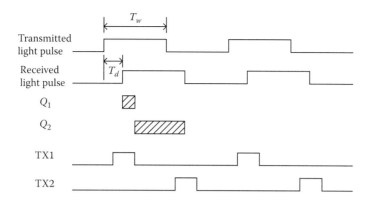

FIGURE 10.12
Operation of two-tap lock-in CCD.

$$Q_1 = I_p\left(T_w/2 - T_d\right) \tag{10.3}$$

$$Q_2 = I_p\left(T_w/2 + T_d\right) \tag{10.4}$$

The delay time can be estimated using Equations (10.3) and (10.4) as

$$T_d = \frac{T_w}{2}\frac{Q_2 - Q_1}{Q_1 + Q_2} \tag{10.5}$$

Using Equation (10.5), the range can be estimated by the charge ratio and a time reference of T_w; it is independent of light intensity.

To read out the two charges, two floating diffusion amplifiers with floating diffusion (floating p–n junction) and a source follower are used. The final voltage outputs V_x, $x = 1$ or 2, are given by

$$V_x = G_{SF}\frac{Q_x}{C_{FD}} \tag{10.6}$$

where G_{SF} is the gain of a source follower amplifier, typically 0.8 to 0.9, and C_{FD} is the floating diffusion capacitance.

10.4.2 Sinusoidal Modulation

Figure 10.13 shows the top view of the four-tap lock-in pixel suitable for TOF range imaging [9]. One pixel consists of a PG, a dump-gate (DG), a dump diffusion (DD), four transfer gates, and a four-phase CCD. Two vertical, four-phase CCD lines are located in each pixel. The four-phase CCD is covered by

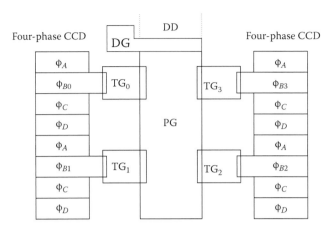

FIGURE 10.13
Four-tap lock-in CCD.

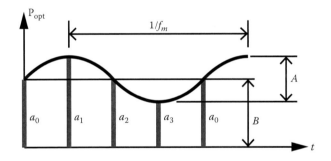

FIGURE 10.14
Timing diagram of the indirect TOF pixel.

a metal layer for light shielding in order to prevent unwanted photo-charge generation during signal readout.

In order to use the four-tap lock-in CCD for TOF range imaging, an LED (light emitting diode) or laser diode is modulated by a sine wave, and the photo charge is sampled in the four taps as shown in Figure 10.14. Using the four sampled signals, the phase delay of the propagated light signal is estimated while reducing the influence of background light.

From the definition of discrete Fourier transform, the amplitude A, offset b, and phase ϕ are given by

$$A = \sqrt{\left(a_0 - a_2\right)^2 + \left(a_1 + a_3\right)^2} \tag{10.7}$$

$$B = \left(\frac{a_0 + a_1 + a_2 + a_3}{4}\right) \tag{10.8}$$

$$\phi = \arctan\left(\frac{a_0 - a_2}{a_1 - a_3}\right) \tag{10.9}$$

The range L is given by

$$L = \frac{c\phi}{4\pi f_m} \tag{10.10}$$

where f_m is the modulation frequency of the light.

10.4.3 Small-Duty-Cycle Pulse Modulation

The four-tap lock-in pixel has a capability of canceling offset charge due to background light. However, strong background light such as direct sunlight may produce extremely large amounts of offset charge and it may exceed the full well capacity of the storage.

In order to reduce the background light charge, a lock-in pixel structure for receiving small-duty-cycle light pulses can be used [14]. The top view of a lock-in pixel for the small duty cycle is shown in Figure 10.15. In addition to two transfer gates at the right and left sides of the photo detector, draining gates are located at the top and bottom for high-speed charge draining. The timing diagram of the transfer gates TX1 and TX2 and draining gate TXD, together with the light pulse received, is shown in Figure 10.16. A small-duty-cycle light pulse is received while TX1 or TX2 is turned on; in the rest of the cycle, TXD is turned on so that unwanted charge components due to ambient light, stray light, and dark noise are reduced.

This technique allows us to obtain higher tolerance to background light because, for a given average power of the light source, the signal-light power

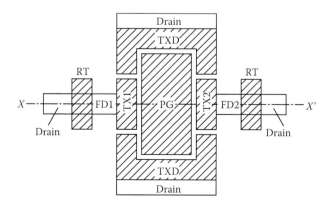

FIGURE 10.15
TOF ranging pixel for small-duty-cycle light pulse.

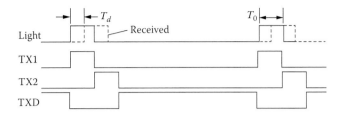

FIGURE 10.16
Timing diagram of the lock-in pixel for small-duty-cycle light pulse.

can be concentrated in a short duration and the signal-light charge is intensified compared with the background-light charge. Because of the property that the maximum rating of the LED (light-emitting diode) can be increased if a small duty cycle is used, this operation also leads to a cost-effective use of LEDs.

10.4.4 CCD and CMOS Implementations

Lock-in pixels for high-modulation frequency need high-speed carrier transfer due to a fringing electric field in a depleted semiconductor region. This idea originated in the CCD. Solid-state image sensors have been attracted by CMOS technology since the mid-1990s; thus, range imagers with lock-in pixels are also often implemented based on CMOS or mixed CCD/CMOS technologies.

The four-tap lock-in pixels have been realized with CCDs [9,11,12] and are most suitable for demodulating the sinusoidally modulated light. However, because each sample needs large storage space for attaining sufficient signal dynamic range, the four-tap approach has a drawback of a low fill factor (2.2% for a wheel-shaped four-tap pixel [12]).

The sensor of the first TOF range camera was based on the one-tap lock-in pixel using mixed CCD/CMOS technology [13]. The cross-sectional view of the one-tap lock-in pixel is shown in Figure 10.17. This pixel has a size of 65×21 μm^2 and an optical fill factor of 22% using 2 μm CCD/CMOS technology and is used for a 64×25-pixel TOF range imager.

Using the one-tap lock-in pixel, the phase measurement of sinusoidally modulated light is based on a serial acquisition of the four samples shown in Figure 10.14. This leads to a reduced frame rate and may cause errors in fast moving objects. Despite these drawbacks, it was worth demonstrating the first all-electronic solid-state TOF range camera.

The preferred choice of the pixel architecture in practical TOF range imagers is often the two-tap lock-in pixel [15]. It has a symmetrical architecture consisting of photo gates (PGR and PGL), storage gates (IG), output gates (OUTG), and sense diffusions at both sides of the middle photo gate (PGM), instead of using dump diffusion at the left end in the one-tap lock-in pixel. Using advanced CCD/CMOS technology of 0.6 μm rules, the number of pixels of the TOF range imager with two-tap lock-in pixels has been increased to 176×144 pixels.

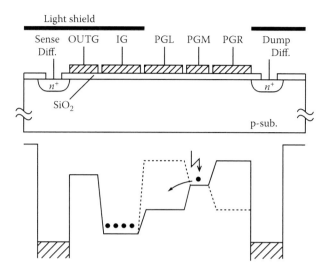

FIGURE 10.17
One-tap lock-in pixel using CCD/CMOS technology.

The architecture shown in Figure 10.17 using a CCD-type charge transfer is preferable for reduced readout noise because reset noise (kTC noise) at the sensing diffusion is canceled by a truly correlated double sampling (CDS) operation. However, such a charge transfer structure requires CCD or mixed CCD/CMOS technology.

To implement lock-in pixels using CMOS technology, a simplified lock-in pixel architecture is often used. Figure 10.18 shows a CMOS-based lock-in pixel using gates on field oxide structure [14]. The pixel uses single-layer polysilicon gates on relatively thick field oxide. To create an n-type buried layer that prevents the interface traps from causing charge transfer delay, one additional mask and an ion implantation process step are used. Because of the thick oxide and buried channel structure on a lightly doped p-type epitaxial layer, a relatively large fringing field is realized by a small voltage swing of 3 V, and a barrierless potential profile at the gap between two gates is also created by the single-layer polysilicon gates. Photo-generated and demodulated charges are directly stored in either of two sensing diffusions. Signal charge storage in the sensing diffusions suffers from reset noise and dark current noise.

However, if a sufficient amount of signal charge is available, the noise level is dominated by the photon shot noise, as discussed in the next section. The lock-in pixel shown in Figure 10.18 is used to implement the pixel whose top view is shown in Figure 10.16 for small-duty light pulse modulation. Thanks to the simplified pixel architecture, a TOF range imager with high spatial resolution of 336 × 252 pixels has been implemented using 0.35 μm CMOS technology. The pixel size is 15 × 15 μm².

A further simplified lock-in pixel architecture is shown in Figure 10.19 [16]. It uses only two gates for modulating a photo-generated charge. The lock-in

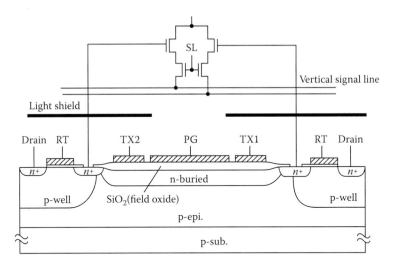

FIGURE 10.18
CMOS-based TOF pixel using gates on field oxide structure.

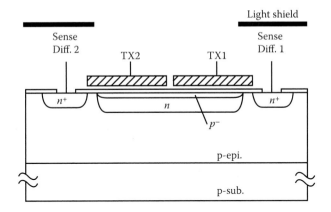

FIGURE 10.19
CMOS-based TOF pixel using two gates on active area and a buried channel.

pixel using two photo gates is also called a photo-mixing device (PMD) [17]. In Figure 10.19, a buried channel is created under the two photo gates and has offsets from the edges of sensing diffusions. This structure is useful for a better demodulation factor at higher modulation frequency.

A pinned photodiode has become a dominant technology for CMOS image sensors and is also useful for lock-in pixels in the TOF range imagers based on standard CMOS image sensor technology. Figure 10.20 shows a lock-in pixel using a pinned photodiode [18]. A photo-generated charge in the pinned photodiode is transferred to sensing diffusion or drained by applying high voltage to the TX and TXD gates, respectively. A prototype TOF range image sensor with 64 × 16 pixels, each of which

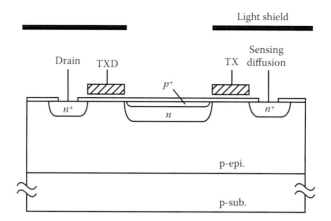

FIGURE 10.20
CMOS-based TOF pixel using two gates on active area and a buried channel.

has the size of 12×12 μm², has been implemented. Using the modulation frequency of 5 MHz, the range accuracy is 2% to 4% in the distance range of 1 to 4 m.

Lock-in pixels based on CMOS technology described before use a floating diffusion as a charge storage as well as a sensing diffusion and suffer from the reset noise during the readout operation. To overcome this problem, two-stage charge transfer architectures using a pinned photodiode and pinned storage diode can be used [19,20].

10.4.5 Range Resolution and Influence of Background Light and Noise

The range resolution of the indirect TOF range imagers is determined by the different types of noise. In the four-tap lock-in CCD using sinusoidal modulation, the range resolution $\Delta\varphi$ is expressed as

$$\Delta\varphi = \sqrt{\sum_{i=0}^{3} \left(\frac{\partial\varphi}{\partial a_i}\right)^2 a_i} \tag{10.11}$$

Using Equations (10.7)–(10.10), the range resolution ΔL is expressed as [13]

$$\Delta L = \frac{L}{2\pi}\Delta\varphi = \frac{L}{\sqrt{8}} \cdot \frac{\sqrt{B}}{2A} \tag{10.12}$$

In this equation, $\sqrt{B}$ means the amplitude of noise fluctuation. Under the influence of read noise and shot noise due to background light, as well as the shot noise of the signal light itself, B in Equation (10.12) is expressed as

$$B = N_s + N_b + N_r^2 \tag{10.13}$$

where N_s is the number of signal electrons, N_b is the number of electrons generated by background light, and N_r is the equivalent number of electrons due to read noise. Equation (10.13) is based on the fact that the square of shot noise due to photo-generated electrons (variance in Poisson distribution) is equal to the mean number of photo-generated electrons. The signal amplitude A depends on the modulation contrast of the light source C_{mod} and the demodulation contrast of the lock-in pixel device C_{demod} and is expressed as

$$A = C_{mod} C_{demod} N_s \tag{10.14}$$

Thus, Equation (10.12) can be rewritten as

$$\Delta L = \frac{L}{\sqrt{8}} \cdot \frac{\sqrt{N_s + N_b + N_r^2}}{2 C_{mod} C_{demod} N_s} \tag{10.15}$$

The equivalent number of reset noise electrons is given by

$$N_{reset} = \sqrt{2 \frac{V_{th}}{V_s} N_{well}} \tag{10.16}$$

where V_{th} is the thermal voltage (26 mV at room temperature) and V_s is the maximum signal voltage amplitude. The factor of 2 in Equation (10.16) is due to the noise power doubling effect in the readout operation. Typically, N_{well} ranges from several thousands to one million. For $V_s = 1$ V and $N_{well} = 200{,}000$, N_{reset} is calculated to be approximately 100.

Figure 10.21 shows the range resolution as a function of the signal amplitude for different conditions of background light and read noise.

For $N_s = 100{,}000$ and $N_b = 100{,}000$, the range resolution $\Delta L/L$ is less than 0.1%. In this condition, for example, the resolution of 5 mm is obtained, or a 5 m range. For a small signal, the resolution is highly dependent on the background light and readout noise. For $N_s = 1000$, $\Delta L/L$ is 0.56% and 5.6% for $N_b = 100$ and $N_b = 100{,}000$, respectively. If the readout noise N_r is increased to 100 because of the reset noise, $\Delta L/L$ is increased to 1.5% for $N_s = 1000$ and $N_b = 100$. Range imaging with relatively low signal light, a lock-in pixel architecture for reset noise canceling, offers a better performance in range resolution.

The background light also limits the available well capacity for a signal charge. If N_b is smaller than the well capacity of the charge storage N_{well}, the available well capacity for the signal charge is given by $N_{well} - N_b$. Under a very strong background light, N_b may exceed N_{well} and no residual well capacity is available for the signal.

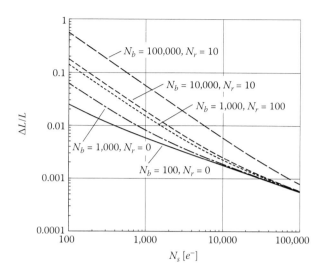

FIGURE 10.21
CMOS-based TOF pixel using two gates on active area and a buried channel.

The well capacity is also important for signal dynamic range and the resulting distance range because the signal intensity of back-reflected light at the focal plane of the image sensor is inversely proportional to the distance to the fourth power. On the other hand, the large well capacity will cause large reset noise. The lock-in pixel architecture with reset noise canceling is a better choice for wide distance range and total high-range resolution.

References

1. http://www.ecoscan.jp/
2. M. Kawakita, K. Iizuka, H. Nakmura, I. Mizuno, T. Kurita, T. Aida, Y. Yamanouchi, et al. High-definition real-time depth-mapping TV camera: HDTV axi-vision camera. *Optics Express* 12 (12): 2781–2794, 2004.
3. A. Rochas, A. R. Pauchard, P.-A. Besse, D. Pantic, Z. Prijic, R. S. Popovic. Low-noise silicon avalanche photodiodes fabricated in conventional CMOS technologies. *IEEE Transactions Electron Devices* 49 (3): 387–394, 2002.
4. A. Rochas, M. Gosch, A. Serov, P.-A. Besse, R. S. Popovic, T. Lasser. R. Rigler. First fully integrated 2-D array of single-photon detectors in standard CMOS technology. *IEEE Photonics Technology Letters* 15 (7): 963–965, 2003.
5. C. Niclass, A. Rochas, P.-A. Besse, E. Charbon. Toward a 3-D camera based on single photon avalanche diodes. *IEEE Journal Selected Topics Quantum Electronics* 10 (4): 796–802, 2004.

6. C. Niclass, A. Rochas, P.-A. Besse, E. Charbon. Design and characterization of a CMOS 3-D image sensor based on single photon avalanche diodes. *IEEE Journal Solid-State Circuits* 40 (9): 1847–1854, 2005.

7. C. Niclass, C. Favi, T. Kluter, M. Gersbach, E. Charbon. A 128 × 128 single-photon image sensor with column-level 10-bit time-to-digital converter array. *IEEE Journal Solid-State Circuits* 43 (12): 2977–2989, 2008.

8. H. Povel, H. Aebersold, J. O. Stenflo. Charge-coupled device image sensor as a demodulator in a 2-D polarimeter with a piezoelastic modulator. *Applied Optics* 29 (8): 1186–1190, 1990.

9. T. Spirig, P. Seitz, O. Vietze, F. Heiger. The lock-in CCD—Two-dimensional synchronous detection of light. *IEEE Journal Quantum Electronics* 31 (9): 1705–1708, 1995.

10. R. Miyagawa, T. Kanade. CCD-based range-finding sensor. *IEEE Transactions Electron Devices* 44 (10): l648–1652, 1997.

11. T. Spirig, M. Marley, P. Seitz. Multitap lock-in CCD with offset subtraction. *IEEE Transactions Electron Devices* 44 (10): 1643–1647, 1997.

12. P. Seitz, A. Biber, S. Lauxtermann. Demodulation pixels in CCD and CMOS technologies for time-of-flight ranging. *Proceedings SPIE* 3965:177–188, 2000.

13. R. Lang, P. Seitz. Solid-state time-of-flight camera. *IEEE Journal Quantum Electronics* 37 (3): 390–397, 2001.

14. S. Kawahito, I. A. Halin, T. Ushinaga, T. Sawada, M. Homma, Y. Maeda. A CMOS time-of-flight range image sensor with gates-on-field-oxide structure. *IEEE Sensors Journal* 7 (12): 1578–1586, 2007.

15. B. Buttgen, T. Oggier, M. Lehmann, R. Kaufmann, S. Neukom, M. Richter, M. Schweizer, et al. High-speed and high-sensitive demodulation pixel for 3D imaging. *Proceedings SPIE* 6056: 22–33, 2006.

16. D. Stoppa, N. Massari, L. Pancheri, M. Malfatti, M. Perenzoni, L. Gonzo. A range image sensor based on 10 μm lock-in pixels in 0.18 μm CMOS imaging technology. *IEEE Journal Solid-State Circuits* 46 (1): 248–258, 2011.

17. T. Ringbeck, T. Moller, B. Hagebeuker. Multidimensional measurement by using 3-D PMD sensors. *Advances Radio Science* 5: 135–146, 2007.

18. S. J. Kim, S. W. Han, B. Kang, K. Lee, D. K. Kim, C. Y. Kim. A three-dimensional time-of-flight CMOS image sensor with pinned-photodiode pixel structure. *IEEE Electron Device Letters* 31 (11): 1272–1274, 2010.

19. H. J. Yoon, S. Itoh, S. Kawahito. A CMOS image sensor with in-pixel two-stage charge transfer for fluorescence lifetime imaging. *IEEE Transactions Electron Devices* 56 (2): 214–221, 2009.

20. K. Yasutomi, S. Itoh, S. Kawahito. A two-stage charge transfer active pixel CMOS image sensor with low-noise global shuttering and a dual-shuttering mode. *IEEE Transactions Electron Devices* 58 (3): 740–747, 2011.

11

Uniaxial 3D Shape Measurement

Yukitoshi Otani

CONTENTS

11.1 Introduction

Recently, noncontact three-dimensional (3D) profilometry has been required in the field of manufacturing inspection for industry, biometrics, and robot vision. Table 11.1 shows typical 3D methods that have already been proposed for this purpose. We can classify the type of optical axis based on measuring depth data by a stereoscopic or uniaxis method. The stereoscopic method is the most popular for measuring 3D shapes. It uses a parallax where two or more optical axes are imaging, or one is for illuminating specified patterns and the other is imaging. The stereo method has two axes of an optical system with projection and observation and originated as aerophotography. A 3D measurement by the stereo machine is proposed multi-eye which is employed multioptical axes and differential illumination. A moiré topography is an elegant method and an easy-to-understand way to produce a contour map by superimposing gratings [1,2]. Optical sectioning in the pattern projection is the oldest method for noncontact 3D measurement [3]. It has been expanded for different methods and production, such as spot scanning, Gray code, random-dots color pattern, and grating projection [4]. However, it has a drawback in measuring deep holes, steep height, and shadow portions (shown in Figure 11.1). Figure 11.1(a) shows a conventional method. It cannot capture the projected pattern because of different axes.

To overcome this problem, a uniaxis profilometry has been proposed [5–14]. It is necessary to capture the projected patterns coaxially, as shown in Figure 11.1(b). We can categorize methods by focus, interferometry, time of

TABLE 11.1

Typical Methods for Three-Dimensional Measurement

	Specific Method	Specific Examples
Stereoscopic method	Optical triangulation	Stereo: multieye:
		Differential illumination
		Moiré topography
		Pattern projection:
		Light section
		Spot scanning
		Gray code
		Random dots
		Color pattern
		Grating pattern
Uniaxis method	Focus	Shape from focus (defocus)
		Confocal
	Interferometry	Subfringe:
		Phase-shifting method
		Optical heterodyne
		Fourier transform method
		Common path
		Oblique incidence
		Low-coherence
		Wave length scanning
	Time of flight	Short plus
		Modulation
	Polarization	Talbot effect
		Conscopic holography

flight, and polarization. The interferometry is a powerful tool to use to measure with high sensitivity less than wavelength. However, there are some limitations in the size of the measurement area and tilt angle because of optical components. The time of flight is identified as future potential areas of 3D metrology. However, it is still necessary and costly equipment. The polarization method (such as a conoscopic holography) has already been in commercial production but it is still a point measurement. The contrast distribution or intensity change along the focus direction to a sample is detected for distance information. One of the famous methods is the shape form focus of 1994 [5]. It is already used in a variety of robotics areas. The first trial of focus detection by grating projection was proposed in 1997 [6]. There was a limitation of sample size because of measuring on the microscope. A focus method by grating projection for the large objects was proposed and succeeded in measuring large objects [7–14].

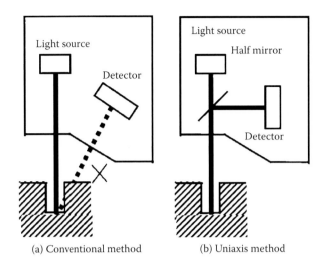

(a) Conventional method   (b) Uniaxis method

FIGURE 11.1
Requirement of uniaxis measurement.

In this section, a uniaxis measurement of a 3D surface is explained within a wide range. Two key components for practical applications are to design a compact measuring system with a telecentric optical system and to detect the contrast with high accuracy by the phase-shifting method. We discuss how to increase accuracy of the grating projection method by detecting the contrast variation by a phase-shifting technique. Its key device is a liquid crystal grating (LCG). The contrast distribution can be changed depending on the pitch, such as frequency of grating, of the LCG. This technique is almost the same as the optical sectioning of the confocal microscope to detect a 3D profile. Finally, we discuss designing a new lens system with a wide measurement area and mounting on the arm robot with six degrees of freedom. We introduce a principle of the focus method using the phase-shifting technique by an LCG and demonstrate measuring the surface profile of samples with a steep profile.

11.2 Principle of the Uniaxial Measurement Method for 3D Profiles

Figure 11.2 shows an optical setup for a uniaxial measurement method 3D shape by the focus method using LCG. We can analyze 3D data from the contrast value. A grating pattern made of liquid grating is projected onto a

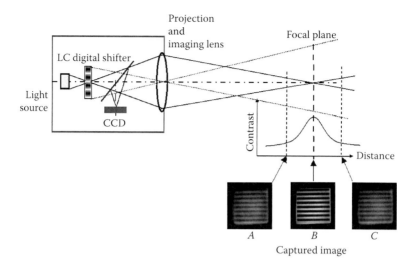

FIGURE 11.2
Optical setup for focus method.

sample and is applied to the phase shifting. In this case, there is no phase change because of one-dimensional measurement.

The contrast variation along an optical axis is theoretically similar to the Bessel function. The maximum contrast means the focus position of the projection lens. Its distribution varies depending not only on the projection and imaging lens but also on the period of the LCG, which is a useful tool for changing the frequency. A measurement range of the system is limited to the dynamic range of the projection lens. The contrast distribution along an optical axis is the relation of the Bessel function along the z-direction. It is written as follows:

$$F(z) = \left\{ \frac{2J_1[\pi z \ fF]}{\pi z fF} \right\} \tag{11.1}$$

where f is grating pitch and F is f-number.

This means the contrast cure is variable by spatial frequency of grating or projection grating pitch. We employ two methods for determining the peak point of the contrast curve. One is a Gaussian fitting, shown in Figure 11.2, and the other is focus changing by a varifocus lens or moving by a robot.

The captured intensity distribution I_i at (x,y) is expressed as

$$I_i = I_0 \left(1 + V(x,y) \cdot \cos\{\varphi(x,y) + \delta_i\} \right) \tag{11.2}$$

where I_0, $V(x,y)$, and $\varphi(x,y)$ are an illuminated intensity, a contrast, and initial phase, respectively. δ_i is the shifted phase with $90° \cdot i/4$ ($i = 0, 1, 2, 3$).

The contrast $V(x,y)$ is determined by the four-step phase-shifting principle as

$$V(x,y) = \frac{2\sqrt{\left(I_{0°} - I_{180°}\right)^2 + \left(I_{90°} - I_{270°}\right)^2}}{I_{0°} + I_{90°} + I_{180°} + I_{270°}} \qquad (11.3)$$

The relation between a focus length and the contrast $V(x,y)$ is a linear function.

11.3 Experimental Setup and Results

Figure 11.3 shows an experimental setup for the 3D surface profile measurement using liquid crystal grating. The measuring system is designed with a telecentric optical system. The measuring area is expanded to 50 × 50 mm² by using a wide objective lens with 92 mm of diameter. The size of this system is less than notebook size: 275 × 153 mm². The appearance of the system on the arm robot with six degrees of freedom is shown in Figure 11.2(b). The robot is used not only for calibration but also for expanding the working distance. Theoretically, we can measure in the area of 500 × 500 mm² by the arm robot.

The sample can be set on either side, as shown in Figure 11.2. A contrast on the sample is captured by a charge-coupled device camera through a half mirror between the projection part and the detection part (shown in Figure 11.2). The contrast distribution is detected by a four-step phase-shifting technique by LCG without mechanical movement. We checked the accuracy of this method with the grating period of 0.6 mm/line on the sample. Figure 11.4 shows a measured result of calibrated distance versus a measured one. The repeatability is determined as 39 μm with 2 s of measuring time.

Figure 11.5 shows the result of a surface profile of a mechanical part with holes and steep height. The background is a hard board. We succeeded in measuring the steep height comparing the gear surface and background.

Figure 11.6 is a measured result of a cylinder part in midair. The square part at the right-hand bottom is a bump with 10 mm height. Moreover, we can detect the characters with less than 1 mm.

11.4 Conclusions

Measurement of a three-dimensional surface by the profiled uniaxial method by projecting a grating pattern was discussed. The optical axis coincides

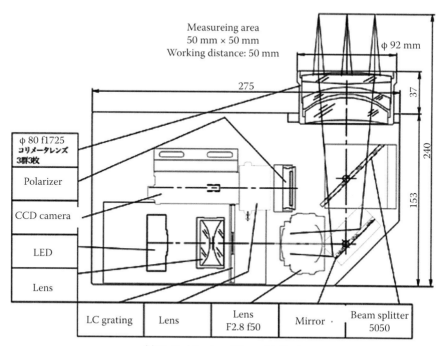

Measureing area
50 mm × 50 mm
Working distance: 50 mm

φ 92 mm

275

37

240

153

φ 80 f1725
コリメータレンズ
3群3枚

Polarizer

CCD camera

LED

Lens

| LC grating | Lens | Lens F2.8 f50 | Mirror · | Beam splitter 5050 |

(a) Optical Design of Focus Method

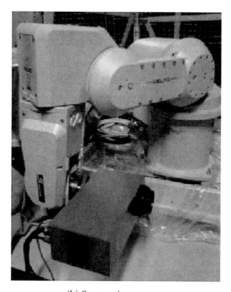

(b) System Appearance

FIGURE 11.3
Experimental setup for the 3D profile measurement on the arm robot.

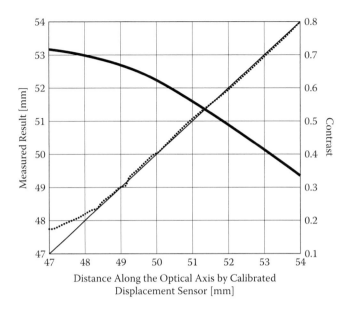

FIGURE 11.4
Accuracy check for distance measurement.

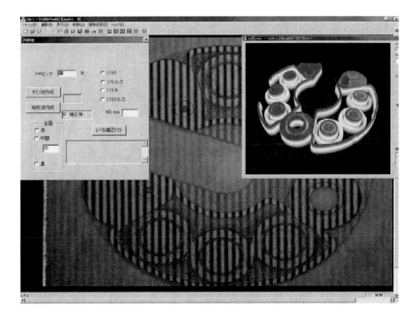

FIGURE 11.5
Measured result of gear sample.

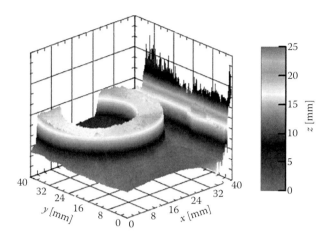

FIGURE 11.6
Measured result of cylinder part.

with a projection and an observation axis. A distance from a projection lens and an object can be determined by a contrast distribution. An advantage of the uniaxial method is measuring for steep shapes and deep holes. Accuracy has been achieved to 36 μm.

Acknowledgment

This work was supported by the Grant for Practical Application of University R&D Results under the matching fund method, New Energy and Industrial Technology Development Organization, Japan. The author would like to express his gratitude to Dr. Mizutani and Dr. Kobayashi.

References

1. D. M. Meadows, W. O. Johnson, J. B. Allen. Generation of surface contours by moiré patterns. *Applied Optics* 9 (4): 942–947 (1970).
2. H. Takasaki. Moiré topography. *Applied Optics* 9 (4): 1467–1472 (1970).
3. G. Schmaltz. *Technische oberflächenkunde*. Berlin: Springer Verlag (1936).
4. T. Yoshizawa, Y. Ohtsuka, et al. *Surface topography and spinal deformity*, 403–410. Stuttgart: Gustav Fischer (1987)
5. S. K. Nayar, Y. Nakagawa. Shape from focus. *IEEE Transactions on Pattern Analysis and Machine Intelligence* 16 (8): 824–831 (1994).

6. M. A. A. Neil, R. Juskaitis, T. Wilson. Method of obtaining optical sectioning by using structured light in a conventional microscope. *Optics Letters* 22 (24): 1905–1907 (1997).
7. M. Ishihara, Y. Nakazato, H. Sasaki, M. Tonooka, M. Yamamoto, Y. Otani, T. Yoshizawa. Three-dimensional surface measurement using grating projection method by detecting phase and contrast. *Proceedings SPIE* 3740: 114–117 (1999).
8. M. Takeda, T. Aoki, Y. Miyamoto, H. Tanaka, R. W. Gu, Z. B. Zhang. Absolute three-dimensional shape measurements using coaxial and coimage plane optical systems and Fourier fringe analysis for focus detection. *Optical Engineering* 39 (1): 61–68 (2000).
9. A. Ishii. 3-D shape measurement using a focused-section method. *Pattern Recognition* 4 (4): 4828 (2000).
10. T. Yoshizawa, T. Shinoda, Y. Otani. Uniaxis range finder using contrast detection of a projected pattern. *Proceedings SPIE* 4190: 115–122 (2001).
11. Y. Mizutani, R. Kuwano, Y. Otani, N. Umeda, T. Yoshizawa. Three-dimensional shape measurement using focus method by using liquid crystal grating and liquid varifocus lens. *Proceedings SPIE* 6000: 6000J-1-7 (2005).
12. R. Kuwano T. Tokunaga, Y. Otani, N. Umeda. Liquid pressure varifocus lens. *Optical Review* 12 (5): 405–408 (2005).
13. K. Yamatani, M. Yamamoto, Y. Otani, T. Yoshizawa, H. Fujita, A. Suguro, S. Morokawa. Three-dimensional surface profilometry using structured liquid crystal grating. *Proceedings SPIE* 3782: 291–296 (1999).
14. Y. Otani, F. Kobayashi, Y. Mizutani, T. Yoshizawa. Three-dimensional profilometry based on focus method by projecting LC grating pattern. *Proceedings SPIE* 7432: 743210 (2009).

12

Three-Dimensional Ultrasound Imaging

Aaron Fenster, Grace Parraga, Bernard Chiu, and Jeff Bax

CONTENTS

12.1 Introduction

X-ray beams used to generate two-dimensional (2D) projection images of the human body have been used since they were studied by Roentgen in 1895. Since 2D x-ray imaging provides only a projection image, complete information of an organ or pathology necessary to diagnose or treat pathology may not be available. In the early 1970s, the development of computed tomography (CT) revolutionized diagnostic radiology. The contiguous tomographic images generated by CT scanners could be assembled into three-dimensional (3D) images and viewed with the aid of computer visualization software. Three-dimensional magnetic resonance imaging (MRI), positron emission tomography (PET), and multislice and cone beam CT imaging have further stimulated the field of 3D medical imaging, stimulating the development of a wide variety of applications in diagnostic and interventional medicine.

In addition to the development of 3D imaging using CT, PET, and MR images, ultrasound (US) imaging has also been extended to 3D imaging [1]. The majority of ultrasound-based diagnostic and interventional procedures are currently performed using 2D imaging; 3D ultrasound (3D US) technology is being improved and applications utilizing that technology has been growing in demand [2–4]. Although 3D US technology is still being improved, it already provides high-quality 3D images of complex anatomical structures and pathology to be used in diagnostic, interventional, and surgical procedures. Researchers and commercial companies are actively pursuing integration of 3D visualization techniques into ultrasound instrumentation as well as specialized biopsy and therapy systems [4–8].

In this chapter, we introduce the different methods for obtaining 3D US images and describe their use in two applications: guided prostate biopsy and monitoring carotid atherosclerosis.

12.2 Benefits of 3D Ultrasound Imaging

Conventional 2D US imaging systems are highly flexible, allowing users to manipulate handheld ultrasound transducers freely over the body in order to generate real-time 2D images of organs and pathology. However, these 2D US systems suffer from the following disadvantages, which 3D US imaging systems attempt to overcome:

- Conventional 2D US imaging systems require users to integrate many 2D images mentally to form an impression of the 3D anatomy and pathology. While this approach can be effective at times, it leads to longer procedures and variability in diagnosis and guidance in interventional procedures.

- Since the 2D US imaging transducer is held and controlled manually, it is difficult to relocate the 2D US image at the exact location and orientation in the body when imaging a patient. Thus, monitoring progression and regression of pathology in response to therapy can be suboptimal, as accurate monitoring requires a physician to reposition the transducer to view the same image of the pathology.
- 2D US imaging does not permit viewing of planes parallel to the skin. Diagnostic and interventional procedures sometimes require an arbitrary selection of the image plane for optimal viewing of the pathology.
- Diagnostic procedures, therapy/surgery planning, and therapy monitoring often require accurate volume delineation and measurements. However, the use of conventional 2D US imaging for measurements of organ or lesion volume is variable and at times inaccurate. Thus, a 3D imaging approach is required to allow accurate and precise volume measurements.

Along with the development of 3D viewing with CT images, 3D US images were demonstrated in the 1970s. The development of commercial 3D US systems took longer to be sold as the first commercial system became available in 1989 by Kretz. However, many researchers and commercial companies have been active over the past two decades and have developed efficient 3D US imaging systems for use in a wide variety of applications [9–15]. While development of commercial CT and MR systems has been rapid, progress in the development of 3D US systems and their routine use has been slow because 3D US systems require significant computational speed for acquiring, reconstructing, and viewing 3D US information in real or near real time on inexpensive systems. Advances in low-cost computer and visualization technology have now made 3D US imaging a viable technology that can be used in a wide range of applications. Most of the major ultrasound system manufacturers are now providing 3D US transducers and viewing software on their imaging platforms.

In the following sections, we describe various approaches used in 3D US imaging systems, with an emphasis on the geometric accuracy of the generation of 3D images as well as the use of this technology in interventional and quantitative monitoring applications.

12.3 3D US Scanning Techniques

Over the past two decades, investigators have explored a wide variety of approaches for generating 3D US images. These include the use of US transducers generating real-time 2D US images in mechanical and freehand scanning and the use of 2D arrays for generating real-time 3D images. Use of conventional transducers to produce 3D US images requires methods to determine the position and orientation of the 2D images within the 3D image volume, while 2D arrays require a 3D scan converter to build the 3D image from the sequence of transmit/receive acoustic signals. In both types of systems, production of high-quality 3D US images without any distortions requires that three factors be optimized:

- The scanning technique must be either rapid (i.e., real time or near real time) or gated to avoid image artifacts due to involuntary, respiratory, or cardiac motion.

- The locations and orientations of the acquired 2D US images must be accurately known to avoid geometric distortions in the generation of the 3D US image. Any geometric errors will lead to geometric distortions and errors of measurement and guidance of needles to targets.

- The scanning apparatus must be simple and convenient to use; therefore, the scanning must be easily added to the examination or interventional procedure.

Although many approaches for production of 3D US images have been explored, current systems make use of one of the following approaches: mechanical scanning, freehand scanning with position sensing, freehand scanning without position sensing, or 2D array scanning for dynamic 3D ultrasound (or four-dimensional [4D] US) imaging.

12.3.1 Mechanical 3D US Scanning Systems

Three-dimensional US systems based on mechanical scanning mechanisms make use of motorized components to translate, tilt, or rotate a conventional 2D US transducer while a sequential series of real-time 2D US images is rapidly acquired by a computer. In this approach the scanning geometry is predefined and precisely controlled by a motor/encoder, allowing the relative positions and orientations of the acquired 2D US images to be known accurately. Thus, the acquired 2D US images and their predefined relative locations and orientation can then be used to reconstruct the 3D US image in real time (i.e., as the 2D images are acquired). These mechanical scanning systems provide great flexibility, allowing the user to adjust the speed of the motor, the number of 2D images to be acquired, and the angular or spatial interval between each acquired 2D image. This flexibility allows the user to optimize the scan time and resolution in the 3D US image through adjustments made to interval spacing between images [16].

Mechanical scanning mechanisms have been particularly successful and have been developed and used in a variety of clinical applications. These include integrated 3D US transducers that house the scanning mechanism within the transducer housing, and external mechanical fixtures that hold the housing of conventional transducers that generate 2D US images.

Many ultrasound manufacturers now offer integrated 3D US transducers that are based on a mechanically swept probe or "wobbler." In these systems an ultrasound transducer is wobbled or swept back and forth inside the handheld housing, while the 2D US-generated planes are reconstructed into the 3D US image. Since these types of 3D transducers are integrated with the ultrasound scanner, the collection and reconstruction of the images can be optimized, allowing generation of two or three 3D images per second. These types of 3D transducers are typically larger than conventional 2D US transducers since they have to accommodate "wobbling" of the transducer. However, they are easier to use than 3D US systems using external fixtures with conventional 2D US transducers.

Since these types of 3D transducers are integrated into the manufacturer's ultrasound system, they require a special US machine that can control them and reconstruct the acquired 2D images into a 3D image. While external mechanical 3D scanning fixtures (discussed later) are generally bulkier than integrated transducers, they can be adapted to hold any conventional US transducer, obviating the need to purchase a special 3D US machine. Since 3D US scanning external fixtures can accommodate any manufacturer's transducer, they can take advantage of improvements in the US machine (e.g., image compounding, contrast agent imaging) and flow information (e.g., Doppler imaging) without any changes in the scanning mechanism.

Three-dimensional mechanical scanning offers the following advantages: short imaging times, ranging from about 3 to 0.2 volumes per second; high-quality 3D images, including B-mode and Doppler; and real-time 3D reconstruction times allowing viewing of the 3D image as it is being acquired. However, 3D mechanical scanners can be bulky and their weight sometimes makes them inconvenient to use. Figure 12.1 shows three basic types of mechanical scanners that are used: linear scanners, tilt scanners, and rotational scanners.

12.3.1.1 Wobbling or Tilt Mechanical 3D US Scanners

This approach uses a motorized drive mechanism to tilt or wobble a conventional one-dimensional (1D) US transducer over an angle that can be as large as 60° about an axis parallel to the face of the transducer. The 2D US images that are generated are arranged as a page in an open book with an angular spacing between the 2D images, such as 1.0°. In both the integrated 3D US probe and the external fixture approach, the housing of the transducer remains fixed on the skin of the patient while the US transducer is angulated or wobbled. Since the reconstruction of the 3D image can be performed as the 2D images are acquired, the time to acquire a 3D US image depends on the 2D US image update rate and the number of 2D images used to generate the 3D image.

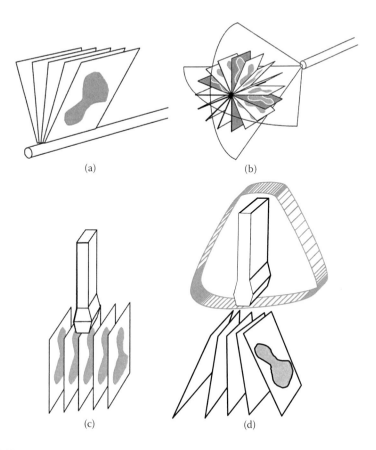

FIGURE 12.1
Schematic diagrams of 3D US mechanical scanning methods. (a) A side-firing TRUS transducer is mechanically rotated and the acquired images have equal angular spacing. The same approach is used in a mechanically wobbled transducer. (b) A rotational scanning mechanism using an end-firing transducer, typically used in 3D TRUS guided prostate biopsy. The acquired images have equal angular spacing. (c) A linear mechanical scanning mechanism in which the acquired images have equal spacing. (d) The mechanically tilting mechanism, but integrated into a 3D US transducer. The US transducer is "wobbled" inside the housing of the transducer.

The 2D US update rate depends on the US machine settings, such as the depth setting and number of focal zones. The number of acquired 2D images is controlled by the choice of the angular separation between the acquired images needed to yield a desired image quality. Typically, these parameters can be adjusted to optimize scanning time, image quality, and the size of the volume imaged [17–22]. The most common integrated 3D probes require special 3D US systems or upgrades that are used for abdominal and obstetrical imaging [15,23–26].

Since the geometry of the acquired 2D US images is predefined, the 3D image reconstruction can be performed in real time while the 2D images are acquired. In this 3D scanning geometry, the resolution in the 3D US image will not be isotropic because it will degrade as the distance from the axis of rotation is increased due to US beam spread in the lateral and elevational directions of the acquired 2D US images. In addition, since the geometry of the acquired 2D images is fan-like, the distance between the acquired US images increases with increasing depth, resulting in a decrease in the spatial sampling and spatial resolution of the reconstructed 3D image [27].

12.3.1.2 Linear Mechanical 3D Scanners

Mechanical 3D scanners, which move conventional transducers in a linear manner, use a motorized drive mechanism to translate the transducer across the skin of the patient. A mechanical housing can hold the transducer perpendicular to the surface of the skin or at an angle for acquiring Doppler images. The 2D images are acquired at a regular but adjustable distance interval so that they are parallel and uniformly spaced (Figure 12.1b). The temporal sampling interval can be chosen to match the 2D US frame rate for the US machine and the translating speed can be chosen to match the required spatial sampling interval, which should be at least half of the elevational resolution of the transducer [16].

Since the acquired 2D US images have a predefined and regular geometry, the 3D image can be reconstructed in real time as the set of 2D US images is acquired. In this scanning approach, the resolution in the 3D image will also not be isotropic. In the direction parallel to the acquired 2D US images (i.e., axial and lateral directions in the 2D images), the resolution of the restructured 3D US image will be the same as the original 2D US images. However, the resolution of the reconstructed 3D image will be equal (if spatial sampling is appropriate) to the elevational resolution of the transducer in the direction perpendicular to the acquired 2D US images. Thus, the resolution of the 3D US image will be poorest in the 3D scanning direction, and a transducer with good elevational resolution should be used for optimal results [28].

The linear scanning is typically used when long scanning distances are needed. Thus, this approach has been successfully used in many vascular B-mode and Doppler imaging applications, particularly for carotid arteries, where scanning distances of about 5 cm are needed (see discussion in Section 12.5) [21,29–37]. Figure 12.2 shows two examples of linearly scanned 3D US images of the carotid arteries made with an external fixture.

12.3.1.3 Endocavity Rotational 3D Scanners

This scanning approach uses an external fixture or internal mechanism to rotate or "wobble" an endocavity probe (e.g., a transrectal ultrasound [TRUS] probe, see Figure 12.1). For endocavity probes using an end-firing transducer, the set of acquired 2D images may be arranged as a fan (Figure 12.1c), intersecting in the center of the 3D US image (see resulting image in Figure 12.3).

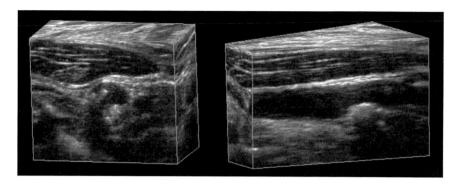

FIGURE 12.2
Examples of 3D carotid US images obtained with the mechanical linear 3D scanning approach. The 3D US images are displayed using the cube-view approach with the faces sliced to reveal the details of the atherosclerotic plaque in the carotid arteries.

For endocavity probes using a side-firing 1D transducer (as used in prostate brachytherapy), the acquired images will also be arranged as a fan, but intersect at the axis of rotation of the transducer (see Figure 12.1a). In this scanning approach, the side-firing probe is typically rotated from 80° to 110° and an end-firing transducer is typically rotated by 180° [22,38,39]. Figure 12.3 shows images generated with an end-firing TRUS transducer using the endocavity 3D scanning approach [21,31,32,38], and guiding 3D US biopsy and therapy [6,8,22,40–43].

In this 3D scanning approach, the resolution of the 3D image will also not be isotropic. The spatial sampling is highest near the axis of the transducer (i.e., axis of rotation) and the poorest away from the axis of the transducer; thus, the resolution of the 3D US image will degrade as the distance from the rotational axis of the transducer increases. Similarly, the axial and elevational resolution will decrease as the distance from the transducer increases. The combination of these effects will cause the 3D US image resolution to vary—highest near the transducer and the rotational axis and poorest away from the transducer and rotational axis.

Three-dimensional rotational scanning with an end-firing transducer is most sensitive to the motion of the transducer and patient during the scanning. Because the acquired 2D images intersect along the rotational axis of the transducer, any motion during the scan will cause a mismatch in the acquired planes, resulting in the production of artifacts in the center of the 3D US image. Artifacts in the center of the 3D US image will also occur if the axis of rotation is not accurately known; however, proper calibrations can remove this source of potential error. Although handheld 3D rotational scanning of the prostate and uterus can produce excellent 3D images (see Figure 12.3), for optimal results in long procedures, such as prostate brachytherapy and biopsy [22,38,44], the transducer and its assembly should be mounted onto a fixture, such as a stabilizer in prostate brachytherapy.

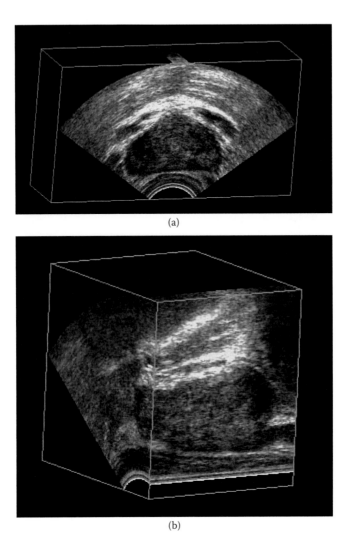

(a)

(b)

FIGURE 12.3
A 3D US image of the prostate acquired using a side-firing endocavity rotational 3D scanning approach (rotation of a TRUS transducer). The transducer was rotated around its long axis, while 3D US images were acquired and reconstructed. The 3D US image is displayed using the cube view approach and has been sliced to reveal (a) a transverse view and (b) a sagittal view.

12.3.2 Freehand Scanning with Position Sensing

Over the past 20 years researchers have been developing freehand 3D scanning approaches that do not require bulky mechanical scanning devices. Since freehand 2D US transducers do not use mechanical mechanisms to control the movement of the transducer, other means must be used to monitor the position and orientation of the transducer. Many approaches in which a small sensor is mounted on the transducer to allow measurement of the transducer's position and orientation as the transducer is moved over the body have been investigated. These approaches include optical, acoustic, mechanical, image-based (speckle decorrelation), and electromagnetic devices used to track the transducer and hence the 2D images and their relative location and orientation, which are then used to reconstruct the 3D image [45]. Since the transducer is moved freely by the user, the locations and orientations of the acquired 2D images are not predefined. Thus, the user must move the transducer at an appropriate speed to ensure that the transducer's position is sampled so that no significant gaps will result in the reconstructed 3D image. The method used most commonly is the magnetic field sensing approach; several companies provide the sensing technology: Ascension: Bird sensor [46], Polhemus: Fastrack sensor [47], and Northern Digital: Aurora sensor [3].

12.3.2.1 Freehand 3D Scanning with Magnetic Field Sensors

The magnetic field sensor approach for freehand 3D US imaging has been used extensively in many applications, including echocardiography, obstetrics, and vascular imaging [3,46–57]. To allow tracking of the conventional transducer as it is moved over the body, a time-varying 3D magnetic field transmitter is placed near the patient, and a small receiver containing three orthogonal coils (with six degrees of freedom) is mounted on the probe and used to sense the strength of the magnetic field in three orthogonal directions. The position and orientation of the transducer are calculated by measuring the strength of the three components of the local magnetic field. This information, together with the acquired 2D images, is used in the 3D reconstruction algorithm.

The main advantage of this approach is that the position sensor can be small and unobtrusive. However, the accuracy of the tracking can be compromised by electromagnetic interference. Since geometric distortions in the 3D US image can occur if ferrous (or highly conductive) metals are located nearby, metal hospital beds in procedure or surgical rooms can cause significant distortions. Modern magnetic field sensors—particularly ones that use a magnetic transmitter placed between the bed and the patient—produce excellent images and are now less susceptible to sources of error. Nonetheless, to minimize these potential sources of error, the position of the sensor relative to the transmitter must be calibrated accurately and precisely using one of the numerous calibration techniques that have been developed [57–65].

12.3.2.2 Freehand 3D US Scanning without Position Sensing

If spatial accuracy of the 3D US image is not required, reconstruction can easily be produced without position sensing. In this approach, the transducer is assumed to be moved over the body in a predefined scanning geometry and speed. As the user moves the transducer, either in a linear manner or by tilting, the 3D US image is reconstructed from the acquired 2D US images using the assumed scanning geometry and speed. Since position or orientation information is not recorded during the motion of the transducer, the operator must move the transducer at a constant linear or angular velocity so that the 2D images are acquired at a regular interval [31,32]. Since this approach does not guarantee that the 3D US image is geometrically accurate, it must not be used for measurements.

12.3.3 Real-Time 3D Ultrasound Imaging (4D Imaging) Using 2D Arrays

The mechanical or freehand 3D scanning approaches limit the volume acquisition update rate to about two or three volumes per second. To increase the volume update rate, transducers with 2D arrays generating 3D images in real time have been developed. In this approach, a transducer containing a 2D phased array allows the transducer to remain stationary. An electronic approach is then used to control the transmission and receiving of a broadly diverging ultrasound beam away from the array, sweeping out a volume shaped like a truncated pyramid [66–72]. The returned echoes detected by the 2D array are processed to display a set of multiple image planes in real time. This approach allows acquisition of a set of 3D images to occur in real time, generating time-dependent 3D images, known as 4D US imaging. Users can interactively control and manipulate these acquired planes to explore the entire volume while they are being updated in real time. This approach is successfully being used in echocardiology, which requires dynamic 3D imaging of the heart and its valves [73–76]. Since the technology for developing a transducer based on a 2D phased array is very difficult, few companies provide this technology.

12.4. Visualization of 3D Ultrasound Images

The wide use of 3D images produced by CT and MRI systems has stimulated the development of many interactive 3D algorithms to help physicians and researchers visualize and manipulate 3D images. Because US images suffer from shadowing, poor tissue–tissue contrast, and image speckle, the display of 3D US images presents significant visualization problems, unlike 3D displays of CT and MR images. Although many 3D display techniques have been developed, two of the most frequently used techniques for 3D US visualization are multiplanar reformatting (MPR) and volume rendering (VR).

12.4.1 Multiplanar Reformatting

The MPR technique is the most commonly used in 3D and 4D US viewing approaches. In this technique, 2D US planes are extracted from the 3D

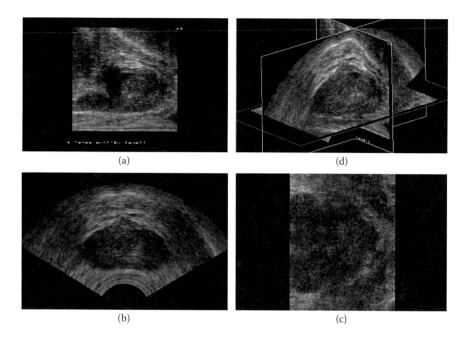

(a)   (d)

(b)   (c)

FIGURE 12.4
The 3D US of the prostate displayed using the multiplanar reformatting approach: (a–c) three orthogonal planes from a 3D image, and (d) the crossed-planes approach. The 3D TRUS image being displayed was acquired using a side-firing transducer using the mechanical rotation approach.

US images and displayed to the user with 3D cubes. Users interact with the images by moving the planes to view the desired anatomy. Three MPR approaches are commonly used to display 3D US images. Figure 12.4(a) illustrates the *crossed-planes* approach, in which multiple planes (typically two or three) are presented in a view that shows their correct relative orientations in 3D. These planes intersect with each other and can be moved in a parallel or an oblique manner to any other plane to reveal the desired anatomy.

A second approach displays the 3D US image using the *cube-view* approach illustrated in Figure 12.4(b). In this approach, an extracted set of 2D US images is texture mapped onto the faces of a polyhedron representing the volume being viewed. Users can select any face of the polyhedron and move it (parallel or obliquely) to any other plane, while the appropriate 2D US image is extracted in real time and texture mapped on the new face. The appearance of a "solid" polyhedron provides users with 3D image-based cubes, which relates the manipulated plane to the other planes [13,77–79]. In a third approach, three *orthogonal planes* are displayed in separate panels with 3D cubes, such as lines on each extracted plane, to designate the intersection with the other planes (Figure 12.4c). These lines can be moved in order to extract and display the desired planes [80,81].

12.4.2 Volume Rendering Techniques

Volume rendering techniques are used to view 3D objects as well as the anatomy using 3D CT and MRI images. However, US images do not typically produce sufficient tissue–tissue contrast to allow easy segmentation or rendering needed for VR rendering, making this approach appropriate only in a few applications of 3D US. Since ultrasound imaging does produce excellent contrast between tissue and fluids (i.e., blood and amniotic fluid), the VR approach is used extensively to view 3D US fetal images [73,75] and 4D US cardiac images [25,74,82]. The VR approach uses ray-casting techniques to project a 2D array of lines (rays) through a 3D image [83–86]. The volume elements (voxels) intersecting each ray are then determined and are weighted, summed, and colored in a number of ways to produce various effects.

The VR technique projects 3D information onto a 2D plane, and thus many VR techniques are not well suited for viewing the details of soft tissues in 3D B-mode ultrasound images. VR techniques are best suited for viewing anatomical surfaces that are distinguishable in 3D B-mode US images, including limbs and fetal face surrounded by amniotic fluid (see Figure 12.5) [80,81], tissue–blood interfaces such as endocardiac surfaces and inner vascular surfaces (see Figure 12.6), and structures where B-mode clutter has been removed from power or color Doppler 3D images [31,32].

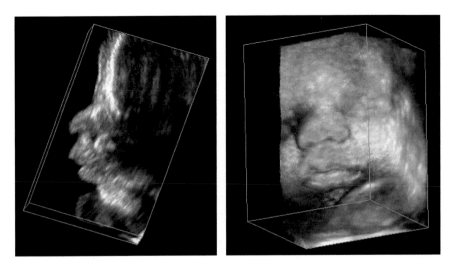

FIGURE 12.5
Two views of a 3D US image of a fetal face that have been volume rendered.

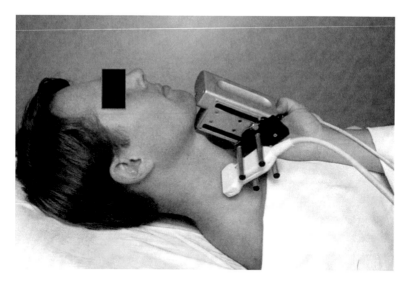

FIGURE 12.6
Mechanically linear scanning approach to produce a 3D US image of the carotid arteries by translating the transducer along the neck by a distance of about 4 cm.

12.5. 3D Ultrasound Imaging of the Carotid Arteries

12.5.1 Motivation

In this section we describe a quickly developing application using 3D US imaging as an example of the use of 3D US. Three-dimensional US imaging of the carotid arteries is used to analyze, quantify, and monitor carotid atherosclerosis. Atherosclerosis is an inflammatory disease in which the inner layer of arteries progressively accumulates low-density lipoproteins and macrophages over a period of several decades, forming plaques [87]. The carotid arteries in the neck supply oxygenated blood to the brain and face and are major sites for developing atherosclerosis. Unstable plaques may suddenly rupture, forming a thrombus causing an embolism, which may ultimately lead to an ischemic stroke by blocking the oxygenated blood supply to parts of the brain. Since carotid arteries are close to the skin, it is relatively easy to acquire high-quality 3D US images of carotid arteries.

The use of 3D US to image the carotid arteries has allowed the development of two sensitive biomarkers of carotid atherosclerosis. Over the past decade *total plaque volume* (*TPV*) and *vessel wall volume* (*VWV*) [33–35,88–91] have emerged as useful US phenotypes of carotid atherosclerosis that quantify plaque burden in 3D. Total plaque volume has been shown to be useful in measuring changes in plaque burden [33–35,92,93] and evaluating the effects of drug therapy [36,94]. More recently, VWV has been shown to provide a more reproducible measure of atherosclerosis burden changes in the carotid arteries. The VWV technique quantifies vessel wall thickness plus plaque within carotid arteries by measuring the volume between the vessel intima and media. It can be implemented more easily in a semiautomated computer segmentation algorithm, resulting in shorter quantification times and greater reliability.

Three-dimensional US-based TPV requires users to distinguish plaque–lumen and plaque–outer vessel wall boundaries, but the measurement of 3D US-based VWV requires the user to outline the lumen–intima/plaque and media–adventitia boundaries, similarly to the measurement of the intima–media (IMT) phenotype [95]. These boundaries are easier to interpret than plaque–lumen and wall boundaries in 3D carotid US images. In this section, we review the method used to acquire 3D carotid US images and discuss its use in the measurement of TPV and VWV.

12.5.2 3D Carotid Ultrasound Scanning Technique

Since imaging of the carotid arteries requires scanning at least a 4 cm of length of the neck, real-time 3D (i.e., 4D) systems using 2D phased arrays optimized for cardiac imaging cannot be used effectively. Thus, investigators have focused on the use of mechanical scanning mechanisms with external fixtures and freehand scanning systems with magnetic field sensors to generate 3D images of the carotid arteries.

We have developed and used a mechanical scanning mechanism with an external fixture (see Figure 12.6) to translate the transducer linearly along the neck of the patient, while transverse 2D US images are acquired at regular spatial intervals [35]. The length of the scan is adjustable and is typically 4 to 6 cm; speed of the scan can be adjusted to minimize the scan time and optimize the spatial resolution. Typically, we acquire 2D US images at 30 frames per second with a spatial interval of 0.2 mm. Thus, a 4 cm scan length will require 200 two-dimensional US images, which can be collected in 6.7 seconds without cardiac gating.

12.5.3 Display of 3D US Carotid Images

The most commonly used method to display 3D carotid US images is the *cube-view* approach, as it is best suited for segmentation of the desired anatomy to quantify TPV and VWV. The user can view transverse images of the carotid arteries with optimal resolution and yet be able to view the vessel and plaque in a longitudinal view to obtain 3D anatomical context [28,78,96]. An example of this approach is shown in Figure 12.7.

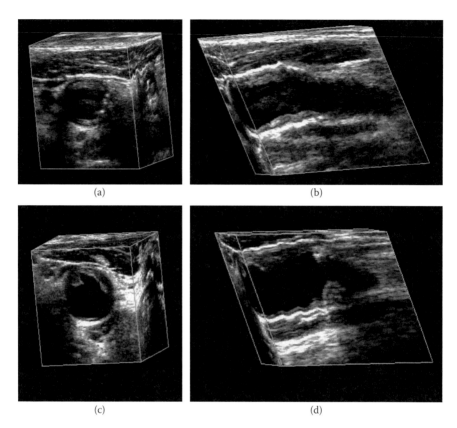

FIGURE 12.7

Multiplaner views of 3D US images of the carotid arteries. (a) Transverse view of a 3D US image of carotid artery for a subject with moderate stenosis; (b) longitudinal view of a 3D US image of carotid artery.

12.5.4 Use of 3D US to Quantify Carotid Atherosclerosis

12.5.4.1 Total Plaque Volume

Quantification of carotid atherosclerosis burden using the *total plaque volume* requires the operator to manipulate the 3D US image by "slicing" it transverse to the vessel axis, starting from one end of the plaque using an inter-slice distance (ISD) of 1.0 mm. In each cross-sectional image, the plaque is contoured and the result is displaced in the 3D image. The area enclosed by each contour is calculated, and the sequential areas are summed and multiplied by the ISD to obtain the plaque volume. The sum of all plaque volumes in the 3D image provides the TPV.

12.5.4.2 Vessel Wall Volume

This vessel wall volume technique proceeds in a similar way to the measurement of TPV. Each 3D US carotid image is "sliced" transverse to the vessel

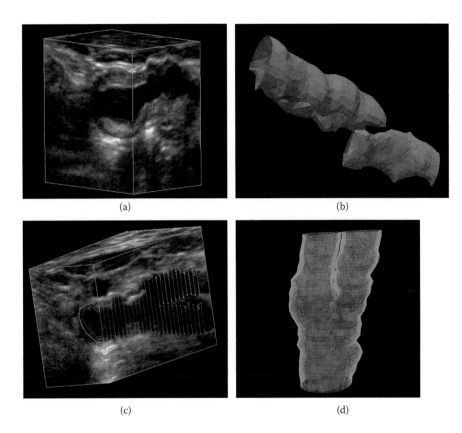

(a) (b)

(c) (d)

FIGURE 12.8
(See color insert.) Manual segmentation of the CCA, ICA, and ECA lumen and outer wall boundaries from 3D US images are used to calculate the vessel wall volume (VWV). The reconstructed surfaces for the lumen and outer wall boundaries from the manual segmentations are shown.

axis, starting from one end of the 3D US image using an ISD of 1.0 mm. However, unlike the TPV technique, in this approach, the lumen (blood–intima boundary) and the vessel wall (media–adventitia boundary) are segmented in each slice. The area between the lumen and vessel wall is calculated giving the vessel wall area. Sequential areas are summed and multiplied by the ISD to give the VWV (see Figure 12.8).

The VWV measurements are global measurements; however, in monitoring of plaque and vessel wall thickness changes, it is important to analyze these on a spatial point-by-point basis. Chiu et al. [97] proposed extending the VWV phenotype to generating vessel wall thickness (VWT) maps and VWT change maps to quantify and visualize local changes in plaque morphology on a point-by-point basis. Identification of the locations of change in plaque burden may assist in developing treatment strategies for patients. To facilitate the visualization and interpretation of these maps for clinicians,

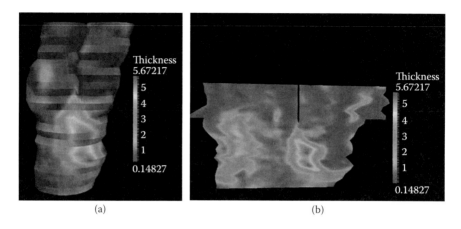

(a)   (b)

FIGURE 12.9
(See color insert.) (a) Vessel wall thickness map for a patient with moderate stenosis. Manual segmentations of the lumen and outer wall boundaries were used to generate the thickness maps (indicated in millimeters). (b) Corresponding flattened thickness map for better visualization.

Chiu et al. [98] proposed a technique to flatten the 3D VWT maps and VWT change maps to 2D (see Figure 12.9a and b).

12.6 3D Ultrasound Guided Prostate Biopsy

12.6.1 Background

Digital rectal exams and prostate-specific antigen (PSA) blood tests are the most common prostate cancer screening methods in men who have no symptoms of the disease. However, a PSA test does not provide sufficient information to provide a definitive diagnosis of prostate cancer. Thus, a histological assessment of tissue cores drawn from the prostate during a biopsy procedure are required to provide a definitive diagnosis. Currently, a physician uses a 2D transrectal ultrasound (TRUS) transducer to guide the needle into the prostate to obtain a biopsy sample. Unfortunately, early stage cancer is not usually visible in ultrasound images. Thus, physicians obtain biopsy samples from predetermined regions of the prostate with a high probability of harboring cancer rather than targeting the lesions directly. Although this approach is commonly used, it has a false negative rate as high as 34%, requiring repeat biopsies to locate the cancer site. Depending on the pathological results, the urologist must either avoid a previously targeted region in the prostate or aim near these sites to obtain additional cores during a repeat biopsy.

Since 2D US images do not provide sufficient information about the 3D location of the biopsy sample, it is difficult for physicians to plan and guide a repeat biopsy procedure. A 3D ultrasound-based navigation system would provide a reproducible record of the 3D locations of the biopsy targets throughout the procedure, allowing guidance of the biopsy to the desired locations in the prostate.

We have developed a mechanical 3D US-guided biopsy system that overcomes the current limitations of a 2D TRUS biopsy procedure while maintaining the procedural work flow, thus minimizing costs and physician retraining [22]. This mechanical system has four degrees of freedom and an adaptable cradle that supports commercially available TRUS transducers. It also allows the acquisition of 3D TRUS images and real-time tracking and recording of the 3D position and orientation of the biopsy needle relative to the 3D TRUS image as the physician manipulates the US transducer.

12.6.2 Mechanically Tracked 3D US System

Our 3D TRUS-guided prostate biopsy system consists of an integrated personal computer, conventional US system with an end-firing transducer, and a mechanical guidance system. Two-dimensional TRUS images are acquired and reconstructed into a 3D TRUS image in real time using the rotational scanning approach (Figure 12.1). The details of the mechanical system have been described elsewhere and are only summarized here [22]. The system uses [99]

- Passive mechanical components for guiding, tracking, and stabilizing the position of a commercially available end-firing TRUS transducer
- Software components for acquiring, storing, and 3D reconstructing (in real time) a series of 2D TRUS images into a 3D TRUS image
- Software that displays a model of the prostate to guide and record the biopsy core locations in 3D

The mechanical assembly consists of a passive four degrees-of-freedom tracking device, an adaptable cradle to accommodate any conventional end-firing TRUS transducer (Figure 12.10). To produce a 3D US image, the transducer is rotated around its long axis (mechanical rotational scanning) to acquire a 3D image, which is then shown in a cube-view display (Figure 12.11).

12.6.3 Biopsy Work Flow

The physician begins the procedure by inserting the TRUS probe into the patient's rectum and aligning the prostate to the center of the 2D TRUS image. A 3D TRUS image of the prostate is acquired by rotating the transducer 180° about its long axis. The boundary of the prostate is then segmented and used

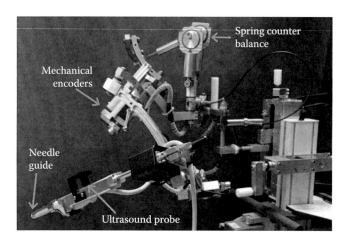

FIGURE 12.10

Photograph of the mechanical tracking system to be used for 3D TRUS guided prostate biopsy. The system is mounted at the base of a stabilizer while the linkage allows the TRUS transducer to be manipulated manually about a remote center of motion (RCM) to which the center of the probe tip is aligned. The spring-loaded counterbalance is used to support the weight of the system fully throughout its full range of motion about the RCM. (Bax, J. et al., *Medical Physics* 35 (12): 5397–5410, 2008. With permission.)

to produce a model of the prostate, which is used in the biopsy navigation display [100–103]. After the model has been displayed, the physician can manipulate the 3D TRUS image on the computer screen and select locations to biopsy. After all of the targets have been selected, the system then displays the 3D biopsy needle guidance interface, which facilitates the systematic targeting of each biopsy location previously selected (Figure 12.12).

Figure 12.12 illustrates the biopsy interface, which is composed of four windows: the live 2D TRUS video stream, the 3D TRUS image, and two 3D model views. The 2D TRUS window displays the real-time 2D TRUS images from the US machine. The 3D TRUS window contains a 2D slice of the 3D prostate image in real time to reflect the expected orientation and position of the TRUS probe. This correspondence allows the physician to compare the 3D image with the real-time 2D image to determine if any prostate motion or deformation has occurred. Finally, the two 3D graphical model windows show orthogonal views (sagittal and coronal) of the 3D prostate model, the real-time position of the 2D TRUS image plane, and the expected trajectory of the biopsy needle.

12.7 Discussion and Conclusions

The utility of 3D US imaging has already been demonstrated in obstetrics, cardiology, and image guidance of interventional procedures. Current 3D US

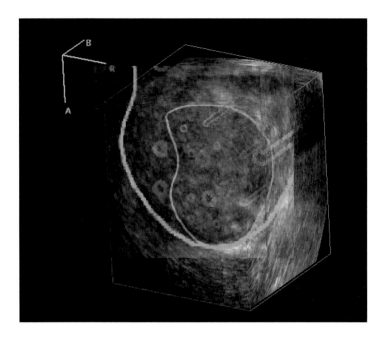

FIGURE 12.11
(See color insert.) A 3D US image of the prostate acquired using an end-firing endocavity rotational 3D scanning approach (rotation of a TRUS transducer), which was used to guide a biopsy using the system shown in Figure 12.10. The transducer was rotated around its long axis, while 3D US images were acquired and reconstructed. The image also showed the segmented boundary of the prostate and the locations of the biopsy cores within the prostate.

imaging technology is sufficiently advanced to allow real-time 3D imaging using 2D array transducers and near real-time 3D imaging with mechanically conventional transducers. Investigators are continuing to establish the utility of 3D US in additional clinical applications using improved image analysis techniques and quantitative measurement approaches. Improved software tools for image analysis are promising to make 3D US a routinely used tool on ultrasound machines. The following are some possible improvements in 3D US imaging that may accelerate its use in routine clinical procedures.

12.7.1 Improved Visualization Tools

Many 3D US imaging applications require interactive visualization tools. However, many approaches are complicated and difficult to use by physicians during busy clinical procedures, particularly interventional procedures. For 3D US to become widely accepted for interventional guidance applications, intuitive tools are required to manipulate the 3D US image and display the result with the appropriate background using the appropriate rendering method. Currently, segmentation, guidance, and volume-rendering approaches using 3D US images require multiple parameters to be

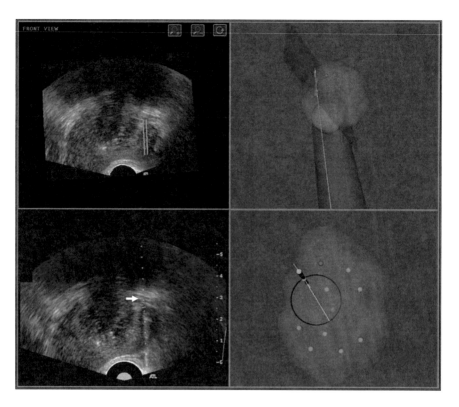

FIGURE 12.12

(See color insert.) The 3D TRUS-guide biopsy system's interface is composed of four windows: (top left) the 3D TRUS image dynamically sliced to match the real-time TRUS transducer 3D orientation; (bottom left) the live 2D TRUS video stream; (right side) the 3D location of the biopsy core is displayed within the 3D prostate models. The targeting ring in the bottom right window shows all the possible needle paths that intersect the preplanned target by rotating the TRUS about its long axis. This allows the physician to maneuver the TRUS transducer to the target (highlighted by the red dot) in the shortest possible distance. The biopsy needle (arrow) is visible within the real-time 2D TRUS image.

optimized and manipulated. Techniques are needed that provide immediate optimal selection of parameters for segmentation, volume rendering, and guidance without significant user intervention.

12.7.2 3D US Segmentation

Over the past decade, many semiautomated and automated segmentation algorithms have been developed for use with 3D US images. Since US images suffer from shadowing and poor tissue contrast, many segmentation approaches are not yet sufficiently robust to be used in routine clinical procedures. Semiautomated segmentation approaches typically require the user to identify the organ or pathology to be segmented (e.g., plaque or prostate), and an algorithm then performs the segmentation. While this approach is easier than manual segmentation, it still requires user interaction. Thus, improvements in segmentation of organs and pathology using robust, accurate, and reproducible algorithms would be highly welcome.

12.7.3 3D US Guided Interventional Procedures

Three-dimensional US imaging has been demonstrated to be useful during interventional procedures, such as biopsy, therapy, and surgery. For example, the 3D TRUS-guided prostate biopsy approach improves the physician's ability to guide the biopsy needle accurately to selected targets and record the biopsy location in 3D. The use of 3D TRUS images allows the physician to view the patient's prostate in views currently not possible in 2D TRUS-based procedures. However, significant work and testing are still required to allow physicians to integrate 3D US imaging efficiently into the interventional procedure. Some required improvements are better 3D registration tools allowing integration of 3D US with images from other modalities, efficient and robust 3D US-based segmentation tools, hands-free methods to control and manipulate the 3D US image and 3D visualization tools to help the physician guide tools within the body.

Acknowledgments

The authors gratefully acknowledge the financial support of the Canadian Institutes of Health Research, the Ontario Institute for Cancer Research, the Ontario Research Fund, the National Science and Engineering Research Council, and the Canada Research Chair program.

References

1. Elliott, S. T. (2008). Volume ultrasound: The next big thing? *British Journal Radiology* 81 (961): 8–9.
2. Downey, D. B., A. Fenster and J. C. Williams (2000). Clinical utility of three-dimensional US. *Radiographics* 20 (2): 559–571.
3. Hummel, J., M. Figl, M. Bax, H. Bergmann and W. Birkfellner (2008). 2D/3D registration of endoscopic ultrasound to CT volume data. *Physics in Medicine and Biology* 53 (16): 4303–4316.
4. Carson, P. L. and A. Fenster (2009). Anniversary paper: Evolution of ultrasound physics and the role of medical physicists and the AAPM and its journal in that evolution. *Medical Physics* 36 (2): 411–428.

5. Chin, J. L., D. B. Downey, G. Onik and A. Fenster (1996). Three-dimensional prostate ultrasound and its application to cryosurgery. *Techniques in Urology* 2 (4): 187–193.

6. Chin, J. L., D. B. Downey, M. Mulligan and A. Fenster (1998). Three-dimensional transrectal ultrasound guided cryoablation for localized prostate cancer in nonsurgical candidates: A feasibility study and report of early results. *Journal Urology* 159(3): 910-914.

7. Smith, W. L., K. Surry, G. Mills, D. Downey and A. Fenster (2001). Three-dimensional ultrasound-guided core needle breast biopsy. *Ultrasound Medicine and Biology* 27 (8): 1025–1034.

8. Wei, Z., G. Wan, L. Gardi, G. Mills, D. Downey and A. Fenster (2004). Robot-assisted 3D-TRUS guided prostate brachytherapy: System integration and validation. *Medical Physics* 31 (3): 539–548.

9. Nelson, T. R. and D. H. Pretorius (1992). Three-dimensional ultrasound of fetal surface features. *Ultrasound Obstetrics and Gynecology* 2: 166-174.

10. Greenleaf, J. F., M. Belohlavek, T. C. Gerber, D. A. Foley and J. B. Seward (1993). Multidimensional visualization in echocardiography: An introduction [see comments]. *Mayo Clinic Proceedings* 68 (3): 213–220.

11. King, D. L., A. S. Gopal, P. M. Sapin, K. M. Schroder and A. N. Demaria (1993). Three-dimensional echocardiography. *American Journal Cardiology Imaging* 7 (3): 209–220.

12. Rankin, R. N., A. Fenster, D. B. Downey, P. L. Munk, M. F. Levin and A. D. Vellet (1993). Three-dimensional sonographic reconstruction: Techniques and diagnostic applications. *American Journal Roentgenology* 161 (4): 695–702.

13. Fenster, A. and D. B. Downey (1996). 3-dimensional ultrasound imaging: A review. *IEEE Engineering in Medicine and Biology* 15: 41–51.

14. Sklansky, M. (2003). New dimensions and directions in fetal cardiology. *Current Opinion Pediatrics* 15 (5): 463–471.

15. Peralta, C. F., P. Cavoretto, B. Csapo, O. Falcon and K. H. Nicolaides (2006). Lung and heart volumes by three-dimensional ultrasound in normal fetuses at 12-32 weeks' gestation. *Ultrasound Obstetrics and Gynecology* 27 (2): 128–133.

16. Smith, W. L. and A. Fenster (2000). Optimum scan spacing for three-dimensional ultrasound by speckle statistics. *Ultrasound Medicine and Biology* 26 (4): 551–562.

17. Gilja, O. H., N. Thune, K. Matre, T. Hausken, S. Odegaard and A. Berstad (1994). In vitro evaluation of three-dimensional ultrasonography in volume estimation of abdominal organs. *Ultrasound Medicine and Biology* 20 (2): 157–165.

18. Delabays, A., N. G. Pandian, Q. L. Cao, L. Sugeng, G. Marx, A. Ludomirski and S. L. Schwartz (1995). Transthoracic real-time three-dimensional echocardiography using a fan-like scanning approach for data acquisition: Methods, strengths, problems, and initial clinical experience. *Echocardiography* 12 (1): 49–59.

19. Downey, D. B., D. A. Nicolle and A. Fenster (1995). Three-dimensional orbital ultrasonography. *Canadian Journal Ophthalmology* 30 (7): 395–398.

20. Downey, D. B., D. A. Nicolle and A. Fenster (1995). Three-dimensional ultrasound of the eye. *Administrative Radiology Journal* 14: 46–50.

21. Fenster, A., S. Tong, S. Sherebrin, D. B. Downey and R. N. Rankin (1995). Three-dimensional ultrasound imaging. *SPIE Physics Medicine Image* 2432: 176–184.

22. Bax, J., D. Cool, L. Gardi, K. Knight, D. Smith, J. Montreuil, S. Sherebrin, C. Romagnoli and A. Fenster (2008). Mechanically assisted 3D ultrasound guided prostate biopsy system. *Medical Physics* 35 (12): 5397–5410.

23. Benacerraf, B. R., C. B. Benson, A. Z. Abuhamad, J. A. Copel, J. S. Abramowicz, G. R. Devore, P. M. Doubilet, et al. (2005). Three- and 4-dimensional ultrasound in obstetrics and gynecology. Proceedings of the American Institute of Ultrasound in Medicine Consensus Conference. *Journal Ultrasound Medicine* 24 (12): 1587–1597.

24. Dolkart, L., M. Harter and M. Snyder (2005). Four-dimensional ultrasonographic guidance for invasive obstetric procedures. *Journal Ultrasound Medicine* 24 (9): 1261–1266.

25. Goncalves, L., J. Nien, J. Espinoza, J. Kusanovic, W. Lee, B. Swope and E. Soto (2005). Two-dimensional (2D) versus three- and four-dimensional (3D/4D) US in obstetrical practice: Does the new technology add anything? *American Journal Obstetrics and Gynecology* 193 (6): S150.

26. Kurjak, A., B. Miskovic, W. Andonotopo, M. Stanojevic, G. Azumendi and H. Vrcic (2007). How useful is 3D and 4D ultrasound in perinatal medicine? *Journal Perinatal Medicine* 35(1): 10-27.

27. Blake, C. C., T. L. Elliot, P. J. Slomka, D. B. Downey and A. Fenster (2000). Variability and accuracy of measurements of prostate brachytherapy seed position in vitro using three-dimensional ultrasound: an intra- and inter-observer study. *Medical Physics* 27 (12): 2788–2795.

28. Fenster, A., D. B. Downey and H. N. Cardinal (2001). Three-dimensional ultrasound imaging. *Physics in Medicine and Biology* 46 (5): R67–99.

29. Pretorius, D. H., T. R. Nelson and J. S. Jaffe (1992). 3-dimensional sonographic analysis based on color flow Doppler and grayscale image data: A preliminary report. *Journal Ultrasound Medicine* 11(5): 225–232.

30. Picot, P. A., D. W. Rickey, R. Mitchell, R. N. Rankin and A. Fenster (1993). Three-dimensional color Doppler imaging. *Ultrasound Medicine and Biology* 19 (2): 95–104.

31. Downey, D. B. and A. Fenster (1995). Three-dimensional power Doppler detection of prostate cancer [letter]. *American Journal of Roentgenelogy* 165 (3): 741.

32. Downey, D. B. and A. Fenster (1995). Vascular imaging with a three-dimensional power Doppler system. *American Journal Roentgenology* 165 (3): 665–668.

33. Landry, A. and A. Fenster (2002). Theoretical and experimental quantification of carotid plaque volume measurements made by 3D ultrasound using test phantoms. *Medical Physics* 10: 2319–2327.

34. Landry, A., J. D. Spence and A. Fenster (2004). Measurement of carotid plaque volume by 3-dimensional ultrasound. *Stroke* 35 (4): 864–869.

35. Landry, A., J. D. Spence and A. Fenster (2005). Quantification of carotid plaque volume measurements using 3D ultrasound imaging. *Ultrasound Medicine and Biology* 31 (6): 751–762.

36. Ainsworth, C. D., C. C. Blake, A. Tamayo, V. Beletsky, A. Fenster and J. D. Spence (2005). 3D ultrasound measurement of change in carotid plaque volume; a tool for rapid evaluation of new therapies. *Stroke* 35: 1904–1909.

37. Krasinski, A., B. Chiu, J. D. Spence, A. Fenster and G. Parraga (2009). Three-dimensional ultrasound quantification of intensive statin treatment of carotid atherosclerosis. *Ultrasound Medicine and Biology* 35 (11): 1763–1772.

38. Tong, S., D. B. Downey, H. N. Cardinal and A. Fenster (1996). A three-dimensional ultrasound prostate imaging system. *Ultrasound Medicine and Biology* 22 (6): 735–746.

39. Tong, S., H. N. Cardinal, R. F. McLoughlin, D. B. Downey and A. Fenster (1998). Intra- and inter-observer variability and reliability of prostate volume measurement via two-dimensional and three-dimensional ultrasound imaging. *Ultrasound Medicine and Biology* 24 (5): 673–681.

40. Downey, D. B., J. L. Chin and A. Fenster (1995). Three-dimensional US-guided cryosurgery. *Radiology* 197 (P): 539.

41. Onik, G. M., D. B. Downey and A. Fenster (1996). Three-dimensional sonographically monitored cryosurgery in a prostate phantom. *Journal Ultrasound Medicine* 15 (3): 267–270.

42. Chin, J. L., D. B. Downey, T. L. Elliot, S. Tong, C. A. McLean, M. Fortier and A. Fenster (1999). Three dimensional transrectal ultrasound imaging of the prostate: Clinical validation. *Canadian Journal Urology* 6 (2): 720–726.

43. Wei, Z., L. Gardi, D. B. Downey and A. Fenster (2005). Oblique needle segmentation and tracking for 3D TRUS guided prostate brachytherapy. *Medical Physics* 32 (9): 2928–2941.

44. Cool, D., S. Sherebrin, J. Izawa, J. Chin and A. Fenster (2008). Design and evaluation of a 3D transrectal ultrasound prostate biopsy system. *Medical Physics* 35 (10): 4695–4707.

45. Pagoulatos, N., D. R. Haynor and Y. Kim (2001). A fast calibration method for 3-D tracking of ultrasound images using a spatial localizer. *Ultrasound Medicine and Biology* 27 (9): 1219–1229.

46. Boctor, E. M., M. A. Choti, E. C. Burdette and R. J. Webster III (2008). Three-dimensional ultrasound-guided robotic needle placement: An experimental evaluation. *International Journal Medical Robotics* 4 (2): 180–191.

47. Treece, G., R. Prager, A. Gee and L. Berman (2001). 3D ultrasound measurement of large organ volume. *Medical Image Analysis* 5 (1): 41–54.

48. Raab, F. H., E. B. Blood, T. O. Steiner and H. R. Jones (1979). Magnetic position and orientation tracking system. *IEEE Transactions Aerospace Electronic Systems* AES-15: 709–717.

49. Detmer, P. R., G. Bashein, T. Hodges, K. W. Beach, E. P. Filer, D. H. Burns and D. E. Strandness, Jr. (1994). 3D ultrasonic image feature localization based on magnetic scanhead tracking: In vitro calibration and validation. *Ultrasound Medicine and Biology* 20 (9): 923–936.

50. Hodges, T. C., P. R. Detmer, D. H. Burns, K. W. Beach and D. E. J. Strandness (1994). Ultrasonic three-dimensional reconstruction: In vitro and in vivo volume and area measurement. *Ultrasound Medicine and Biology* 20 (8): 719–729.

51. Pretorius, D. H. and T. R. Nelson (1994). Prenatal visualization of cranial sutures and fontanelles with three-dimensional ultrasonography. *Journal Ultrasound Medicine* 13 (11): 871–876.

52. Nelson, T. R. and D. H. Pretorius (1995). Visualization of the fetal thoracic skeleton with three-dimensional sonography: A preliminary report. *American Journal Roentgenology* 164 (6): 1485–1488.

53. Riccabona, M., T. R. Nelson, D. H. Pretorius and T. E. Davidson (1995). Distance and volume measurement using three-dimensional ultrasonography. *Journal Ultrasound Medicine* 14 (12): 881–886.

54. Hughes, S. W., T. J. D'Arcy, D. J. Maxwell, W. Chiu, A. Milner, J. E. Saunders and R. J. Sheppard (1996). Volume estimation from multiplanar 2D ultrasound images using a remote electromagnetic position and orientation sensor. *Ultrasound Medicine and Biology* 22 (5): 561–572.

55. Gilja, O. H., P. R. Detmer, J. M. Jong, D. F. Leotta, X. N. Li, K. W. Beach, R. Martin and D. E. J. Strandness (1997). Intragastric distribution and gastric emptying assessed by three-dimensional ultrasonography. *Gastroenterology* 113 (1): 38–49.

56. Leotta, D. F., P. R. Detmer and R. W. Martin (1997). Performance of a miniature magnetic position sensor for three-dimensional ultrasound imaging. *Ultrasound Medicine and Biology* 23(4): 597-609.
57. Hsu, P. W., R. W. Prager, A. H. Gee and G. M. Treece (2008). Real-time freehand 3D ultrasound calibration. *Ultrasound Medicine and Biology* 34 (2): 239–251.
58. Lindseth, F., G. A. Tangen, T. Lango and J. Bang (2003). Probe calibration for freehand 3-D ultrasound. *Ultrasound Medicine and Biology* 29 (11): 1607–1623.
59. Leotta, D. F. (2004). An efficient calibration method for freehand 3-D ultrasound imaging systems. *Ultrasound Medicine and Biology* 30 (7): 999–1008.
60. Dandekar, S., Y. Li, J. Molloy and J. Hossack (2005). A phantom with reduced complexity for spatial 3-D ultrasound calibration. *Ultrasound Medicine and Biology* 31 (8): 1083–1093.
61. Gee, A. H., N. E. Houghton, G. M. Treece and R. W. Prager (2005). A mechanical instrument for 3D ultrasound probe calibration. *Ultrasound Medicine and Biology* 31 (4): 505–518.
62. Gooding, M. J., S. H. Kennedy and J. A. Noble (2005). Temporal calibration of freehand three-dimensional ultrasound using image alignment. *Ultrasound Medicine and Biology* 31 (7): 919–927.
63. Mercier, L., T. Lango, F. Lindseth and D. L. Collins (2005). A review of calibration techniques for freehand 3-D ultrasound systems. *Ultrasound Medicine and Biology* 31 (4): 449–471.
64. Poon, T. C. and R. N. Rohling (2005). Comparison of calibration methods for spatial tracking of a 3-D ultrasound probe. *Ultrasound Medicine and Biology* 31 (8): 1095–1108.
65. Rousseau, F., P. Hellier and C. Barillot (2005). Confhusius: A robust and fully automatic calibration method for 3D freehand ultrasound. *Medical Image Analysis* 9 (1): 25–38.
66. Shattuck, D. P., M. D. Weinshenker, S. W. Smith and O. T. von Ramm (1984). Explososcan: A parallel processing technique for high speed ultrasound imaging with linear phased arrays. *Journal Acoustic Society America* 75 (4): 1273–1282.
67. von Ramm, O. T. and S. W. Smith (1990). Real time volumetric ultrasound imaging system. *SPIE* 1231:15–22.
68. Smith, S. W., H. G. Pavy, Jr. and O. T. von Ramm (1991). High-speed ultrasound volumetric imaging system. Part I. Transducer design and beam steering. *IEEE Transactions Ultrasonics Ferroelectrics Frequency Control* 38: 100–108.
69. Turnbull, D. H. and F. S. Foster (1991). Beam steering with pulsed two-dimensional transducer arrays. *IEEE Transactions Ultrasonics Ferroelectrics Frequency Control* 38: 320–333.
70. von Ramm, O. T., S. W. Smith and H. G. Pavy, Jr. (1991). High-speed ultrasound volumetric imaging system. Part II. Parallel processing and image display. *IEEE Transactions Ultrasonics Ferroelectrics Frequency Control* 38: 109–115.
71. Smith, S. W., G. E. Trahey and O. T. von Ramm (1992). Two-dimensional arrays for medical ultrasound. *Ultrasonic Imaging* 14(3): 213–233.
72. Oralkan, O., A. S. Ergun, C. H. Cheng, J. A. Johnson, M. Karaman, T. H. Lee and B. T. Khuri-Yakub (2003). Volumetric ultrasound imaging using 2-D CMUT arrays. *IEEE Transactions Ultrasonics Ferroelectrics Frequency Control* 50 (11): 1581–1594.
73. Sklansky, M. (2004). Specialty review issue: Fetal cardiology—Introduction. *Pediatric Cardiology* 25 (3): 189–190.

74. Devore, G. R. and B. Polanko (2005). Tomographic ultrasound imaging of the fetal heart: A new technique for identifying normal and abnormal cardiac anatomy. *Journal Ultrasound Medicine* 24 (12): 1685–1696.

75. Xie, M. X., X. F. Wang, T. O. Cheng, Q. Lu, L. Yuan and X. Liu (2005). Real-time 3-dimensional echocardiography: A review of the development of the technology and its clinical application. *Progress Cardiovascular Disease* 48 (3): 209–225.

76. Prakasa, K. R., D. Dalal, J. Wang, C. Bomma, H. Tandri, J. Dong, C. James, et al. (2006). Feasibility and variability of three-dimensional echocardiography in arrhythmogenic right ventricular dysplasia/cardiomyopathy. *American Journal Cardiology* 97 (5): 703–709.

77. Nelson, T. R., D. B. Downey, D. H. Pretorius and A. Fenster (1999). *Three-dimensional ultrasound*. Philadelphia PA: Lippincott-Raven.

78. Fenster, A. and D. Downey (2000). Three-dimensional ultrasound imaging. In *Handbook of medical imaging, volume 1, physics and psychophysics. In text* (J. Beutel, H. Kundel and R. Van Metter. Bellingham, WA, SPIE Press.) 1: 433–509.

79. Fenster, A. and D. Downey (2001). Three-dimensional ultrasound imaging. *Proceedings SPIE* 4549: 1–10.

80. Pretorius, D. H. and T. R. Nelson (1995). Fetal face visualization using three-dimensional ultrasonography. *Journal Ultrasound Medicine* 14 (5): 349–356.

81. Nelson, T. R., D. H. Pretorius, M. Sklansky and S. Hagen-Ansert (1996). Three-dimensional echocardiographic evaluation of fetal heart anatomy and function: Acquisition, analysis, and display. *Journal Ultrasound Medicine* 15 (1): 1–9, quiz 11–12.

82. Deng, J. and C. H. Rodeck (2006). Current applications of fetal cardiac imaging technology. *Current Opinion Obstetrics and Gynecology* 18 (2): 177–184.

83. Levoy, M. (1990). Volume rendering, a hybrid ray tracer for rendering polygon and volume data. *IEEE Computer Graphics Applications* 10: 33–40.

84. Fruhauf, T. (1996). Raycasting vector fields. *Visualization '96. Proceedings* 115–120.

85. Sun, Y. and D. L. Parker (1999). Performance analysis of maximum intensity projection algorithm for display of MRA images. *IEEE Transactions Medical Imaging* 18 (12): 1154–1169.

86. Kniss, J., G. Kindlmann and C. Hansen (2002). Multidimensional transfer functions for interactive volume rendering. *IEEE Transactions on Visualization and Computer Graphics* 8 (3): 270–285.

87. Lusis, A. J. (2000). Atherosclerosis. *Nature* 407 (6801): 233–241.

88. Delcker, A. and H. C. Diener (1994). 3D ultrasound measurement of atherosclerotic plaque volume in carotid arteries. *Bildgebung* 61 (2): 116–121.

89. Delcker, A. and H. C. Diener (1994). Quantification of atherosclerotic plaques in carotid arteries by three-dimensional ultrasound. *British Journal Radiology* 67 (799): 672–678.

90. Delcker, A. and C. Tegeler (1998). Influence of ECG-triggered data acquisition on reliability for carotid plaque volume measurements with a magnetic sensor three-dimensional ultrasound system. *Ultrasound Medicine and Biology* 24 (4): 601–605.

91. Palombo, C., M. Kozakova, C. Morizzo, F. Andreuccetti, A. Tondini, P. Palchetti, G. Mirra, G. Parenti and N. G. Pandian (1998). Ultrafast three-dimensional ultrasound: Application to carotid artery imaging. *Stroke* 29 (8): 1631–1637.

92. Delcker, A., H. C. Diener and H. Wilhelm (1995). Influence of vascular risk factors for atherosclerotic carotid artery plaque progression. *Stroke* 26 (11): 2016–2022.
93. Schminke, U., L. Motsch, B. Griewing, M. Gaull and C. Kessler (2000). Three-dimensional power-mode ultrasound for quantification of the progression of carotid artery atherosclerosis. *Journal Neurology* 247 (2): 106–111.
94. Zhao, X. Q., C. Yuan, T. S. Hatsukami, E. H. Frechette, X. J. Kang, K. R. Maravilla and B. G. Brown (2001). Effects of prolonged intensive lipid-lowering therapy on the characteristics of carotid atherosclerotic plaques in vivo by MRI: A case-control study. *Arteriosclerosis Thrombosis Vascular Biology* 21 (10): 1623–1629.
95. O'Leary, D. H. and M. L. Bots (2010). Imaging of atherosclerosis: Carotid intima–media thickness. *European Heart Journal* 31 (14): 1682–1689.
96. Fenster, A. and D. B. Downey (2000). Three-dimensional ultrasound imaging. *Annual Review Biomedical Engineering* 2: 457–475.
97. Chiu, B., M. Egger, J. D. Spence, G. Parraga and A. Fenster (2008). Quantification of carotid vessel wall and plaque thickness change using 3D ultrasound images. *Medical Physics* 35 (8): 3691–3710.
98. Chiu, B., M. Egger, J. Spence, G. Parraga and A. Fenster (2008). Development of 3D ultrasound techniques for carotid artery disease assessment and monitoring. *International Journal Computer Assisted Radiology Surgery* 3 (1): 1–10.
99. Ding, M. and A. Fenster (2004). Projection-based needle segmentation in 3D ultrasound images. *Computer Aided Surgery* 9 (5): 193–201.
100. Hu, N., D. B. Downey, A. Fenster and H. M. Ladak (2003). Prostate boundary segmentation from 3D ultrasound images. *Medical Physics* 30 (7): 1648–1659.
101. Ladak, H. M., Y. Wang, D. B. Downey and A. Fenster (2003). Testing and optimization of a semiautomatic prostate boundary segmentation algorithm using virtual operators. *Medical Physics* 30 (7): 1637–1647.
102. Wang, Y., H. N. Cardinal, D. B. Downey and A. Fenster (2003). Semiautomatic three-dimensional segmentation of the prostate using two-dimensional ultrasound images. *Medical Physics* 30 (5): 887–897.
103. Ding, M., B. Chiu, I. Gyacskov, X. Yuan, M. Drangova, D. B. Downey and A. Fenster (2007). Fast prostate segmentation in 3D TRUS images based on continuity constraint using an autoregressive model. *Medical Physics* 34 (11): 4109–4125.

13

Optical Coherence Tomography for Imaging Biological Tissue

Michael K. K. Leung and Beau A. Standish

CONTENTS

Clinicians and medical scientists are faced with an ever increasing number of choices for the early detection of life-threatening medical conditions, such as heart disease or cancer. Several of these pathologies require invasive biopsy procedures to verify the presence and stage of disease progression. Once these factors are identified, the patient must then undergo treatment where existing modalities used in the medical industry lack the resolution for treatment monitoring such that in vivo measurements can be correlated to the gold standard of disease-free survival—namely, histology. Optical coherence tomography (OCT) is an exciting, high-resolution, noninvasive, volumetric imaging modality that may provide both structural and functional information, such as blood flow detection, to aid in the early detection and treatment monitoring of diseases. As this optically based technology continues to improve, there is great potential for OCT to become a widely accepted tool to aid in the clinical decision-making process, where emphasis is placed on the ability to provide proper patient risk stratification and subsequent appropriate therapy or therapies. In this chapter, an introduction to the working principles of OCT technologies and the extension of OCT for blood flow detection are presented with respect to the field of biological tissue imaging.

13.1 Introduction

Optical coherence tomography (OCT) has been developed as an imaging modality to visualize subsurface tissue morphology at a resolution (~10 μm in tissue) approaching histology [1]. Current state-of-the-art, ultrahigh resolution systems have achieved submicron spatial resolutions in both in vivo and ex vivo imaging scenarios [2–4], allowing for the detection of biological subcellular features. In vivo endoscopic OCT systems with 10–20 μm spatial resolution have been sufficient in resolving larger structures such as muscular layers of bulk tissue in the gastrointestinal wall [5], along with extension into intraluminal imaging to detect vessel wall structures as a way to stratify patient risk or monitor the placement of stents in coronary arteries [6].

For imaging moving targets, such as blood flow, Doppler OCT has been used to investigate fluid dynamics, interstitial blood flow, hereditary hemorrhagic telangiectasia, retinal blood flow, and the vascular response of photodynamic therapy [7–11]. Modern frequency domain (FD) OCT systems are capable of acquiring and processing three-dimensional (3D) subsurface (~3 mm in depth) images of tissue at multiple volumes per second [12,13]. This increase in system speed has been crucial for the early adoption of OCT as a clinical tool because large anatomical regions can now be imaged in seconds, such as the previously mentioned coronary artery or gastrointestinal tract [3,14].

The technique of OCT itself is considered analogous to ultrasound. However, instead of measuring backscattered pressure waves, OCT measures backscattered photons using an interferometer, avoiding the difficult task of directly measuring the time of flight of the backscattered light. The following sections present an introduction to the working principles of OCT technologies and the extension of OCT for blood flow detection for use in imaging biological tissue.

13.2 Frequency Domain OCT

Optical coherence tomography is an emerging noninvasive imaging modality that can yield real-time, three-dimensional subsurface (~3 mm in depth) images of tissue structure at a spatial resolution (~10 μm) that approaches histology. No preparation of the sample is required, and current state-of-the-art ultra-high resolution systems have achieved submicron spatial resolutions, allowing for the detection of subcellular features [15]. Optical coherence tomography images are similar in appearance to those obtained with ultrasound. Figure 13.1(a)–(c) show sample OCT images of different regions of the finger. To interrogate a sample, instead of measuring backscattered pressure waves, OCT measures backscattered photons using an interferometer, avoiding the difficult task of directly measuring the time-of-flight of the backscattered light. In turbid media and other scattering tissue, the penetration depth of imaging is limited primarily by scattering. The maximum imaging depth of OCT is approximately 1–3 mm at wavelengths between 800 and 1300 nm for a variety of tissues [16]. The axial and transverse resolutions of OCT are decoupled. The axial resolution depends on light source characteristics such as spectral bandwidth and central wavelength, whereas the transverse resolution is dependent on the focusing optics.

To date, two main embodiments of OCT exist. The first generation of systems was based on time-domain OCT (TD-OCT), and these systems can be routinely found in commercial Food and Drug Administration (FDA)-approved systems, used primarily in ophthalmology. The second-generation frequency domain OCT (FD-OCT) systems are more prevalent in research, although commercial interests are gaining momentum.

In TD-OCT, one-dimensional depth ranging is achieved through low-coherence interferometry [17]. Using broadband light, interference fringes are observed only when the reference and sample arm optical path length are matched within the coherence length of the light source. Refractive indices vary between interfaces within a sample manifest as intensity peaks in the detected interference pattern [18]. A reference arm is translated linearly in time during imaging to capture entire depth profiles (an A-scan). By acquiring multiple A-scans at different spatial locations, two- (B-mode) or three-dimensional images can be obtained similar to those of Figure 13.1(a)–(c). The imaging speed of TD-OCT is generally restricted due to the requirement to translate the reference arm mechanically. This limits its use in screening large tissue volumes, such as significant portions of a tumor or the walls of an artery [6].

In FD-OCT, depth ranging is performed by means of spectral interferometry [19]. The reference arm is kept stationary, and depth information is acquired by an inverse Fourier transform of the spectrally resolved interference fringes [20]. Two configurations of FD-OCT exist. One, called spectral domain OCT (SD-OCT), separates individual spectra components using a spectrometer and a charge-coupled device array, and generally uses a superluminescent diode as light source. The other is called swept-source OCT (SS-OCT), which uses a wavelength-swept tunable laser source and a single photodetector (together acting as a "time-resolved" spectrometer). The depth ranging principles for both techniques are the same and differ only in method in acquiring the interference signal.

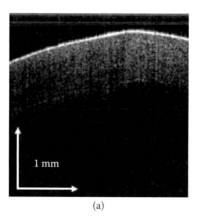

(a)

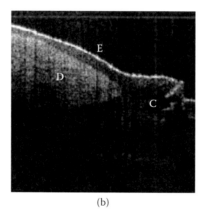

(b)

FIGURE 13.1
Structural B-mode OCT images of the different parts of the human finger: (a) the nail, (b) the nail root, and (c) the fingerprint. (d) Schematic explaining the notation used to describe the dimensions of a 3D OCT image data set. A B-mode image is located on the x–z plane. An A-scan extends in the z, or depth, direction. The x–y direction is also called the en-face plane. Labels: D, dermis; E, epidermis; C, cuticle; F, fingerprint ridges; J, epidermal–dermal junction; S, sweat ducts.

13.2.1 Signal Formation in Frequency Domain OCT

This section describes a derivation of the principle of FD-OCT, which is based on spectral interferometry. To understand depth ranging, assume that the light backscattered from the imaged object consists of reflected elementary plane waves U from different depths z:

$$U(z) = U_0 e^{-ik_0 nz} \qquad (13.1)$$

where
U_0 is the amplitude

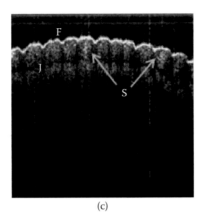

(c)

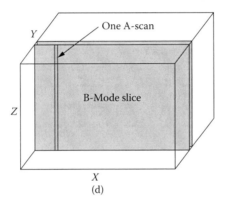

(d)

FIGURE 13.1 *(Continued)*

$k_0 = 2\pi/\lambda_0$ is the wave number

λ_0 is the wavelength

n is the refractive index (≈ 1.3 in tissue)

i is the imaginary number

For a monochromatic source with wave number k_0, a single interface reflector that is detected via interferometry is given by

$$
\begin{aligned}
I_{\Delta z}(k_0) &= \left| (U_R + U_S) \right|^2 \\
&= \left| U_0 \left(e^{-ik_0 nz} + e^{-ik_0 n(z+2\Delta z)} \right) \right|^2 \\
&= U_0^2 \left[2 + e^{ik_0 n 2\Delta z} + e^{-ik_0 n 2\Delta z} \right] \\
&= 2I_0 \left[1 + \cos(2k_0 n \Delta z) \right]
\end{aligned}
\tag{13.2}
$$

where the additional $2\Delta z$ accounts for the round-trip path length difference. The plane waves originating from the reference and sample arms are U_R and U_S, respectively. Ignoring dispersion, for a material with depth-dependent reflectivity $a(z)$, the intensity from many different depths can be summed as

$$I(k_0) = 2I_0\left[1 + \int_0^\infty a(z)\cos(2k_0 nz)dz\right] + \text{mutual inteference terms (MIT)} \quad (13.3)$$

The MIT, which describes the mutual interference of the elementary waves, has the expression

$$MIT = \int_0^\infty \int_0^\infty a(z')a(z)e^{-i2k_0n(z-z')}dzdz' \quad (13.4)$$

This term lies in baseband frequencies (i.e., at $z = 0$) and is generally much weaker than the signal of interest. Therefore, the MIT can be removed by filter as the object signal only needs to be a small offset away from $z = 0$, which is usually the case during OCT imaging. When measurements are performed with many difference frequencies, the interference signal $I(k)$ takes on the following form:

$$I(k) = B(k)\left(1 + \int_0^\infty a(z)\cos(2knz)dz + MIT\right) \quad (13.5)$$

where $B(k)$ is spectral intensity distribution of the laser source (swept source spectrum, or bandwidth of the superluminescent diode in the case of SD-OCT). It can be seen that the depth z of scattering amplitudes is encoded in the argument (frequency, $2nz$) of the cosine function.

To achieve a suitable form with which a Fourier transform can be performed, $a(z)$ can be replaced by the symmetrical expansion $\hat{a}(z) = a(z) + a(-z)$, since $a(z)$ is zero on the opposite side of the reference plane [19]. The expression then becomes

$$I(k) = B(k)\left(1 + \int_{-\infty}^\infty \hat{a}(z)e^{-i2knz}dz + MIT\right) \quad (13.6)$$

Finally, by the definition of the Fourier transform and ignoring the MIT,

$$FT^-[I(k)] \approx FT^-[B(k)] \otimes [\delta(z) + \hat{a}(z)] \quad (13.7)$$

where FT$^-$ is the inverse Fourier transform and $\otimes$ denotes the convolution operator. The Dirac delta function is represented by $\delta(z)$. The depth-dependent reflectivity $\hat{a}(z)$ of the sample is thereby recovered. It can also be seen that the light source spectrum $B(k)$ determines the axial resolution of the system, where the general expression for the axial resolution $\bar{z}$ of the FD-OCT system is [21]:

$$\bar{z} = \frac{2\sqrt{2\ln 2}}{\pi} \frac{\lambda_0^2}{\Delta\lambda} \tag{13.8}$$

where λ_0 represents the center wavelength of a Gaussian light source and $\Delta\lambda$ is the $1/e^2$ width of the spectrum.

As an example, given a $\Delta\lambda$ of 110 nm and λ_0 of 1310 nm, the axial resolution that can be achieved is ~12 μm in air. It has been shown that the FD-OCT system offers superior sensitivity compared to TD-OCT systems, which improves the signal-to-noise ratio (SNR) [22]. In addition, faster scanning speeds can be obtained by FD-OCT as mechanical translation of the reference mirror is no longer required. These benefits are crucial for wide field imaging of biological tissues.

13.2.2 Choice of Wavelength for OCT Imaging

This section briefly discusses the wavelength of the light source used in OCT imaging. OCT is based on detection of light that is backscattered only once (to a first-order approximation) by the imaging target. Therefore, to achieve high penetration depth and sensitivity, scattering or absorption by the sample must be minimized. Equation (13.8) states that the central wavelength λ_0 of the light source should be low to obtain good axial resolution. However, tissue scattering tends to increase with decreasing wavelengths. As well, tissue absorption is high for the wavelength range of 200–600 nm due to the inherent absorption by hemoglobin and above 1000 nm due to water [23].

For ophthalmometry (the current main application of OCT), light is required to penetrate the vitreous humor of the eye, which has a volume of 98%–99% water, to reach the retina. Based on the preceding description, a good compromise has been found by using light sources in the wavelength range of ~800 nm. Recently, there has been a shift to the 1000 nm regime for imaging of the eye. This allows ~200 nm of additional penetration depth compared to 800 nm, which suffers from high absorption and scattering from the retinal pigment epithelium [24]. Although water absorption at this wavelength is significantly higher, the International Council on Non-Ionizing Radiation Protection (ICNIRP) guidelines for maximal permissible light exposure in the eye increases with wavelength (from ~1.7 mW at 800 nm to ~5 mW at 1060 nm) [24]. Hence, loss due to absorption may be compensated by using a laser that can provide higher incident power.

For imaging biological tissue, there are advantages to imaging in the 1300 nm range for several reasons. First, due to the explosion of the telecom industry, there is widespread availability of low-cost fiber optic components to develop complete OCT systems in the 1300 nm range. Second, tissue (e.g., gastrointestinal tract, interstitial, epithelium) is not concealed under a thick layer of water, as in the eye; thus, water absorption is not a significant problem. Therefore, to enhance penetration depth and minimize scattering effects, a higher wavelength such as 1300 nm is typically chosen to image biological tissue.

13.2.3 Swept-Source System Description

A schematic of a standard polygon-based OCT system is shown in Figure 13.2 [25]. The laser is in a fiber-ring-cavity configuration, with a semiconductor optical amplifier (BOA1017, Covega, Washington, DC) as the gain medium. Wavelength tuning is accomplished by a polygon mirror [26] with typical output sweeping ranges ($\Delta\lambda$) of ~100 nm centered at 1310 nm. These system characteristics result in axial resolutions of ~10 μm in tissue and average output power of 20–40 mW. Due to the speed with which the polygon can be spun, A-scan repetition rates can reach over 50 kHz.

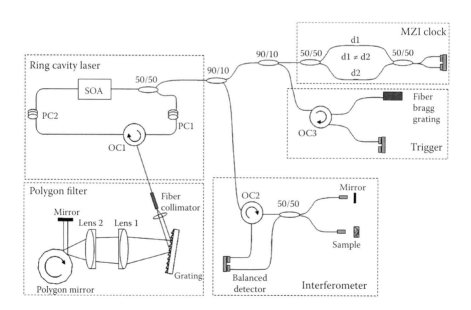

FIGURE 13.2
System schematic of a polygon-based swept source OCT imaging system used for biological tissue imaging. The system consists of multiple modules. (1) The light source is a ring cavity laser. (2) Wavelength tuning used for swept source OCT is achieved by a polygon filter. (3) Signals from the sample arm and reference arm are combined in the interferometer. (4) A-scan triggering and signal recalibration using a Mach–Zehnder interferometer clock are required for data acquisition. Labels: PC, polarization controller; OC, optical circulator; SOA, semiconductor optical amplifier. Ratios denote the routine splitting ratios of fiber optic couplers.

In the current optical filter design, a spinning polygon mirror is used to sweep a beam of light across a diffraction grating. The sweeping motion causes a change in incident angle to the grating that is linear in time. Since the data acquisition sampling rate is constant, this means that the interference signal is acquired linearly in wavelength. To recalibrate wavelength-space data to k-space (as used in the derivations) such that a standard fast Fourier transform can be used to retrieve the depth-intensity profile, a clock signal is generated using a Mach–Zehnder interferometer (MZI) clock [27]. A fiber Bragg grating (FBG) can provide the A-scan trigger. The FBG reflects a particular wavelength of light and transmits all others and it is used to provide a trigger signal that indicates the beginning of a particular wavelength sweep. For instance, given a sweeping range of ~100 nm centered at 1310 nm, the FBG would be designed to reflect wavelengths near 1260 nm.

13.2.4 Structural OCT

In the structural imaging mode, OCT signal intensity indicates the local reflectivity of a particular depth-dependent volume of tissue. For a given two-dimensional (2D) image frame, after inverse Fourier transform, the value of a given pixel S is calculated as [28]

$$S = \frac{1}{MN} \sum_{m=1}^{M} \sum_{n=1}^{N} \left[I_{m,n}^2 + Q_{m,n}^2 \right] \qquad (13.9)$$

where I and Q are the real and imaginary components of the complex inverse Fourier transformed interference fringe signal. M and N represent the number of voxels to average spatially in the axial and lateral directions (i.e., within the same B-mode frame), respectively, to improve SNR. The number of voxels to average is hence $M \times N$.

13.3 Methods of Imaging Blood Vessels with OCT

Two functional extensions of OCT offer the ability to visualize and quantify changes in the tumor vasculature after a therapeutic treatment. Doppler OCT (DOCT), which utilizes similar principles as Doppler ultrasound, permits quantitative measures of flow velocities within vessels ~100 μm in diameter to be measured and quantified [28]. Speckle variance OCT (SVOCT) is a newly developed technique that, while discarding flow information, can identify vessels whose velocity and/or dimension are too small to be detected by DOCT (~10–15 μm diameter) [29]. The strengths and weaknesses of both techniques as applied to biological tissue are discussed next.

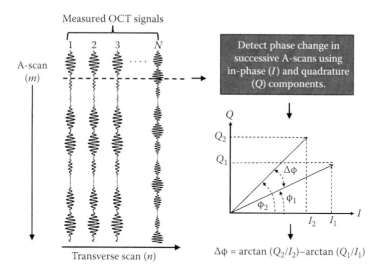

FIGURE 13.3

Doppler frequency detection from moving elements (i.e., blood) using DOCT. A change in phase between two successive A-scans can be calculated using the in-phase (I) and quadrature (Q) components to calculate the phase shift between two successive A-scans.

13.3.1 Doppler OCT

Doppler optical coherence tomography has been used in a number of clinical applications to investigate fluid dynamics in the gastrointestinal tract, interstitial blood flow, and retinal blood flow and for therapeutic monitoring [5,30–32]. The principle of DOCT is borrowed from ultrasound, where flow can be calculated by comparing the phase shifts in consecutive A-scans. A complex signal is readily available from the inverse Fourier transform. To compute Doppler information, the complex signal is separated into its in-phase (I) and quadrature (Q) components. Figure 13.3 shows a flow chart where the mean phase shift can be calculated by evaluating the phase angle of individual OCT signals, consisting of their in-phase and quadrature components, followed by the computation of the phase difference between axial scans. This value can then be related to the Kasai autocorrelation velocity estimator [33] through a trigonometric identity [34], yielding the following equation to calculate the mean Doppler frequency shift (f_D) used to display regions of blood flow. Using the Kasai autocorrelation function, the mean Doppler frequency shift can be calculated as follows [35]:

$$f_D = \frac{f_a}{2\pi} \arctan\left\{ \frac{\sum_{m=1}^{M}\sum_{n=1}^{N-1}(I_{m,n+1}Q_{m,n} - Q_{m,n+1}I_{m,n})}{\sum_{m=1}^{M}\sum_{n=1}^{N-1}(Q_{m,n+1}Q_{m,n} - I_{m,n+1}I_{m,n})} \right\} \qquad (13.10)$$

where f_D is the Doppler frequency shift, f_a is the axial scan frequency, m represents the indices in the depth direction (M = total number of indices), and n represents the indices in the lateral direction (N = total number of indices), resulting in a 2D color Doppler image. The arctangent component gives the Doppler phase shift. From f_D, the velocity v of the moving target can be calculated as

$$v = \frac{\lambda_o f_D}{2n_t \cos(\theta_D)} \tag{13.11}$$

where θ_D is the angle between the optical axis and the direction of flow, λ_0 is the center wavelength of the laser, and n_t is the index of refraction of the sample. In the case of imaging biological tissue, θ_D is assumed to be the angle between the tissue surface and scan beam, which is typically 70°–80°.

A trade-off exists in DOCT, where improved sensitivity to low blood velocities results in a reduction in the maximum detectable velocity. To achieve great sensitivity to slow-moving flow, f_a should be low to allow sufficient phase buildup to exceed the noise floor of the imaging system. However, in such a condition, aliasing may occur if the imaged blood flow is high, due to sampling rate limitations. Conversely, increasing the A-scan sampling rate increases the dynamic range of flow velocities that can be detected, but lowers the detection threshold for slow-moving flow.

This trade-off is one of the factors limiting the broad application of DOCT for acquiring accurate velocity maps in biological tissue. The range of blood flow velocities, which is inversely proportional to the vascular cross-sectional area, between arteries, veins, and capillaries is very broad (centimeters per second to submillimeters per second) [36,37]. For a given A-scan repetition rate f_a optimized for the measurement of the blood flow in the venules (which have relatively low flow compared to arteries), significant aliasing will be observed within the arterial blood vessels. For this reason, it is difficult to capture a snapshot of the complete vasculature in a 3D tissue volume, with each vessel having nonaliased velocity information. Therefore, DOCT's utility in imaging and quantifying biological tissue has been limited to several niche areas or for only acquiring architectural information of the vessels [38]. That is, the velocity information is discarded, and the Doppler signal is used as a binary metric to identify regions of blood flow or no blood flow.

Another limitation of DOCT is its relatively long acquisition time, since oversampling is required [28]. To image a useable volume of tissue (such as the dorsal skin slab in a mouse model, 5×5 mm²), the acquisition time is of the order of tens of minutes. Note that the pulsatile nature of flow requires a sufficient number of repeated acquisitions to obtain a mean value for velocity. For human imaging, heart rates are ~60 beats per minute. This means many repeated acquisitions would be required, further lengthening the acquisition time to obtain an average blood velocity map. Therefore, the benefit of quantitative DOCT in tissue may be limited to monitoring specific vessels, where repeated acquisition over a small spatial volume can be accomplished rapidly [39].

13.3.2 Speckle Variance OCT

Recently, a new OCT technique called speckle variance, which calculates interframe speckle modulation (changes in pixel intensity as a function of time), has been developed [29]. This technique is angle and flow-velocity independent and is more sensitive in detecting microvasculature compared to DOCT, at the expense of lacking quantification of blood velocity information. The intrinsic contrast of SVOCT is based on time-dependent scattering properties of fluids and solids. Speckle variance OCT is calculated from the following equation [29]:

$$SV_{ijk} = \frac{1}{N_f} \sum_{i=1}^{N} (S_{ijk} - S_{mean})^2 \qquad (13.12)$$

where
N_f is the number of frames used for variance calculation
i is the frame number
j,k are indices of corresponding pixels in the transverse and axial directions, respectively
S is the structural OCT intensity

Essentially, it is a variance calculation of a given pixel at N_f different time points, each separated by the time required to acquire a frame, or B-mode image. It has been shown that, at the imaging frame rate of the system used for typical experiments (~40 frames per second), complete decorrelation of structural intensity occurs even in stationary scattering fluids; the origin comes from intrinsic Brownian motion [40]. This property allows SVOCT to detect vasculature even if there is no flow, unlike phase-sensitive techniques.

Figure 13.4 shows fluorescence and white light pictures of a mouse, which highlight the imaged regions (box), via OCT, to demonstrate Doppler and speckle variance imaging. These regions are shown in Figure 13.5, which highlights the ability of OCT to provide functional imaging of biological tissue. The SVOCT image (Figure 13.5a) is what is referred to as a "first-generation" picture, acquired in mid-2008. The second generation is discussed in the next section. This image is a projection in the en-face direction (x–y plane). The corresponding DOCT images at the indicated slice locations (Figure 13.5d–h) are shown (x–z plane). A number of observations can be made here to compare and contrast the DOCT and SVOCT techniques.

First, the number of detected blood vessels in Figure 13.5(a), using SVOCT, is clearly much greater than that of Figure 13.5(b), acquired using DOCT. The

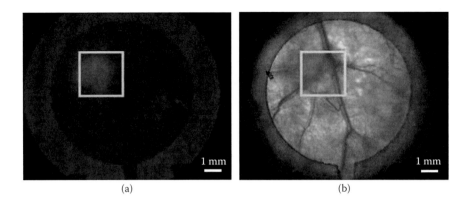

FIGURE 13.4
(See color insert.) Wide-field (a) fluorescence and (b) corresponding white light image of a tumor-bearing (ME-180) nude mouse implanted within a mouse window chamber. The square region is imaged with Doppler and speckle variance optical coherence tomography (shown in Figure 13.5).

lower sensitivity of DOCT requires vessels to have sufficient diameter and flow velocity in order for them to be detected. Second, Figure 13.5(b) and (c) show a reconstruction of many DOCT image slices into 3D, where a threshold was applied to show only the direction of flow. The angle dependence of OCT can be observed here, where the flow direction seems to reverse in approximately the middle of the image due to orientation of how the beam is used to scan the tissue. This angle dependence complicates flow measurement over a large 3D volume. Third, aliasing can be observed in Figure 13.5(d) and (e), where artery–vein pairs are present. Flow velocity may be calculated from the Doppler shift in the vein, but is obscured in the artery due to aliasing.

Although, SVOCT has the ability to image down to the capillary level, it is also very prone to motion artifacts. In DOCT, moderate bulk tissue motion may be removed by analyzing the mean phase of the image. It is assumed that, for a given A-scan, the diameters of blood vessels are much smaller than the region, which consists of bulk tissue (easily satisfied in most cases). Then, the detected mean phase, which would come mostly from the bulk tissue because of the previous assumption, can be subtracted. This is currently not possible for SVOCT, where correlation between variance and magnitude of motion has yet to be established as the speckle variance intensity is identical whether the imaged fluid is stationary or moving. Therefore, it is not possible to subtract the bulk tissue motion by methods used in phase-sensitive techniques.

13.3.3 Optimized Speckle Variance OCT

An optimized version of SVOCT imaging has been developed by Mariampillai et al. [40]. Previously, acquisition of an entire 3D volume was performed by raster scanning, as shown in Figure 13.7(a). In this scan pattern, two consecutive

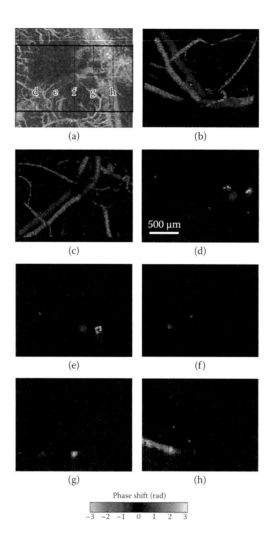

FIGURE 13.5
(See color insert.) Functional imaging of a tumor-bearing nude mouse 1 week after ME-180 cell implantation. (a) Speckle-variance OCT image (3 × 3 mm²) of a tumor and surrounding blood vessels. A necrotic core is present, which manifests as a region devoid of vessels. The labels indicate the locations of DOCT image slices. (b, c) Three-dimensional reconstruction of the detected Doppler signal viewed from an angle and parallel (x–z plane) and perpendicular (x–y plane) to the imaging beam, respectively. (d–h) Corresponding DOCT image slices labeled in (a). Each image is 2.0 mm across and has a 1.7 mm depth of view.

frames, even though they are very close, are still in spatially different positions. In the improved SVOCT scheme, the scan pattern is altered such that the same region is imaged multiple times, as shown in Figure 13.6(b). This reduces the spatial decorrelation that occurred with the original method and improves the rejection of stationary and nonfluid objects [40].

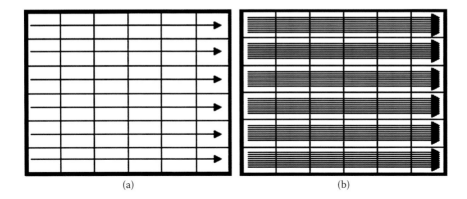

(a)   (b)

FIGURE 13.6
Speckle variance optical coherence tomography scan patterns. In the schematic, each row in the grid represents a single frame, and each cell represents a voxel. (a) First-generation, raster scan pattern. (b) Second-generation, repeated acquisitions in a given frame.

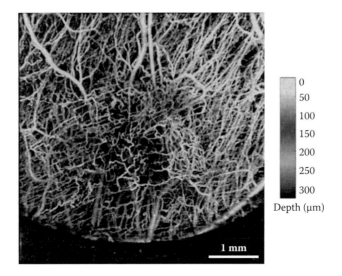

FIGURE 13.7
(See color insert.) Second-generation speckle variance optical coherence tomography image of the mouse dorsal skin. The color indicates the relative depth of the vessels. A tumor mass (ME-180) is in the center of the image. The black border at the bottom is the edge of the plastic fastener of the tissue window.

Figure 13.7 shows an exemplary second-generation SVOCT image of biological in vivo mouse tissue with an implanted tumor. Vasculature down to the capillary level may be detected. In addition, it has been color coded to indicate the relative depth of the vessels. The location of tumors may be identified by two features, and this detection technique may lend itself to the clinical environment to detect tumors or be used as a quantitative method to

TABLE 13.1

Strengths and Weaknesses of Doppler and Speckle Variance Optical Coherence Tomography

Technique	Strengths	Weaknesses
Doppler OCT	1. Can quantify flow velocities 2. Three dimensional	1. Angle dependence 2. Prone to aliasing 3. Limited sensitivity to small/low-flow vessels
Speckle variance OCT	1. High sensitivity to blood vessels regardless of flow 2. Angle independence	1. Binary metric 2. Two-dimensional projection image 3. Prone to motion artifacts

monitor therapeutic responses. The first feature includes the physical vessel architecture as tumorous vessels are different when compared to normal blood vessels. The vessels near the center of Figure 13.7 are more tortuous (tumor), compared to normal vessels, which do not change in direction or tortuosity as frequently. The second feature includes vessel density. Although there is angiogenesis within the tumor, the vessel density tends to be less where a tumor mass is present. This is observed in the center of Figure 13.7.

As stated, the Doppler and speckle variance techniques have their own advantages and disadvantages as outlined in Table 13.1. It is important for the researcher or clinician to understand their unique requirements when imaging biological tissue and to choose the appropriate method to obtain relevant and useful data.

13.4 Current Limitations, Future Solutions, and Conclusions

Optical coherence tomography is a promising noninvasive, real-time, high-resolution, volumetric imaging modality that has found initial clinical utility for imaging biological tissue such as atherosclerotic detection in the coronary artery, gastrointestinal imaging, and ophthalmic applications. Although imaging speeds and biomedical OCT research have increased drastically over the last several years, there are still substantial hurdles for the widespread clinical adoption of OCT.

Essential requirements of OCT as a realistic platform for routine biomedical and clinical imaging necessitate a system that is cost effective and yields high-resolution and/or functional imaging along with extensive scientific study to demonstrate clinical utility. As the field of OCT is rapidly expanding, there is a degree of optimism that, in the foreseeable future, OCT will become a standard clinical tool. It has already found a foothold in ophthalmic imaging and is in the early stages of clinical use for intraluminal images such as of the coronary artery. However, several hurdles exist before this technique becomes a standard imaging option for widespread biomedical or clinical environments. Although the high resolution of OCT is extremely beneficial for identifying microstructural and microvascular architecture, there are issues with the manipulation, evaluation, and storage of the acquired data. A typical pullback data set, commonly produced for gastrointestinal imaging, can approach 40 GB and consist of ~1200 individual cross-sectional images [41]. This overload of data must be simplified, as it would take an extensive amount of human capital by highly qualified personnel to analyze each of the large 3D data sets thoroughly. A potential solution to this problem includes the incorporation of smart algorithms that have the ability to identify regions of interest via pattern recognition algorithms to filter these large data sets down to a manageable set of "high-risk" images that necessitate additional clinical review.

With continued hardware and software OCT advancements, this platform technology has the potential to provide evidence-based disease detection along with an ability to track disease progression longitudinally. Therefore, these inherent system characteristics have resulted in a new cost-effective solution to understanding biological processes as a first step in ultimately providing superior health care to patients.

References

1. D. Huang, E. A. Swanson, C. P. Lin, J. S. Schuman, W. G. Stinson, W. Chang, M. R. Hee, et al. Optical coherence tomography. *Science* 254:1178–1181 (1991).
2. K. Bizheva, A. Unterhuber, B. Hermann, B. Povaûay, H. Sattmann, A. Fercher, W. Drexler, M. Preusser, H. Budka, and A. Stingl. Imaging ex vivo healthy and pathological human brain tissue with ultra-high-resolution optical coherence tomography. *Journal of Biomedical Optics* 10:011006 (2005).
3. B. J. Vakoc, M. Shishko, S. H. Yun, W. Y. Oh, M. J. Suter, A. E. Desjardins, J. A. Evans, et al. Comprehensive esophageal microscopy by using optical frequency-domain imaging (with video) (a figure is presented). *Gastrointestinal Endoscopy* 65:898–905 (2007).
4. S. Yazdanfar, M. D. Kulkarni, and J. A. Izatt. High resolution imaging of in vivo cardiac dynamics using color Doppler optical coherence tomography. *Optics Express* 1:424–431 (1997).
5. V. X. D. Yang, S. J. Tang, M. L. Gordon, B. Qi, G. Gardiner, M. Cirocco, P. Kortan, et al. Endoscopic Doppler optical coherence tomography in the human GI tract: Initial experience. *Gastrointestinal Endoscopy* 61:879–890 (2005).
6. G. J. Tearney, S. Waxman, M. Shishkov, B. J. Vakoc, M. J. Suter, M. I. Freilich, A. E. Desjardins, et al. Three-dimensional coronary artery microscopy by intracoronary optical frequency domain imaging. *JACC: Cardiovascular Imaging* 1:752–761 (2008).

7. B. A. Standish, X. Jin, J. Smolen, A. Mariampillai, N. R. Munce, B. C. Wilson, I. A. Vitkin, and V. X. D. Yang. Interstitial Doppler optical coherence tomography monitors microvascular changes during photodynamic therapy in a Dunning prostate model under varying treatment conditions. *Journal of Biomedical Optics* 12:034022 (2007).

8. J. Moger, S. Matcher, C. Winlove, and A. Shore. Measuring red blood cell flow dynamics in a glass capillary using Doppler optical coherence tomography and Doppler amplitude optical coherence tomography. *Journal of Biomedical Optics* 9:982 (2004).

9. B. White, M. Pierce, N. Nassif, B. Cense, B. Park, G. Tearney, B. Bouma, T. Chen, and J. de Boer. In vivo dynamic human retinal blood flow imaging using ultra-high-speed spectral domain optical coherence tomography. *Optics Express* 11:3490–3497 (2003).

10. S. Tang, M. Gordon, V. Yang, M. Faughnan, M. Cirocco, B. Qi, E. Yue, G. Gardiner, G. Haber, and G. Kandel. In vivo Doppler optical coherence tomography of mucocutaneous telangiectases in hereditary hemorrhagic telangiectasia. *Gastrointestinal Endoscopy* 58:591–598 (2003).

11. B. Standish, K. Lee, X. Jin, A. Mariampillai, N. Munce, M. Wood, B. Wilson, I. Vitkin, and V. Yang. Interstitial doppler optical coherence tomography as a local tumor necrosis predictor in photodynamic therapy of prostatic carcinoma: An in vivo study. *Cancer Research* 68:9987 (2008).

12. R. Huber, D. Adler, and J. Fujimoto. Buffered Fourier domain mode locking: Unidirectional swept laser sources for optical coherence tomography imaging at 370,000 lines/s. *Optics Letters* 31:2975–2977 (2006).

13. D. Adler, Y. Chen, R. Huber, J. Schmitt, J. Connolly, and J. Fujimoto. Three-dimensional endomicroscopy using optical coherence tomography. *Nature Photonics* 1:709–716 (2007).

14. G. van Soest, T. Goderie, E. Regar, S. Koljenovi, G. van Leenders, N. Gonzalo, S. van Noorden, T. Okamura, B. Bouma, and G. Tearney. Atherosclerotic tissue characterization in vivo by optical coherence tomography attenuation imaging. *Journal of Biomedical Optics* 15:011105.

15. K. Bizheva, A. Unterhuber, B. Hermann, B. Povay, H. Sattmann, A. F. Fercher, W. Drexler, et al. Imaging ex vivo healthy and pathological human brain tissue with ultra-high-resolution optical coherence tomography. *Journal of Biomedical Optics* 10:1–7 (2005).

16. M. E. Brezinski and J. G. Fujimoto. Optical coherence tomography: High-resolution imaging in nontransparent tissue. *IEEE Journal on Selected Topics in Quantum Electronics* 5:1185–1192 (1999).

17. A. F. Fercher, W. Drexler, C. K. Hitzenberger, and T. Lasser. Optical coherence tomography—Principles and applications. *Reports on Progress in Physics* 66:239–303 (2003).

18. P. H. Tomlins and R. K. Wang. Theory, developments and applications of optical coherence tomography. *Journal of Physics D: Applied Physics* 38:2519–2535 (2005).

19. M. W. Lindner, P. Andretzky, F. Kiesewetter, and G. Häusler. In *Handbook of optical coherence tomography,* ed. B. E. Bouma, and G. J. Tearney, 335–357. New York: Marcel Dekker, Inc. (2002).

20. A. F. Fercher, C. K. Hitzenberger, G. Kamp, and S. Y. El-Zaiat. Measurement of intraocular distances by backscattering spectral interferometry. *Optics Communications* 117:43–48 (1995).

21. S. H. Yun, and B. E. Bouma. Wavelength swept lasers. In *Optical coherence tomography: Technology and applications,* ed. W. Drexler, and J. G. Fujimoto, 359–377. New York: Springer (2008).

22. M. A. Choma, M. V. Sarunic, C. Yang, and J. A. Izatt. Sensitivity advantage of swept source and Fourier domain optical coherence tomography. *Optics Express* 11:2183–2189 (2003).

23. M. E. J. van Velthoven, D. J. Faber, F. D. Verbraak, T. G. van Leeuwen, and M. D. de Smet. Recent developments in optical coherence tomography for imaging the retina. *Progress in Retinal and Eye Research* 26:57–77 (2007).

24. A. Unterhuber, B. Povay, B. Hermann, H. Sattmann, A. Chavez-Pirson, and W. Drexler. In vivo retinal optical coherence tomography at 1040 nm—Enhanced penetration into the choroid. *Optics Express* 13:3252–3258 (2005).

25. G. Y. Liu, A. Mariampillai, B. A. Standish, N. R. Munce, X. Gu, and I. A. Vitkin. High power wavelength linearly swept mode locked fiber laser for OCT imaging. *Optics Express* 16:14095–14105 (2008).

26. W. Y. Oh, S. H. Yun, G. J. Tearney, and B. E. Bouma. 115 kHz tuning repetition rate ultrahigh-speed wavelength-swept semiconductor laser. *Optics Letters* 30:3159–3161 (2005).

27. R. Huber, M. Wojtkowski, J. G. Fujimoto, J. Y. Jiang, and A. E. Cable. Three-dimensional and C-mode OCT imaging with a compact, frequency swept laser source at 1300 nm. *Optics Express* 13:10523–10551 (2005).

28. V. X. D. Yang, M. L. Gordon, B. Qi, J. Pekar, S. Lo, E. Seng-Yue, A. Mok, B. C. Wilson, and I. A. Vitkin. High speed, wide velocity dynamic range Doppler optical coherence tomography (part I): System design, signal processing, and performance. *Optics Express* 11:794–809 (2003).

29. A. Mariampillai, B. A. Standish, E. H. Moriyama, M. Khurana, N. R. Munce, M. K. K. Leung, J. Jiang, et al. Speckle variance detection of microvasculature using swept-source optical coherence tomography. *Optics Letters* 33:1530–1532 (2008).

30. H. Li, B. A. Standish, A. Mariampillai, N. R. Munce, Y. Mao, S. Chiu, N. E. Marcon, B. C. Wilson, A. Vitkin, and V. X. D. Yang. Feasibility of interstitial Doppler optical coherence tomography for in vivo detection of microvascular changes during photodynamic therapy. *Lasers in Surgery and Medicine* 38:754–761 (2006).

31. Y. Wang, B. A. Bower, J. A. Izatt, O. Tan, and D. Huang. In vivo total retinal blood flow measurement by Fourier domain Doppler optical coherence tomography. *Journal of Biomedical Optics* 12:041215 (2007).

32. B. A. Standish, K. K. C. Lee, X. Jin, A. Mariampillai, N. R. Munce, M. F. G. Wood, B. C. Wilson, I. A. Vitkin, and V. X. D. Yang. Interstitial Doppler optical coherence tomography as a local tumor necrosis predictor in photodynamic therapy of prostatic carcinoma: An in vivo study. *Cancer Research* 68:9987–9995 (2008).

33. C. Kasai, and K. Namekawa. Real-time two-dimensional blood flow imaging using an autocorrelation technique. *Ultrasonics Symposium Proceedings,* pp. 953–958, (1985).

34. W. D. Barber, J. W. Eberhard, and S. G. Karr. A new time domain technique for velocity measurements using Doppler ultrasound. *IEEE Transactions on Biomedical Engineering* 32:213–229 (1985).

35. R. S. C. Cobbold. Pulsed methods for flow velocity estimation and imaging. In *Foundations of biomedical ultrasound,* ed. R. S. C. Cobbold. Oxford, England: Oxford University Press (2005).

36. C. J. Hartley, L. H. Michael, and M. L. Entman. Noninvasive measurement of ascending aortic blood velocity in mice. *American Journal of Physiology—Heart and Circulatory Physiology* 268:H499–H505 (1995).

37. A. C. Guyton, and J. E. Hall. Overview of the circulation; medical physics of pressure, flow, and resistance. In *Textbook of medical physiology,* 11th ed., ed. A. C. Guyton, and J. E. Hall, 161–170. New York: Elsevier (2006).

38. B. J. Vakoc, R. M. Lanning, J. A. Tyrrell, T. P. Padera, L. A. Bartlett, T. Stylianopoulos, L. L. Munn, et al. Three-dimensional microscopy of the tumor microenvironment in vivo using optical frequency domain imaging. *Nature Medicine* 15:1219–1223 (2009).

39. H. A. Collins, M. Khurana, E. H. Moriyama, A. Mariampillai, E. Dahlstedt, M. Balaz, M. K. Kuimova, et al. Blood-vessel closure using photosensitizers engineered for two-photon excitation. *Nature Photonics* 2:420–424 (2008).

40. A. Mariampillai, M. K. K. Leung, M. Jarvi, B. A. Standish, K. Lee, B. C. Wilson, A. Vitkin, and V. X. D. Yang. Optimized speckle variance OCT imaging of microvasculature. *Optics Letters* 35:1257–1259 (2010).

41. M. J. Suter, B. J. Vakoc, P. S. Yachimski, M. Shishkov, G. Y. Lauwers, M. Mino-Kenudson, B. E. Bouma, N. S. Nishioka, and G. J. Tearney. Comprehensive microscopy of the esophagus in human patients with optical frequency domain imaging. *Gastrointestinal Endoscopy* 68:745–753 (2008).

14

Three-Dimensional Endoscopic Surface Imaging Techniques

Jason Geng

CONTENTS

This chapter provides an overview of state-of-the-art three-dimensional (3D) endoscopic imaging technologies. Physical objects in the world are 3D, yet traditional endoscopes can only acquire two-dimensional (2D) images that lack depth information. This fundamental restriction greatly limits our ability to perceive and to measure quantitatively the complexity of real-world objects, as well as to understand the spatial relationship among them. In both medical imaging and industrial inspection applications, 3D surface imaging capability would add one more dimension, literally and figuratively, to the existing imaging technologies. Over the past decades, tremendous new technologies and methods have emerged in the 3D surface imaging field. We provide a classification of these technologies first and then describe each category in detail, with representative designs and examples. This overview would be useful to researchers in the field since it provides a snapshot of the current state of the art, from which subsequent research in meaningful directions is encouraged. This overview also contributes to the efficiency of research by preventing unnecessary duplication of already performed research.

14.1 Introduction

The term "endoscopy" is coined from the Greek words *endo,* meaning "inside," and *skopeein,* meaning "to see." Endoscopy is a broad term used to describe any examination of the inside of the body or a physical structure with the help of an endoscope. A flexible, rigid, or semirigid endoscope consists of a camera and an illuminator or set of illuminators at the tip or at the base. Transmission through flexible endoscopes is generally through fiber optic cables or video cables, which are connected to the host computer, controller, and light-generation mechanism. In general, endoscopic imaging systems use a fiber optic illumination bundle that is disposed side by side with the imaging sensor optics at the end of a small insertion tube. The miniature diameter of the insertion tube facilitates easy insertion into cavities of the human body or physical structure for viewing, acquiring images, and certain interventions (such as taking samples, making diagnoses, delivering drugs, etc.). Though generally there is uniformity to endoscope design, manufacturers do vary the specifications of instruments in any given category of endoscopes. Design variables include insertion tube length, diameter, and stiffness characteristics; imaging sensor resolution and signal-to-noise ratio; imaging optics configurations, quality, aberration, and distortion; instrument channel size and number; and configuration of the distal end of the insertion tube [1–4].

The physical world around us is three dimensional; yet, traditional video endoscopes are only able to acquire 2D images that lack depth information. This fundamental restriction has greatly limited our ability to perceive and to measure the complexity of real-world objects.

In medical endoscopic imaging applications, most existing endoscopes used in minimally invasive surgery (MIS) for diagnosis and treatment are monocular based; therefore, they provide only 2D images for in vivo visualization. The lack of depth perception inherent in the current endoscopic imaging technology significantly affects a surgeon's ability to determine the size, shape, and precise location of anatomical structures and to maneuver safely and efficiently in a confined space, thus often impairing the efficiency and outcomes of diagnoses and operations.

In industrial inspection applications, lack of quantitative 3D measurement capabilities of existing endoscopic instruments limits the effectiveness and efficiency of nondestructive inspection tasks. Such systems are often used to inspect inaccessible locations for damage or wear that exceeds an operational limit or to check whether a manufactured part or assembly meets its specifications. It would be desirable to produce a 3D model or surface map for comparison to a reference, 3D viewing, reverse engineering, or detailed surface analysis. However, traditional endoscopic devices are not designed to perform such tasks.

14.1.1 Extant Endoscope Technologies and Limits

Medical endoscopes, originated in the early nineteenth century, are optical instruments used for viewing internal organs through the human body's natural openings (mouth, ear, throat, rectum, etc.) or through a small incision in the skin. Classical rigid endoscopes have a number of periscopic and field lenses in order to convey the image from the distal end to the eyepiece or image sensor. In some cases, as many as 40–50 lenses may be used that can cause considerable optical aberration, light loss, and ghost images. These instruments' rigid bodies greatly limit their ability to access deeper sites of body lumina [1–4].

In 1957, Hirschowitz, at the University of Michigan, invented the first flexible endoscope using a bundle of precisely aligned (i.e., coherent) flexible optical fibers to transmit images. Flexible fiber endoscopes permit visualization of normally inaccessible areas (by rigid endoscopes) within the body without discomfort for patients. Endoscopes help medical procedures to be less invasive, thereby reducing tissue trauma, risk of complications, operation costs, and recovery times. By allowing physicians to see and operate inside a human body, the invention of endoscopes forever changed the face of medical practice.

Conventional endoscopes use either bundles of coherent optical fibers to transmit a 2D image (we herein refer to these as fiber bundle endoscopes) or solid-state, charge-coupled device (CCD) or complementary metal oxide semiconductor (CMOS) sensors at the distal end for superior image quality (we herein refer to these as video endoscopes).

For fiber bundle endoscopes, existing ultrathin endoscopes contain coherent bundles with 2,000 to 30,000 fibers and a millimeter diameter. They can image internal organs via very small channels. However, achieving good image quality for fiber bundle endoscopes with millimeter diameters is very challenging, due to these drawbacks:

- *Limited room for further miniaturization:* Each optical fiber, including its cladding, has a finite diameter. Only a limited number of optical fibers (e.g., 10,000 ~ 30,000 pixels) can be packed into a very confined space with ultrasmall diameter. Each fiber has a minimum diameter of 4 µm. This means that a minimum possible diameter of a fiber bundle scope with 800 × 600 pixels is at least 3 mm.
- *Poor image quality:* The cladding itself is also a problem, as it does not transmit image data and typically results in a honeycomb pattern superimposed on the image (Figure 14.3).
- *Expensive:* The manufacturing process of high-resolution miniature fiber bundles is complex and error prone, making them very expensive devices.
- *Fragile/not highly flexible:* Bundles with a high density of fibers and clad materials are fragile and not very flexible.
- *Lack of 3D imaging capability:* More importantly, none of the existing fiber bundle endoscopes provides 3D information, which is crucial for diagnosis and is particularly useful when conducting surgical operations in confined spaces that can only be accessed by small instruments.

Compared with fiber bundle endoscopes, video endoscopes have advantages in many aspects: The image resolution and quality are much higher and video endoscopes can achieve much better flexibility, durability, and lower costs than fiber bundle endoscopes. However, for traditional video endoscopes, no real-time, full-field, high-resolution 3D measurement capability is available.

14.1.2 Traditional Methods of Measuring the Sizes of Targets Using an Endoscope

One traditional method of measuring the size of a target object under endoscope observation is to rely on rough comparisons of the object size with an object with known dimensions from the image on the video viewing. The measurement results are often subjective and error prone, depending heavily on the operator's experience and proper placement of the comparison object.

Another traditional measurement method used by endoscopists relies on providing a physical stand-off over the lens on the end of the probe insertion tube, at which point the magnification is known and the end of the probe is adjusted until it just touches the object to be viewed at the stand-off. With this known magnification, the image can be measured on the screen and the precise size determined [6].

The past several decades have marked tremendous advances in research, development, and commercialization of 3D surface imaging technologies, stimulated by application demands in a variety of market segments, advances in high-resolution and high-speed electronic imaging sensors, and ever increasing computational power. However, the unique features of endoscopic imaging make it a challenging task to apply the 3D surface imaging techniques directly to the endoscopic imaging instruments:

- The miniature size of components makes it extremely difficult and costly to fabricate optical electronic and structural components.
- A narrow baseline due to the small diameter of endoscopes often limits the accuracy of 3D surface reconstruction.

14.1.3 Organization of This Chapter

In this chapter, we provide an overview of recent advances in 3D endoscopic surface imaging technologies. The chapter is organized as follows: Section 14.2 provides a classification framework of various 3D surface imaging technologies available. Section 14.3 focuses on dual sensor stereo techniques. Section 14.4 describes a few representative single sensor stereo techniques.

Both dual sensor and single sensor stereo techniques are 3D surface imaging schemes based on a pair of stereo images of a target to obtain a 3D surface profile of the target. Section 14.5 describes several 3D surface imaging techniques based on structured light that use one sensor and a miniature structured-light projector to obtain 3D surface profiles. If the texture of the target surface is not prominent, the structured-light techniques can often obtain better 3D imaging results. The shape from video (SfV) is another group of 3D modeling techniques that can be applied to the endoscopic 3D imaging. Shape from video is an important aspect and warrants a serious discussion. However, due to space limits, we will leave this topic to a separate article.

Over the past decades, tremendous new technologies and methods have emerged in the 3D surface imaging field. This overview provides a snapshot of the current state of the art, from which subsequent research in meaningful directions is encouraged. This overview also contributes to the efficiency of research by preventing unnecessary duplication of already performed research.

14.2 Classification of 3D Endoscopic Surface Imaging Techniques

The term "3D imaging" refers to techniques that are able to acquire true 3D data (i.e., values of some property of a 3D object, such as the distribution of density) as a function of the 3D coordinates (x, y, z). Examples from the medical imaging field are computed tomography (CT) and magnetic resonance imaging (MRI), which acquire volumetric pixels (or voxels) of the measured target, including its internal structure.

By contrast, surface imaging deals with measurement of the (x,y,z) coordinates of points on the surface of an object. Since the surface is, in general, nonplanar, it is described in a 3D space, and the imaging is called 3D surface imaging. The result of the measurement may be regarded as a map of the depth (or range), z, as a function of the position (x, y) in a Cartesian coordinate system, and it may be expressed in the digital matrix form $\{z_{ij} = (x_i, y_j), i = 1, 2, ..., L, j = 1, 2, ..., M\}$. This process is also referred to as 3D surface measurement, range finding, range sensing, depth mapping, surface scanning, etc. These terms are used in different application fields and usually refer to loosely equivalent basic surface imaging functionality, differing only in details of system design, implementation, and/or data formats [7].

Numerous techniques for 3D endoscopic surface imaging are currently available. In this chapter, we first classify these techniques into four categories: dual chips at the tip, single chip–dual optical channels, structured light, and video-to-3D modeling techniques, as illustrated schematically in Figure 14.1. Once 3D images are obtained, comprehensive 3D display technology has to be developed and utilized to provide users an effective visualization tool to maximize the utilities of 3D image data. This classification

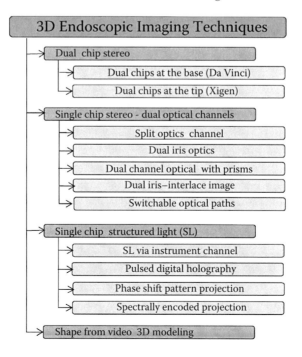

FIGURE 14.1
Classification of 3D endoscopic surface imaging techniques.

framework may serve as a road map for reviewing the 3D endoscopic surface imaging technology.

It would be an impossible task to cover all possible 3D surface imaging techniques in this chapter. Instead, we select a few representative techniques to discuss in order to gain perspective of the entire field and to understand fundamental technical principles and typical system characteristics.

14.3 3D Acquisition Using Dual Sensor Chips

This 3D image acquisition technique is based on a single pair of imaging sensors to acquire binocular stereo images of the target scene in a manner similar to human binocular vision, thus providing the ability to capture 3D information of the target surface (Figure 14.2). The geometric relationship between two image sensors and an object surface point P can be expressed by the triangulation principle as

$$R = B\frac{\sin\beta}{\sin(\alpha+\beta)}$$

where B is the baseline between the two image sensors and R is the distance between the optical center of an image sensor and the surface point P. The (x,y,z) coordinate values of the target point P can then be calculated precisely based on the R, α, β, and geometric parameters.

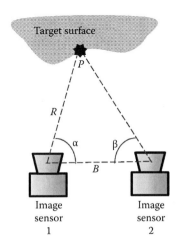

FIGURE 14.2
Concept of stereo 3D surface imaging.

The image sensor may be located at the proximal end of the probe, as with an optical rigid borescope or fiberscope, or at the distal end, as with a video borescope or endoscope. There are essentially two basic design configurations for the dual sensor chip 3D endoscopes:

1. Place two sensor chips at the tip: If the size of sensor chips is sufficiently small, they can be packaged at the tip (distal end) of the endoscope housing. This is an ideal configuration if the performance of small chips meets the 3D imaging requirements in resolution, frame rate, signal-to-noise ratio, etc. Since optics are placed at the tip, the endoscope housing can be made flexible, providing tremendous advantages in miniature flexible endoscope product design and applications.

2. Place two sensor chips at the base: Some high-performance sensor chips often have larger sizes that are not suitable to be placed at the tip, due to restriction of endoscope diameter. In these cases, chips can be placed at the base, and relay optics can be used to lead the acquired optical energy to the sensor chip. However, such system designs usually require rigid outer housings to host the optical train assembly, so this is suited for laparoscopes with rigid housing and shorter length.

We now discuss two representative designs of dual sensor chip endoscopes.

14.3.1 Dual Sensor Chips at the Base

A number of stereo endoscope products adopt the dual sensor chip at the base design configuration. One of the most prominent state-of-the-art medical devices that use stereo endoscopes is the Da Vinci surgical robot, including the one used by the Da Vinci robotic surgery system (Intuitive Surgical System, www.intuitivesurgical.com [8]). It consists of a surgeon's console containing two master manipulators, a patient side cart with up to four robotic arms—three for the slave instrument manipulators, which can be equipped with removable instruments, and a stereo endoscope camera manipulator connected to a high-performance stereo vision system. The endoscopic vision system of the Da Vinci robot delivers high-resolution video to the operator located at the surgeon console. This provides surgeons with visual sharpness that is greater than anything previously available. As the surgeon operates, the Da Vinci vision system displays high-definition video in stereo for true perception of depth. The immersive quality of the 3D vision provides a virtual extension of the surgeon's hands and eyes into the patient's body.

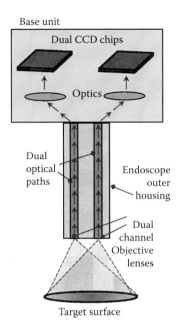

FIGURE 14.3
Stereo endoscope on Da Vinci robotic surgery system (www/intuitivesurgical.com) [8].

However, stereo cameras or endoscopes are only able to offer stereo image pairs; no quantitative 3D measurement and visualization can be performed without advanced 3D reconstruction algorithms. Existing stereo endoscopes/laparoscopes only provide in vivo stereo viewing ability to their operators. Lack of sophisticated image processing algorithms and software for 3D surface image reconstruction prevents these devices from providing in vivo quantitative 3D visualization, 3D measurement (sizing), and 3D registration capability.

Figure 14.3 illustrates the 3D image acquisition mechanism of this design. There are dual optical paths built in on the body of the endoscope, where the optical energy collected by the dual channel objective lenses is relayed by the dual optical paths toward the back end of the endoscope housing. There is a set of specially designed prisms and optical lenses that adjust the dual optical paths such that the large image sensor chips can be hosted at the base unit to receive the collected optical energy for generating a pair of stereo images.

U.S. Patent 5,860,912 [9] shows another example of the dual chips at the base design configuration. Two independent optical channels are used to acquire stereo images and relay them back to the dual image sensor chips. Figure 14.4 provides an example of optical design of a dual chip at the base

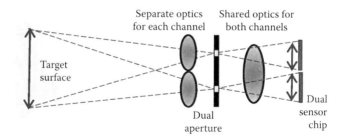

FIGURE 14.4
Example of dual chips at the base optical design: Olympus stereo endoscope (U.S. Patent 5,743,846).

stereo endoscope, designed by Olympus (U.S. Patent 5,743,846) [11]). Note that this design employs dual irises and, in some portions, dual optical trains to facilitate the stereo image optical paths. On the rear portion of the optical train, both channels share the same set of optics. The footprints of stereo channels fall onto two sensor chips, respectively. If the size of sensor chip is large, the optical path needs to be split into two to provide sufficient space for sensor chip installation.

14.3.2 Dual Sensor Chips at the Tip

The recent advance of miniature CMOS sensor modules makes it feasible to build miniature stereo endoscopes where dual sensor modules are placed at the distal end of the endoscope while still keeping the diameter of the stereo endoscope minimal.

Xigen LLC, Gaithersburg, Maryland, has recently developed an ultraminiature flexible stereo endoscope based on the latest advances in miniature CMOS imaging sensor technology [13] [23–28]. Figure 14.5 shows a prototype of an ultraminiature stereo endoscope made by Xigen with an outer diameter of only $\phi = 2.8$ mm. It has a pair of miniature CMOS sensors, each of which is 1 mm × 1 mm × 1.5 mm in physical dimension. The entire endoscope head consists of a stereo pair of CMOS sensors, fiber illumination means, and the outer housing. To the best of our knowledge, this is the smallest stereo endoscope available to date.

14.3.2.1 Unique Features of the Ultrathin 3D Stereo Endoscope

The 3D stereo endoscope technology has several unique capabilities for medical diagnosis, intervention procedures, and minimally invasive surgeries:

- It provides real-time stereo 3D image visualization of the surgical site to surgeons, helping them gain depth sensation on manipulation of instruments in surgery.

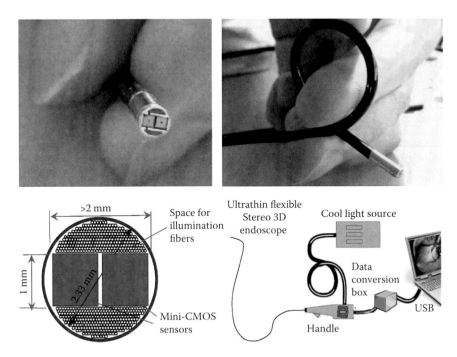

FIGURE 14.5
Xigen's ultraminiature flexible stereo endoscope (j = 2.8 mm). (Geng, J. 2010. Novel ultra-miniature flexible videoscope for on-orbit NDE. Technical report.)

- It provides quantitative 3D measurement (sizing) of targets and surface profile.
- It can be built with a diameter of ~2.8 mm.
- Three-dimensional images provide quantitative measurement and sizing capability for surgeons during operations.
- Three-dimensional images acquired by the miniature 3D stereoscope provide 3D visual feedback for surgeons during manipulation, positioning, and operation.
- The acquired 3D images facilitate real-time true 3D display for surgery.
- The 3D surface data of a surgical site can be generated continuously in real time, or in snapshot fashion triggered by the operator. The 3D surface data enable image-guided interventions by providing self-positioning capability for surgical instruments, via real-time 3D registration of preoperative images (CT, MRI, etc.) and intraoperative (endoscopic) images.
- Flexibility of the cable allows for dexterous control of viewing direction.

14.3.2.2 Minimal Diameter for Reducing Skin Damage

Reducing the diameter could reduce the traumatic damage to the skin or tissue and eliminate surgical scars. The clinical feedback we received suggested that if the diameter of endoscopes is smaller than 2 ~ 3 mm, it would act like a needle and cause minimal damage to skin and tissues and the surgical scar can be totally eliminated. We therefore should design our stereo 3D endoscope with a diameter smaller than 3 mm. The overall diameter of the NextGen 3D design depends on the separation distance (i.e., the baseline) between two sensor modules. The minimal diameter is achieved by leaving no separation space between the two sensors, resulting in a 3D endoscope 2.33 mm in diameter.

14.3.2.3 The Ultrathin 3D Stereo Endoscope: Real-Time 3D Visual Feedback

The real-time 3D imaging capability of the ultraminiature 3D stereo addresses the clinical needs of real-time 3D video in diagnosis, treatment, and MIS for navigation, manipulation, rough sizing, etc. These uses of 3D images often do not need highly accurate 3D surface profile data. Instead, a rough estimate of distance, proximity, and/or direction would suffice for the initial decision making. One of the most important requirements in these applications is the real-time 3D video capability. Clinicians need real-time visual feedback in 3D in order to perform dexterous maneuvers in a speedy and safe fashion.

14.3.2.4 Overall System Configuration

The overall system design of the proposed ultrathin 3D stereo endoscope is shown in Figure 14.5. This novel ultrathin 3D stereo endoscope is able to

- Provide a pair of stereo images
- with a resolution of >250 × 250 pixels in video image
- at 44 frames per second acquisition rate
- with an ultrathin outer diameter of ~2.8 mm
- and integrated sensor chip and optics design, eliminating the need to add components at the tip

14.4 3D Acquisition Using a Single Chip with Dual Optical Channels

14.4.1 Single Chips with Split Channel Optics

Instead of using two sensor chips to acquire a stereo image pair, a number of designs split the sensing area of a single sense chip into two portions by using specially designed split channel optics and acquire a stereo image pair

Split optics attachment

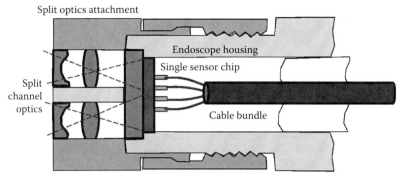

(a) Olympus Stereo Endoscope (U.S. 7,443,488)

Dual iris optics attachment

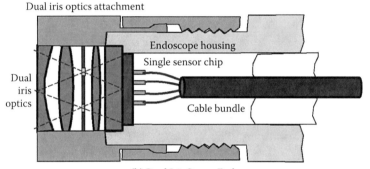

(b) Dual Iris Stereo Endoscope

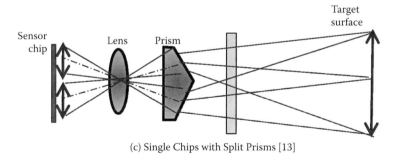

(c) Single Chips with Split Prisms [13]

FIGURE 14.6
Dual iris stereo endoscope.

with reduced image resolution (in comparison with the original resolution of the sensor chip). Figure 14.6(a) shows a split channel optics stereo image acquisition system designed by Olympus (US Patent 7,443,488) [14]. This design facilitates a set of interchangeable optical attachments, offering users multiple options to have different fields of view (FOV) and viewing directions (forward-looking or side-viewing angles). One unit of the attachment set is the split channel optics. It consists of two identical sets of optics, each of which has an effective image footprint on half the area of the single sensor chip. In other words, the entire sensing area of a single sensor chip is split into two halves, acquiring two images simultaneously of a stereo image pair.

The advantages of using split channel optics to acquire stereo images include:

1. No significant changes are made on electronics and sensors.
2. Acquisition of both images in a stereo pair is simultaneous; no time delay problem may occur as on other time-sequential acquisition methods.
3. The potential problem of multichip acquisition with potential discrepancies in image brightness, focus, contrast, etc., is eliminated.

The potential drawbacks of this type of design include:

1. Complex design
2. Difficulty in aligning the optics to have proper image footprints
3. Due to necessity of ensuring sufficient margins between two image footprints on the single chip, frequent occurrences of a black band that may render a significant portion of the sensing area useless
4. Size of the single endoscope outer housing limiting the baseline of the stereo image acquisition

With the small size of endoscopes, this type of design may result in a "weak 3D" effect, meaning that the 3D effect is not quite strong due to the small separation distance between two imaging channels.

14.4.2 Single Chips with Dual Iris Optics

A logic extension of the split channel optics design concept is to employ a dual iris on single channel optics. Figure 14.6(b) illustrates the concept. The design is very similar to a monocular endoscope design, except that there is a specially designed iris diaphragm that has two irises (instead of one) on it. In a normal optical system, replacing a single iris diaphragm with a dual iris diaphragm may result in an image blur because slightly different images coming from different irises may reach the same pixel location, reducing the sharpness of images. However, the optics of a dual iris system are designed such that the footprints of images coming from different irises located in different portions of the sensing area on the single chip, thus facilitating the simultaneous stereo image acquisition using a single chip and single optical channel.

14.4.3 Single Chips with Split Channel Optics Using Prisms

There are many different ways to design split channel optics. For example, one design uses two prisms for left and right sides of a sensing area on the single chip (U.S. Patent application 2005/0141088 [15]). The apparatus comprises three periscope prisms, so the viewing areas of the left and right images are side by side with correct orientation. The optical paths of the left and right channels are separated via the prisms.

Another straightforward way to split the optical channel using a simple prism is shown in Figure 14.6(c) [13]. The refraction of tilted prism surfaces bends optical rays into two optical paths. With careful optical design, the footprints of these two optical paths on the image sensor chip could be separated into two regions, forming a stereo image pair.

14.4.4 Single Chips with Dual Irises and Interlaced Optics

An Israeli company, VisionSense, has introduced a single chip stereo endoscope product, 6.5 mm in diameter, for performing sinus surgeries [16]. The idea of a stereoscopic camera based on a lenticular array was first proposed by the French physicist G. Lippmann. The VisionSense adaptation of this stereoscopic plenoptic camera is shown in Figure 14.7. The 3D object is imaged by a single set of optics with two apertures. This setup generates a telecentric objective in which all the rays passing the center of each aperture emerge as a parallel beam behind the lens. The sensor chip is covered by a sheet of lenticular array—an array of cylindrical microlenses with zero power in the horizontal axis. Each lenticule covers exactly two pixel columns. Rays that pass through a point at the left aperture are emitted as a parallel beam marked in red color. These rays are focused by the lenticular array on the pixels on the right side under the lenslets. Similarly, rays that pass through the

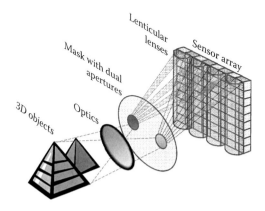

FIGURE 14.7
(See color insert.) VisionSense stereoscopic camera: The camera has a mask with two apertures, a lenticular microlens array in front of the sensor chip. Each lenslet is 2 pixels wide. Rays from the 3D object that pass through the left aperture generate a left view on columns marked with a red color, and rays that pass through the right aperture generate a right view image on neighboring columns marked with a green color. A pair of stereoscopic images is thus generated using a single sensor chip. (Yaron, A. et al. www.visionsense.com.)

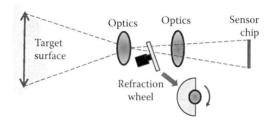

FIGURE 14.8
Single chips with switchable optical paths. (Lee, Y. H. U.S. Patent application 12162482.)

right aperture are focused by the lenslets on the left pixel columns marked in green color.

Thus, a point on the 3D object is imaged twice: once through the left aperture generating an image on the pixel in a red color, and once through the right aperture generating an image on the pixel in a green color. The pair of pixels form a stereoscopic view of the point in the scene. The left and right images can be acquired simultaneously. The distance between a pixel of the left view to that of the right view (disparity) is a function of the distance of the corresponding point from the camera, from which the 3D range can be calculated.

14.4.5 Single Chips with Switchable Optical Paths

U.S. Patent application 2009/0040606 discloses an interesting design of a single chip stereo endoscope [17]. This design (Figure 14.8) utilizes a single optical channel to collect an image from the target and a single image sensor to produce a dynamic stereo image in sequential (even/odd frame) fashion. It places a rotating, transparent half-plate at the entrance pupil position. When the rotating plate is in the position where the transparent plate is not in the optical path of the imaging system, the image sensor acquires a "left" image. When the transparent plate is in the optical path of the imaging system, the optical path is shifted by the refraction of the transparent plate, and the "right" image acquired by the image sensor has parallax disparity from the left image. The rotating angle of the transparent plate is synchronized with the timing of the acquisition frame of the image sensor such that the left and right images of the stereo image pair are acquired in sequential fashion. Using one image sensor, this system is claimed to eliminate the possible dizziness provoked by disparities in focus, zoom, and alignment between two separate images from two sensors.

14.5 3D Acquisition Using a Single Chip with Structured Light

Stereo endoscopes are able to produce stereoscopic views of the target surface for viewing and display. If quantitative 3D (x,y,z) coordinate data of surface points

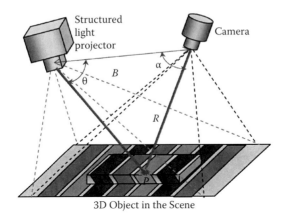

3D Object in the Scene

FIGURE 14.9
Illustration of structured light 3D surface imaging principle. (J. Geng, J. 2011. *Advances in Optics and Photonics* 3:128–160. With permission.)

are needed, stereo images can only provide information where two points on the image can be correlated. This can be problematic when little surface texture exists on the target. The correlation process can also require significant processing, so producing a full 3D surface map can be time consuming. It is more typical to correlate only a small number of points needed for basic measurements.

Laser scanners are a popular approach to acquiring 3D surface profile data from a large object. In a typical commercial 3D laser scanner system, a single profile line is controlled to scan across the entire target surface. The data acquired during the scanning process is used to build a 3D profile model of the target surface. Due to the use of a scanning component, it is generally not practical to adopt a similar approach in an endoscope imaging system because of limitations on the miniature dimension of the entire probe size.

One practical method of full-field 3D surface imaging for endoscopes is based on the use of "structured light" (i.e., active illumination of the scene with a specially designed 2D spatially varying intensity pattern) [7]. As illustrated in Figure 14.9, a spatially varying 2D structured illumination is generated by a special projector or a light source modulated by a spatial light modulator. The intensity of each pixel on the structured-light pattern is represented by the digital signal $\{I_{ij} = (i, j), i = 1, 2, …, I, j = 1, 2, …, J\}$, where (i, j) represents the (x,y) coordinate of projected pattern. The structured-light projection patterns discussed herein are 2D patterns.

An imaging sensor (a video camera, for example) is used to acquire a 2D image of the scene under the structured-light illumination. If the scene is a planar surface without any 3D surface variation, the pattern shown in the acquired image is similar to that of the projected structured-light pattern. However, when the surface in the scene is nonplanar, the geometric shape of the surface "distorts" the projected structured-light pattern as seen from the camera. The principle of structured-light 3D surface imaging techniques is to extract the 3D surface shape based on the information from the "distortion" of the projected structured-light pattern. Accurate 3D surface profiles of objects in the scene can be computed using various structured-light principles and algorithms.

As shown in Figure 14.9, the geometric relationship between an imaging sensor, a structured-light projector, and an object surface point can be expressed by the triangulation principle as

$$R = B\frac{\sin(\theta)}{\sin(\alpha + \theta)}$$

The key for triangulation-based 3D imaging is the technique used to differentiate a single projected light spot from the acquired image under a 2D projection pattern. Various schemes have been proposed for this purpose, and this chapter will provide an overview of various methods based on the structured-light illumination.

Numerous techniques for surface imaging by structured light are currently available. However, due to the miniature size and complexity of optical/structural and electronic designs, there is little published literature on using structured light on 3D endoscopic image acquisition.

14.5.1 Structured Light via Instrument Channel

A relatively easy approach to implement 3D surface endoscopic imaging by applying the structured-light technique without building a special endoscope is to insert a miniature structured-light projector via the instrument channel of a regular endoscope. Schubert and Muller [18] presented a structured-light method for bronchoscopy that can be used for measurement of hollow biological organs. For this purpose, a fiber probe projecting a ring of laser light (laser diode, 675 nm) is inserted into the instrument channel. There is a cone-shaped right angle reflector placed at the end of the fiber such that the fiber light source generates a side-view "sheet of light" around it, forming a "ring of light" on the lumen surface. Using a position registration device, the distance between the tip of the fiber light source and the tip of the camera lens is fixed and known. The 3D profile of target surface is estimated by the deformation of a projected ring of laser light. By viewing the location of the light ring on the image, the diameter of the light ring can be calculated using simple triangulation. By moving the endoscope through the entire length of the lumen section, a complete 3D surface profile of the lumen structure can be obtained. This method enhances the 3D grasp of endoscopically examined lesions.

This method is able to generate only a "ring" of 3D data at one time. To produce 3D surface data, the endoscope has to be moved and its position must be known in order to register a stack of rings into a complete surface profile.

14.5.2 Endoscopic Pulsed Digital Holography for 3D Measurements

Saucedo et al. [19] and Kolenovic et al. [20] have examined the deformation of a 3D scene with an electronic speckle pattern interferometer. A stiff, minimally invasive endoscope is extended with an external holographic camera. Saucedo et al. [19] extended this approach and developed an endoscopic pulsed digital holography approach to endoscopic 3D measurements. A rigid endoscope and three different object illumination source positions are used in pulsed digital holography to measure the three orthogonal displacement components from hidden areas of a harmonically vibrating metallic cylinder. In order to obtain simultaneous 3D information from the optical setup, it is necessary to match the optical paths of each of the reference object beam pairs and to incoherently mismatch the three reference object beam pairs, such that three pulsed digital holograms are incoherently recorded within a single frame of the CCD sensor. The phase difference is obtained using the Fourier method and by subtracting two digital holograms captured for two different object positions.

In the experiment, the high-resolution CCD sensor (1024 × 1024, pixel elements, with an area of $\Delta x \times \Delta x = 9$ μm^2) receives the image. The depth of field of the endoscope is from 3 to 12 mm and the stand-off distance from the endoscope edge to the object surface may vary from 5 to 25 mm. Pulses from a Nd:YAG laser emitting at 532 nm are divided initially into two beams, one that serves as reference and the other as object. Each beam is further divided into three reference and object beams by means of beam splitters. Each object beam is made to converge with a lens into a multimode optical fiber, and the lens-fiber set is attached to a sliding mechanical component that is later used to adjust the optical path length, like the one introduced by the distance from the endoscope edge to the object surface. The remaining part of the fiber length is attached to the endoscope so that all three object beams illuminate the surface area from three different positions by means of a mechanical support at the end of each fiber. Finally, the three reference beams are sent to the CCD sensor using single mode optical fibers.

The target object in the experiment is a metallic cylinder of 13 mm diameter, 19 mm height, and 0.3 mm width. The cylinder is tightly fixed to a mechanical support so that the imaginary origin of the rectangular coordinate system rests always on its center. The cylinder is harmonically excited using a mechanical shaker on a point perpendicular to the optical axis. The excitation frequencies are scanned so that the resonant modes are identified and one is chosen at 2180 Hz, with the laser pulses fired at a separation of 80 μs so that two object positions are acquired during a vibration cycle. Each pulse captures three incoherent digital holograms and their individual phase difference is obtained by simply subtracting the phase information from one pulse to the other. The displacement information may now be calculated.

14.5.3 Phase Shift Structured Light

Phase shift is a well-known fringe projection method for 3D surface imaging. A set of sinusoidal patterns is projected onto the object surface. The intensities for each pixel (x, y) of the three projected fringe patterns are described as

$$I_1(x, y) = I_0(x, y) + I_{mod}(x, y)\cos\big(\phi(x, y) - \theta\big)$$

$$I_2(x, y) = I_0(x, y) + I_{mod}(x, y)\cos\big(\phi(x, y)\big)$$

$$I_3(x, y) = I_0(x, y) + I_{mod}(x, y)\cos\big(\phi(x, y) + \theta\big)$$

where
$I_1(x, y), I_2(x, y)$, and $I_3(x, y)$ are the intensities of three fringe patterns
$I_0(x, y)$ is the DC component (background)
$I_{mod}(x, y)$ is the modulation signal amplitude
$\phi(x,y)$ is the phase
θ is the constant phase shift angle

Since there are multiple lines in the acquired images, which line is which (or absolute phase) must be determined in order to calculate an accurate 3D surface profile. The absolute phase at a given point in the image is defined as the total phase difference (2π times the number of line periods) between a reference point in the projected line pattern and the given point. Phase unwrapping is the process that converts the wrapped phase to the absolute phase. The phase information can be retrieved (i.e., unwrapped) from the intensities in the three fringe patterns:

$$\phi'(x, y) = \arctan\left[\sqrt{3}\,\frac{I_1(x, y) - I_3(x, y)}{2I_2(x, y) - I_1(x, y) - I_3(x, y)}\right]$$

The discontinuity of arc tangent function at 2π can be removed by adding or subtracting multiples of 2π on the $\phi'(x, y)$ value:

$$\phi(x, y) = \phi'(x, y) + 2k\pi$$

where k is an integer representing projection period.
Note that unwrapping methods only provide a relative unwrapping and do not solve for the absolute phase. The 3D (x,y,z) coordinates can be calculated based on the difference between measured phase $\phi(x, y)$ and the phase value from a reference plane [7]. In a simple case,

$$\frac{Z}{L-Z} = \frac{d}{B}, \quad \text{or} \quad Z = \frac{L-Z}{B} d$$

Simplifying the relationship leads to

$$Z \approx \frac{L}{B} d \propto \frac{L}{B}(\phi - \phi_0)$$

Phase-shifting methods have not been practical for use in miniature imaging devices such as borescopes and endoscopes. The components required to produce suitable line pattern projections for phase-shifting methods typically comprise a projector, scanner, piezomirror, etc. Among other things, miniature moving parts are extremely difficult to fabricate and maintain reliably in long-term operation, and the size limitations of miniature probes on endoscopes make the use of these components structurally challenging.

Armbruster and Scheffler [1] developed a prototype of an endoscope that measures the 3D scene by deformation of several projected lines. The accuracy of this structured-light method is increased by a phase shift of the lines. U.S. Patent 7,821,649 B2 [21] proposed a phase shift pattern projection mechanism, which involves no moving parts, that can be built in miniature size.

The illumination device on an endoscopic probe comprises multiple independently controllable light emitters and a fringe pattern mask, as shown in Figure 14.10. When one of the emitters is on, it generates a fringe pattern on the surface of the 3D object in the scene. When another emitter is on, it also generates fringe patterns on the 3D object, but with an offset in the horizontal

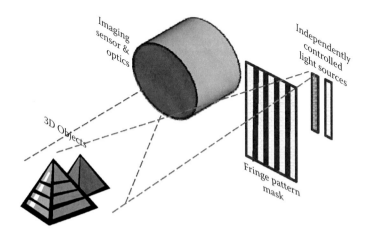

FIGURE 14.10
U.S. Patent 7,821,649 B2 proposed a phase shift pattern projection mechanism that involves no moving parts and miniature size. (Bendall, C. A. et al. U.S. Patent 7,821,649 B2. GE Inspection Technologies.)

position. This mechanism facilitates generation of sequential fringe projection patterns with controllable offset on horizontal positions and thus can be used to produce the phase shift projection patterns for structured-light-based 3D surface imaging.

Each of the fringe sets comprises a structured-light pattern that is projected when one emitter group of light emitters is emitting. The probe further comprises an image sensor for obtaining an image of the surface and a processing unit that is configured to perform phase shift analysis on the image. An imaging sensor is placed on the side of the fringe pattern generator with known geometric parameters with respect to the location of fringe pattern generator. The 3D surface image can be calculated using the phase shift principle.

14.5.4 Spectrally Encoded Structured Light

Researchers at Imperial College London, UK, developed a structured light scheme that uses a flexible fiber bundle to generate multispectral structured illumination that labels each projected point with a specific wavelength. Figure 14.11 illustrates the principle of the spectral illumination generation. A supercontinuum laser is used as a "white light" source. In the visible spectral range (400 ~ 700 nm), variation of the emission spectrum of the supercontinuum laser source is within a reasonable range and can be used as a broadband light source. The broadband laser light from the supercontinuum is dispersed by a prism. The rainbow spectrum light is then coupled into the end of a linear array of a fiber bundle, which consists of 127 cores with 50 μm in diameter. The linear array of the fiber bundle then forms a flexible noncoherent fiber bundle of the required length (up to several meters) to pass the spectral light illumination to the distal end of the probe.

At the distal end, the fibers with different spectral colors are randomly placed. The projected color pattern is a magnified image of the end face of the bundle formed by the projection lens. The identification and segmentation of each light spot is based on recovery of the peak wavelength of each spot, using knowledge of the relationship between the red–green–blue color space and the wavelength of the light.

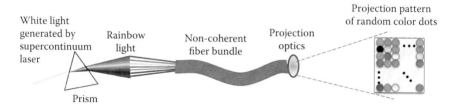

FIGURE 14.11
(See color insert.) Spectrally encoded fiber-based structured lighting probe. (Clancy, N. T. et al. *Biomedical Optics Express* 2:3119–3128. With permission.)

Using the spectrally encoded fiber bundle as the structured light and a video camera (CCD/CMOS) as imaging sensor, a 3D endoscope can be constructed to acquire real-time 3D surface images. Reconstructed 3D data for ovine kidney, porcine liver, and fatty tissue have varying physical features (convex/concave curve, "step" discontinuity, and miscellaneous "fine" structure). The reconstructed surface profiles were observed to match the observed surface profiles of the tissues well, which can be observed in the color photographs presented. These tests demonstrate that the technique may be applied on ex vivo tissue, while the validation with the test objects demonstrates the accuracy that may be obtained.

14.6 Future Directions of Endoscopic 3D Imaging

Three-dimensional endoscopic surface imaging is an exciting new frontier for 3D surface imaging technology development. Numerous methods and approaches have been proposed and tested in laboratories. However, tremendous work still has to be done in order to bridge 3D imaging technology to the practical applications of medical diagnosis and treatment and industrial inspection.

There are several challenges in the future development of 3D surface imaging endoscopes:

1. *Miniaturization of instruments:* Applications always demand smaller, faster, more accurate, and cheaper instrument tools. In terms of size (diameter), existing endoscope products have an array of options ranging from 4 to 15 mm. The endoscpes with these sizes may suffice most application needs. However, there are certain applications for which smaller endoscopes would be ideal—for accessing smaller holes in industrial inspection or for minimizing damage of skin for minimally invasive surgery.

2. *Lack of consistent and reliable surface features for performing 3D reconstruction:* Biological tissues and lumen surfaces are notoriously difficult for performing feature extraction, correlation, and 3D reconstruction, due to lack of consistent and reliable surface features.

3. *Trade-off between miniature size and required baseline for triangulation:* As we know, to achieve a certain accuracy level, 3D surface imaging techniques relying on triangulation require sufficient length of baseline separation. On the other hand, the miniature size of endoscopes places a physical limitation on the length of the baseline. One of the challenges in 3D endoscopes is the trade-off between smaller diameter and sufficient baseline.

4. *Go beyond the surface imaging:* Some applications in both industrial inspection and medical imaging demand subsurface or even volumetric 3D imaging capability. New technologies will have to be developed in order to extend the capability of current 3D surface imaging endoscopes into these new territories.

5. *Seamless integration with existing 2D endoscopes:* Most users of endoscopes prefer a 3D endoscope that can perform all functions of existing 2D endoscopes and, upon demand, perform 3D surface imaging functions. In practical applications, it is unreasonable to ask a user to use a separate 3D endoscope when he or she needs to acquire 3D surface images. The seamless integration of a 3D surface imaging capability into existing 2D endoscopes is important.

Although exciting progress has been made in terms of developing a variety of 3D imaging methods for endoscopes, we are still far away from achieving our ultimate goal, which is to develop viable hardware and software technology that can provide the unique capability of 3D surface imaging with the comparable image quality, speed, and system cost of 2D counterparts.

The field of true 3D surface technology is still quite young compared to its 2D counterpart, which has developed over several decades. It is our hope that this overview chapter provides some stimulation and attraction for more talented people to this fascinating field of research and development.

References

1. Armbruster and Scheffler. 2006. US endoscope markets. F441-54, www.frost.com.
2. Frost and Sullivan. 2004. Endoscopy marketers—Industry overview.
3. Millennium Research Group. 2006. US markets for laparoscopic devices 2006. April.
4. Millennium Research Group. 2004. US markets for arthroscopy, 2005. November.
5. C. M. Lee, C. J. Engelbrecht, T. D. Soper, F. Helmchen, and E. J. Seibel. 2010. Scanning fiber endoscopy with highly flexible, 1-mm catheterscopes for wide-field, full-color imaging. *Journal Biophotonics* 3 (5–6):385–407.
6. C. A. Bendall et al. Fringe projection system and method for a probe using a coherent fiber bundle. U.S. Patent 7812968 (GE Inspection Technologies).
7. J. Geng. 2011. Structured-light 3D surface imaging: A tutorial. *Advances Optics Photonics* 3:128–160.
8. Intuitive Surgical System, www.intuitivesurgical.com.
9. M. Chiba. Stereoscopic-vision endoscope system provided with function of electrically correcting distortion of image. U.S. Patent 5860912 (Olympus Optical Co.).

10. Japanese patent 28278/1993.
11. S. Takahashi et al. Stereoscopic endoscope objective lens system having a plurality of front lens group. U.S. Patent 5743846 (Olympus Optical Co.).
12. Everest XLG3™ VideoProbe® System operating manual by GE Inspection Technologies. www.everestvit.pL/pdf/GEST-65042.xlg3brochure.pdf.
13. J. Geng. 2010. Novel ultra-miniature flexible videoscope for on-orbit NDE. Technical report.
14. K. Ogawa. Endoscope apparatus, method of operating the endoscope apparatus. U.S. Patent 7,443,488 (Olympus Optical Co.)
15. J. A. Christian. Apparatus for the optical manipulation of a pair of landscape stereoscopic images. U.S. Patent application 2005/0141088.
16. A. Yaron et al. Blur spot limitations in distal endoscope sensors. www.visionsense.com.
17. Y. H. Lee. 3-Dimensional moving image photographing device for photographing. U.S. Patent application 12162482.
18. M. Schubert and A. Müller. 1998. Evaluation of endoscopic images for 3-dimensional measurement of hollow biological organs. *Biomedizinische Technik* (Berlin) 43:32–33.
19. A. T. Saucedo, F. Mendoza Santoyo, M. De la Torre-Ibarra, G. Pedrini, and W. Osten. 2006. Endoscopic pulsed digital holography for 3D measurements. *Optics Express* 14 (4):1468–1475.
20. E. Kolenovic, W. Osten, R. Klattenhoff, S. Lai, C. von Kopylow, and W. Jüptner. 2003. Miniaturized digital holography sensor for distal three-dimensional endoscopy. *Applied Optics* 42 (25):5167–5172.
21. Bendall, C. A. et al. Fringe projection system and method for a probe suitable for phase-shift analysis. U.S. Patent 7,821,649 B2 (GE Inspection Technologies).
22. N. T. Clancy, D. Stoyanov, L. Maier-Hein, A. Groch, G.-Z. Yang, and D. S. Elson. 2011. Spectrally encoded fiber-based structured lighting probe for intraoperative 3D imaging. *Biomedical Optics Express* 2:3119–3128.
23. J. Geng. 2009. Video-to-3D: Gastrointestinal tract 3D modeling software for capsule cameras. Technical report.
24. J. Geng. 1996. Rainbow three-dimensional camera: New concept of high-speed three-dimensional vision systems. *Optical Engineering* 35:376–383. doi:10.1117/1.601023.
25. J. Geng. 2008. 3D volumetric display for radiation therapy planning. *IEEE Journal of Display Technology*, Special Issue on Medical Display 4 (4):437–450, invited paper.
26. J. Geng. 2011. Overview of 3D surface measurement technologies and applications: Opportunities for DLP®-based structured illumination. *Emerging Digital Micromirror Device Based Systems and Applications III* (Conference 7932), SPIE Photonics West, CA, Jan., invited paper.
27. R. Y. Tsai. 1987. A versatile camera calibration technique for high-accuracy 3D machine vision metrology using off-the-shelf cameras and lenses. *IEEE Transactions on Robotics and Automation* 3 (4):323–344.
28. R. M. Soetikno, T. Kaltenbach, R. V. Rouse, W. Park, A. Maheshwari, T. Sato, S. Matsui, and S. Friedland. 2008. Prevalence of nonpolypoid (flat and depressed) colorectal neoplasms in asymptomatic and symptomatic adults. *JAMA* 299 (9):1027–1035.

15

Biometrics Using 3D Vision Techniques

Maria De Marsico, Michele Nappi, and Daniel Riccio

CONTENTS

Scientists claim that sometimes just 20 seconds are enough time in which to recognize a person. However, the human ability to recognize a person is quite limited to people who are closer or that have been known or associated with. On the contrary, an automatic identification system may require a longer time for recognition, but is able to recognize people in a larger set of potential identities. Nowadays, the most popular biometric identification system is certainly represented by fingerprints. However, present biometric techniques are numerous and include, for example, those exploiting hand shapes, iris scanning, face and ear feature extraction, handwriting recognition, and voice recognition. Early biometric systems were implemented by exploiting one-dimensional (1D) features such as voice sound, as well as two-dimensional (2D) ones such as images from fingerprints or the face. With technology advances and cost decreases, better and better devices and capture techniques have become available. These have been investigated and adopted in biometric settings too. Image vision techniques have not been excluded from this evolution-supported acceleration. The increase of computational resources has supported the development of stereo vision and multiview reconstruction. Later, three-dimensional (3D) capture devices, such as laser or structured-light scanners as well as ultrasound-based systems, were introduced. The third dimension has therefore represented a significant advancement for the implementation of more and more accurate and efficient recognition systems. However, it also presents new challenges and new limits not yet overcome, which are the main motivation for the research in this context.

15.1　2D, 3D, 2D + 3D Biometrics

Traditional automatic identification systems may be divided into two categories: *knowledge based,* which exploit information that the user must know, such as user name, login, and password, and *token based,* which exploit special objects like smart-cards, magnetic cards, keys, etc. Both approaches present weaknesses, since passwords may be forgotten or guessed by others, and objects may be lost, stolen, or duplicated. *Biometrics* allows recognizing a subject according to physical or behavioral features, and its usage potential is very high: access control for critical areas such as penitentiary structures, an investigation tool for search and identification of missing persons or criminals, subject identification during bank transactions, immigration control, authentication during computer and network use, and local as well as remote procedures. A human feature can be considered a biometric key suitable for recognition if it satisfies the following requirements:

- *Universality:* each person must own the biometric trait.
- *Uniqueness:* two persons must be sufficiently distinguishable according to the biometric trait.
- *Permanence:* the biometric trait must resist aging.
- *Collectability:* the biometric trait must be easy to capture and quantitatively measurable.

It is possible to claim that no biometric key is optimal by itself, but rather there is a wide range of possible choices. Selecting one or another depends on the system at hand and on the controlled context. Different factors play their role in the choice of the biometrics to use (face, fingerprint, iris) and of the capturing and processing modality (infrared, thermic, 2D, 3D). Among them, ambient conditions and user typology are particularly important. The operating setting may heavily influence device performance and biometric trait stability, while a user may be *cooperative* (interested in being recognized) or *noncooperative* (uninterested or unaware of the recognition process).

15.1.1 Some 2D Biometrics and Their Limits

Biometric traits can be further classified into *physiological* and *behavioral*. Biometrics in the first category are related to the way a person is and appears, and they tend to be more stable in time; examples are fingerprints, face appearance, ear morphology, and configuration of the retina's blood vessels. On the other hand, behavioral biometrics reflects an aspect of the person's attitude and therefore depends on psychology and habits. This causes such biometric traits to be more variable in time. Many biometrics have been studied over the years (see Figure 15.1).

- *DNA* (deoxyribonucleic acid) represents the most reliable identification code for a person, despite the number of similarities that can be found among relatives. Though widely exploited in forensics for identification, it does not represent a viable biometric technique since many factors limit its use (most of all, the required time, the required equipment, and its being seen as a form of intrusiveness).

- Deep medical investigations have demonstrated that the *ear* complies with the main principles of biometrics. Its advantages are the limited area, which allows faster processing; and the lack of expression variations, which limits intraclass differences. However, it undergoes strong variations when the subject is not aligned with respect to neck rotations (i.e., when the ear is not perfectly frontal with respect to the capture device).

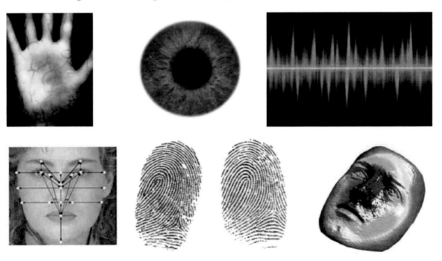

FIGURE 15.1
Some examples of existing biometric traits: vein patterns, iris, voice, 2D face, fingerprints, 3D face.

- The *face* is the most immediate feature through which we recognize a person. However, it presents a number of limitations that are difficult to overcome and limit its reliability. As a matter of fact, though it is faster and more accepted because it is less invasive, this form of identification is far from the recognition rates assured by fingerprints.

- *Fingerprints* are legally recognized as a highly reliable biometric technique and have been long exploited for civil as well as military identification and for incrimination of suspect subjects. Even the FBI has adopted such biometrics as an identification system so that it is surely the most accepted among invasive biometrics. Moreover, since fingerprints were proposed as an identification trait about 50 years ago, we have both wide and deep studies on the topic, as well as many very efficient algorithms that are the core of most present systems. Though especially reliable, fingerprints are not suited for online systems or when the user is not aware or consenting.

- *Gait* is a unique trait and represents a complex spatial biometrics. It is not so distinctive, yet provides sufficient recognition features for a low-security system. Moreover, it provides the same features of acceptability and noninvasiveness of other biometric techniques such as face recognition or thermography. Disadvantages come from the variability in time: A person's way of walking may vary due to a number of factors, such as age, weight, and alcohol use. Moreover, this technique is based on a video capture of steps, and therefore measurement and analysis of the joints' movements require a high computational cost.

- *Hand and fingers geometry* is a consolidated technique based on the capture of the image of an open hand on a guide panel. Hand features like shape, palm width, finger dimensions, etc., are relatively invariant and peculiar, though not very distinctive. Among advantages, we have a good acceptability from the user supported by a relative computational speed and compactness of extracted features. Disadvantages come from the limited distinctiveness, which limits its large-scale use and supports its use as a verification technique.

- The formation of the human *iris* is determined by morphological processes during embryonic development. The iris is part of an annular muscle that allows the amount of light that reaches the eye retina to vary. The layout of radial muscle fibers and their color represent a unique biometric feature for each subject. This feature is captured from the objective of a small color camera while the subject looks at it. This biometrics assures a very high distinctiveness for each eye, even of the same person, and therefore offers accuracy and reliability levels higher than those of other biometrics. The great disadvantage comes from its acceptability and from the necessary user cooperation during the capturing process.

- *Signature* is a quite distinctive feature that is commonly used to verify identity in different settings. It is easy to capture and highly accepted by most users. Disadvantages come from the high variability of the trait due to different factors. This requires a very robust system for effective recognition and does not assure high genuineness levels since it can be easily forged by experts.

- *Voice* recognition is highly accepted, like the signature; the voice can be easily captured and processed. Disadvantages are the great instability of the feature, which can be conditioned by many factors, and the ease of forging it.

15.1.2 The Third Dimension in Biometric Systems

Most research in the field related to biometric systems has long regarded 1D or 2D capture sources, such as voice, face, ear, or fingerprint. However, we recently experimented with a fast technological evolution and a related decrease of the costs of hardware components and devices, like 3D scanners, *motion capture* devices, or multicamera hardware and software. These factors spurred research interest toward solutions that had formerly appeared not to be viable. We can mention gait analysis and recognition, as well as 3D face and ear recognition. The amount of work devoted to 3D face recognition is surely outstanding among biometric 3D techniques. The related range of methodologies is very wide and spans from the adaptation of 2D techniques to the 3D setting to the implementation of specific approaches for this kind of information. The third dimension allows at least limiting, if not overcoming, many problems related to face recognition (e.g., pose and illumination variations). In second place we find the ear, for which the third dimension surely represents a way to address problems related to location of such a trait in an in-profile image. As a matter of fact, the upper part of the ear presents an extremely marked curvature, which is much more easily identified within a 3D model.

The counterpart of this higher richness of information and therefore of the higher achievable recognition accuracy is represented by the huge amount of data to process. This causes some complex problems, which are far easier in 2D, especially for data normalization and preprocessing. One of the main problems in 3D settings is the alignment of the (biometric) models of two objects before classification—for instance, two faces. This is because it is not possible to define a canonical pose, and there are more degrees of freedom. Biometric applications have inherited alignment algorithms commonly used in other 3D settings, most of all the iterative closest point (ICP) technique. This iterative technique requires a reference model as input and a model to align, and it modifies the pose of the latter until a distance criterion between the two is minimized. This algorithm presents two significant disadvantages that mainly derive from the fact that it does not exploit any specific knowledge about the object at hand.

Rather, it operates on three-dimensional shapes in the general sense, and therefore it is slow and does not guarantee convergence to a unique pose.

The types of data operated by a 3D classification system can be of different natures, though in most cases we can reduce them to two standard cases: *point clouds* and *polygon meshes*. Point clouds that are too dense or meshes that are too complex might require a huge amount of computation and therefore significantly slow down the system. A way to get around this problem is *subsampling,* where the number of points in the cloud (or of polygons in the mesh) is pruned to obtain a simpler representation, though as representative as possible. Therefore, such pruning is not performed at random, but is based on discarding those points with less representative results according to a predefined criterion. Another important preprocessing step for data captured through a 3D device is region smoothing, which smoothes parts of the model that are too sharp in order to facilitate the location of possible curvature zones, like in the ear.

15.1.3 3D Capture Systems

At present different technologies exist that can provide a three-dimensional reproduction of the captured object, according to working principles that can be very different. Of course, each has specific features that make it preferable in specific settings as well as unadvisable in others.

One of the earliest among the previously mentioned techniques is stereo vision, where the scene is captured by two or more cameras which are calibrated among them. In other terms, the intrinsic and extrinsic parameters of single cameras are known in advance (through the process known as calibration), and this makes it possible to compute the depth of the scene, as well as of the objects within it, through a triangulation technique. The main critical point in this approach is in searching corresponding scene points in different images. An alternative introduced in relatively recent years is represented by laser scanners. Even in this case, if the relevant object is within a quite limited distance, its three-dimensional reconstruction exploits triangulation, based on laser light. Limiting factors for the diffusion of this technology in biometric settings are elapsed times (from many seconds to some minutes) and still relatively high costs.

A further evolution is represented by structured-light scanners. Though based on a triangulation process, these scanners are different because a specific pattern is projected over the object to be captured to increase the location precision of corresponding points. The main limit of scanners exploiting structured light is just this projection of a visible pattern on the object. In the specific case of face biometrics, this may result particularly in discomfort for the user. In 1998 Raskar et al. [1], at the University of North Carolina, implemented the first imperceptible structured-light scanner. The working principle is based on the idea of projecting a pattern and its inverse in a consecutive way and at an extremely high speed. In this way the subject should perceive only the white light. Despite the proof of concept, many aspects of this technology must still be improved. An alternative solution is the projection of the pattern within the infrared spectrum.

In addition to optical systems, other 3D capturing technologies are spreading (e.g., ultrasound [2]). These present some advantages with respect to other systems. For example, they are not influenced by surface tainting factors like stains or dirt or by skin temperature; moreover, they are able to provide information about the skin surface and about the thin layer underneath, therefore allowing one to obtain a 3D volume of tissues just under the skin (about 10 mm). Such capture devices have been exploited especially for fingerprints and palm prints. Scanning time may be reduced by using scanner arrays, which are able to scan a whole line of points at a time, instead of X–Y scanners, which scan the surface point by point.

Signature recognition requires totally different devices to capture related data. In particular, classical systems exploit devices equipped with accelerometers and gyroscopes, allowing capture of speed and angle of hand movements while a document is signed. In other cases, it is not the pen but rather the signing surface that is equipped with sensors—tablets with different layers with different sensitivity—to measure pen pressure on the paper. Only recently, devices that also record the sound from the point of a ballpoint pen when it moves on the paper during a signature have been introduced. As we can see, even these devices are reaching an extremely high level of complexity, in the same way as technologies adopted for image and volume capture.

15.2 Face

Evaluations such as FERET (face recognition technology) tests and the face recognition vendor test 2002 [3] have underlined that the current state of the art in 2D face recognition is not yet sufficient for biometric applications. As a matter of fact, the performance of the 2D face recognition algorithms is still heavily affected by the pose and illumination conditions of the subject. Much effort has been spent to overcome these drawbacks, but the obtained results in terms of recognition rate confirmed that this is a very hard problem. Nowadays, the 3D capturing process is becoming cheaper and faster and recent work has attempted to solve the problem directly on a 3D model of the face (see Figure 15.2). Few databases of 3D face models are publically available at present; moreover, in most cases the number of subjects is limited and the quality is often very low.

The constant progress of 3D capturing technologies also influenced the kinds of recognition algorithms. The first algorithms directly processed clouds of points [4] after a suitable triangulation, while the more recent ones work on

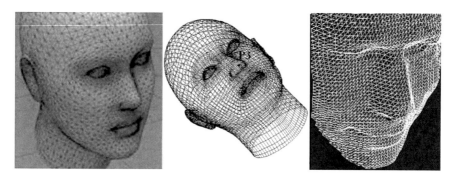

FIGURE 15.2
Examples of 3D meshes of human faces.

a mesh, considering in some cases the information provided by both the 3D shape and the texture. The 3D RMA (Royal Military Academy) [5] is an example of a database of 3D face models represented by clouds of points. It has long been the only publicly available database, even if its quality is quite low. On the other hand, 3D meshes (wrl, 3ds, etc.) are available today from the newer technologies, but in most cases they are just part of proprietary databases.

The vast majority of face recognition research and all of the main commercial face recognition systems use normal intensity images of the face. All these methods are referred to as 2D. On the other hand, 3D techniques consider the whole shape of the head. Since the early work on 3D face recognition was done over 15 years ago and the number of published papers is rapidly growing, a brief description of the most interesting algorithms in the literature follows.

Some of the first approaches to 3D face recognition worked on range data directly obtained by range sensors, due to the lower costs of this hardware compared with the laser scanners used, for example, by Gordon [6]. As a matter of fact, in Achermann, Jiang, and Bunke [7], the range images are captured by means of a structured-light approach. The most important potential disadvantage of this choice is the lack of some data due to occlusions or improperly reflected regions. This problem can be avoided by using two sensors, rather than one, and applying a merging step for integrating the obtained 3D data.

The initial step is sensor calibration so that such parameters as projection matrix, camera direction, etc., are computed. Then, merged images are computed; for every original 3D data point, the coordinates in the merged range image are computed according to the parameters of the virtual sensor. If 3D points of two different surfaces have to be mapped onto the same pixel, a sort of z-buffering is applied to disambiguate the mapping. The template images obtained from this acquisition process are then used as training and testing sets for two different approaches.

The first approach exploits eigenfaces. The dimension of the space of face templates is reduced by applying principal component analysis (PCA) for both training and testing so that, for each testing image, the nearest one in terms of Euclidean distance is searched. Another method, exploiting HMMs (hidden Markov models), is also tested on the template images. As this technique is only applicable on 1D signals, the template images are first transformed into a monodimensional signal by means of a sliding window that moves along the image from the top to the bottom and from the left to the right. The involved HMM has five states. For every person in the database, the parameters of the HMM are calculated in a training phase. When a test image is presented, the probability of producing this image is computed by means of the Viterbi algorithm.

In 2D approaches, the features used in describing faces are still often limited to eyes, nose, mouth, and face boundaries, while neglecting the additional information provided by low-contrast areas, such as jaw boundary, cheeks, and forehead. Therefore, it is clear that an approach based on range and curvature data has several advantages over intensity image approaches thanks to the higher amount of available information. Furthermore, curvature has the valuable characteristic of being *viewpoint invariant*. Gordon [6] proposed a method that defines a set of high-level regions (e.g., eyes, nose, and head), including the following features: *nose bridge* (nasion), *nose base* (base of septum), *nose ridge, eye corner cavities* (inner and outer), *convex center of the eye* (eyeball), *eye socket boundary, boundary surrounding nose,* and *opposing positions on the cheeks.* Each of these regions on the face image is described in terms of a set of relations of depth and curvature values.

Since several regions can fulfill a set of constraints, this set is designed in order to reduce the search to a single definition of the feature. The set of constraints is given by

- Sign of Gaussian and mean curvature
- Absolute extent of a region on the surface
- Distance from the symmetry plane
- Proximity to a target on the surface
- Protrusion from the surrounding surface
- Local configuration of curvature extremes

The high-level features and regions are used to compute a set of low-level features, where the most basic scalar features correspond to measurements of distances. The set of low-level descriptors is given by *left and right eye width; eye separation; total span of eyes; nose height, width, depth; head width; maximum Gaussian curvature on the nose ridge; average minimum curvature on the nose ridge;* and *Gaussian curvature at the nose bridge and base.* For each face image, this set of features is computed, placing it in the space of all possible faces, while the Euclidean distance is used as a measure in the scaled feature space.

Other approaches directly work on clouds of 3D points, like the algorithm proposed by Xu et al. [4], which applies to the face models of the 3D RMA database (models are represented by scattered point clouds). The first problem to be addressed consists of building the mesh from the clouds of points. This is done by means of an iterative algorithm. The nose tip is localized first as the most prominent point in the point cloud. Then a basic mesh is aligned with the point cloud, subdivided, and tuned step by step. Four steps are considered to be enough for the refinement process. Point clouds have different orientations, and resulting meshes preserve this orientation; an average model is computed and all the meshes are aligned with it by tuning six parameters for the rotations and six for the translations. Due to possible noise, some built mesh models cannot describe the geometric shape of the individual. These mesh models are called nonface models. Each mesh contains 545 nodes and is used as a bidimensional intensity image in which the intensity of the pixel is the Z-coordinate of the corresponding node. The eigenfaces technique is applied to these intensity images. After computing the similarity differences between test samples and the training data, the nearest neighbor (NN) classifier and the k-nearest neighbor (KNN) classifier are used for recognition.

The 3D morphable model-based approach [8] has been shown to be one of the most promising ones among all those presented in the literature. This face recognition system combines deformable 3D models with a computer graphics simulation of projection and illumination. Given a single image of a person, the algorithm automatically estimates 3D shape, texture, and all relevant 3D scene parameters. The morphable face model is based on a vector space representation of faces. This space is constructed such that any convex combination of the examples belonging to the space describes a human face. In order to assure that continuous changes on coefficients represent a transition from one face to another, avoiding artifacts, a dense point-to-point correspondence constraint has to be guaranteed. This is done by means of a generalization of the optical flow technique on gray-level images to the three-dimensional surfaces.

Two vectors, $\mathbf{S}$ and $\mathbf{T}$, are directly extracted from the 3D model, where $\mathbf{S}$ is the concatenation of the Cartesian coordinates (x, y, z) of the 3D points and $\mathbf{T}$ is the concatenation of the corresponding texture information (R, G, B). Furthermore, the PCA is applied to the vectors $\mathbf{S_i}$ and $\mathbf{T_i}$ of the example faces $i = 1, 2, …, m$, while Phong's model is used to describe the diffuse and specular reflection of the surface. In this way, an average morphable model is derived from training scans (obtained with a laser scanner). Then, by means of a cost function, the fitting algorithm optimizes a set of shape coefficients and texture coefficients along with 22 rendering parameters concatenated in a feature vector ρ (e.g., pose angles, 3D translations, focal length).

A similar approach, but working on two orthogonal images, has been proposed by Ansari and Mottaleb [9]. This method uses the 3D coordinates of a set of facial feature points, calculated from two images of a subject, in order to deform a generic 3D face model. Images are grabbed by two cameras with perpendicular optical axes. The 3D generic model is centered and aligned by means of Procrustes analysis, which models the global deformation, while local deformations are described by means of 3D spline curves. The front and profile view of a subject are used in order to locate facial features, such as eyes and mouth, by means of a probabilistic approach. This model contains 29 vertices divided into two subsets: 15 principal vertices and 14 additional vertices.

A multimodal approach combining results from both 3D and 2D recognition has been investigated by Wang, Chua, and Ho [10]. The method applies on both range data and texture. In the 3D domain, the point signature is used in order to describe the shape of the face, while the Gabor filters are applied on the texture in order to locate and characterize 10 control points (e.g., corners of the eyes, nose tip, corners of the mouth). The PCA analysis is then applied separately to the obtained 3D and 2D feature vectors, and the resulting vectors are integrated to form an augmented vector that is used to represent each facial image. For a given test facial image, the best match in the galley is identified according to a similarity function or SVM (support vector machine).

One limitation to some existing approaches to 3D face recognition involves sensitivity to size variation. Approaches that use a purely curvature-based representation, such as extended Gaussian images, are not able to distinguish between two faces of similar shape but different size. Approaches based on PCA or ICP avoid this problem, but their performance falls down when changes in expression are present between gallery and probe images. Even multimodal recognition systems require more sophisticated combination strategies. As a matter of fact, in most cases, the score results are computed separately for 2D and 3D and then combined together. It is at least potentially more powerful to exploit possible synergies between the two modalities.

15.3 Ear

The ear represents a rich and stable biometric trait. It can be easily captured at a distance even with unaware subjects, though, in this case, it might be partially occluded by hair and/or earrings. Research has been especially focused on implementing processing and matching techniques based on 2D images, which are, however, strongly influenced by subject pose and by illumination conditions during capture. Addressing such limitations is possible by working on three-dimensional models of the ear captured by range sensors, laser scanners, or structured-light scanners. These captured 3D models are far less sensitive to pose and illumination and allow a significant precision increase during both segmentation and matching. Yan and Bowyer [11] demonstrated that 3D system performance is significantly higher than that from 2D methods.

Even for the ear, the pipeline of a 3D-based system is generally not much different from that of a 2D system, but for the capture and analysis tools. The scene reconstructed from a 3D capture system undergoes a detection process to locate the ear region. The ear is then segmented and processed to extract the features used for matching with other feature vectors from different ear models. However, in some cases, it may happen that the sequence of steps is slightly different, due to a specific capture method or to a nonconventional matching method. For instance, this is the case for techniques that exploit the reconstruction of a model through stereoscopic vision or through shape-from-shading techniques starting from video frames (more details follow).

15.3.1 Ear Capture

Many 3D approaches for ear recognition are based on range data capture (see Figure 15.3). Higher robustness of such systems to pose and illumination is obtained at the cost of often too expensive devices, which are further characterized by the requirement of an almost complete immobility of the subject for a time lapse of some seconds. On the other hand, 2D image capture is quite easy and viable in many scenarios where 3D standard technology is not usable. For these reasons, research has focused on techniques able to reconstruct 3D information from 2D samples (images). As for the ear, the techniques that are most used and have undergone most progressive improvements are multiview reconstruction (MVR), structure from motion (SFM), and shape from shading (SFS).

Liu, Yan, and Zhang [12] proposed a semiautomatic method, where motion analysis and multiview epipolar geometry are the foundations of the 3D model reconstruction process. In a first attempt, the correspondence between homologous points of the ear in different images is identified automatically through Harris corner detection and the RANSAC (random-sample consensus)

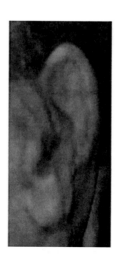

FIGURE 15.3
(a) Range image, (b) visible image, and (c) rendering of the image according to range information of an ear.

algorithm. However, the limited number of obtained correspondences suggested that the authors introduce a further step of manual adjustment, based on an interactive process of ear contour division, to locate correspondences. Though this increases the number of correspondences, the obtained model still presents a relatively limited number of vertices (330–600), which provide a point cloud (or even a mesh) more sparse than the one that can be obtained through a pure 3D capture system (range scanner, structured light).

Zhou, Cadavid, and Abdel-Mottaleb [19] and Cadavid and Abdel-Mottaleb [20] focused their attention on reconstruction techniques from video, along two different lines: structure from motion and shape from shading. In SFM, ear image details are discarded through a smoothing process based on force field. The output of this process is the input for the edge detection implemented through the ridges and ravines detection algorithm. Tracking of characteristic points is performed through the Kanade–Lucas–Tomasi algorithm [21], while the reconstruction of the 3D model is produced by the algorithm by Tomasi and Kanade. Matching between two different ear models is computed according to the partial Hausdorff distance (PHD).

Along the other line, the SFS algorithm extracts a set of frames from a video and independently processes each of them. It first locates the ear region and then approximates its 3D structure through the actual shape-from-shading algorithm. The models obtained are aligned through the ICP algorithm so that recognition can be performed by cross validation. Even this approach presents some disadvantages: first of all, sensitivity to illumination variations since it derives depth information from scene reflectance features.

In order to address some of the limitations of SFM and SFS techniques, Zeng et al. [13] proposed an improvement of the SFM model where images are captured by a pair of calibrated cameras; the scale invariant features transform, together with a coarse to fine policy, is used to identify matches between corresponding points. The 3D ear model is computed through triangulation and refined through a bundle adjustment technique.

15.3.2 Recognition Process

In their first work in 2004, Chen and Bhanu [14] only addressed the ear location problem, without reporting any results regarding recognition rates. The first step of their algorithm applies a step edge detection associated with a threshold to identify the most evident contours (sharp edges) in regions surrounding the ear. This first step is followed by a dilation of contour points and by a search for connected components, whose labeling process aims at determining all candidate regions. Each such region is represented by a rectangle, which is expanded toward the four main directions to minimize the distance from a reference template that has been computed in advance. The region with the minimum distance is then considered the ear region.

In a following work [22], the same authors provided a description of a complete ear recognition system. The detection phase was far more refined than the one in the previous work and enriched with more details. Skin detection was performed according to the statistical model presented by Jones and Rehg [16]. Characteristic points are put in correspondence through a local-to-global registration procedure, instead of training a reference template. Global registration focuses on helix and antihelix regions and is therefore performed through the DARCES [17] algorithm, which solves the problem of aligning rigid structures without any prealignment. A local deformation process is performed after global registration. Matching between two models (probe/gallery) is partly based on information from the helix/antihelix alignment process, which is performed during the detection phase (information about the global rigid transformation that is initially applied), and partly on some local surface descriptors, called local surface patch (LSP), which are computed in those characteristic points that are a minimum or a maximum of shape indices. Distance between two 3D models is then computed by root mean square registration error.

In 2008, Cadavid and Mottaleb presented a system for detection [18], modeling, and recognition of the ear from video, based on a shape-from-shading technique. Successively, they presented a far more refined method together with Zhou in 2010 [19], which only detects the ear in range images. Their aim was to identify features able to characterize the peculiar 3D ear structure that was, at the same time, robust to noise and to pose variations. To address this, they started from the histograms of oriented gradients (HOGs) descriptor [23] used by Dalal and Triggs in pedestrian detection. The descriptor introduced by the authors is called histograms of categorized shapes and is used to represent regions of 3D range images. The detection window is divided into overlapping blocks, for each of which a histogram of shape indices and of curvature is computed over single pixels in the block. The histogram of the detection window is given by chaining the histograms of its blocks.

Yan and Bowyer, who are authors of many works on the topic, have also made a significant contribution to the research about ear-based recognition. Their method is composed of a detection part, based on both 2D and 3D information, and a recognition part based on multibiometrics strategies. Ear detection partly works on the 2D original image through a skin detector. Once the ear pit is identified, a contour detection algorithm exploits both color from the 2D image and depth information from the range image to separate the ear from the remaining part of the range image. Matching is based on ear surface segmentation and on curvature estimation. The recognition process was also studied as a multibiometric one [15] by experimenting with multimodal, multiexpert, and multisample approaches. From the authors' studies, it was determined that all integration methods take advantage of the presence of more samples for the same subject, while it is not possible to find a single fusion rule that overcomes all the others in all cases; on the contrary, a different fusion rule exists for different settings that is the best for the specific case.

15.4 Hand-, Palm-, and Fingerprints

Hand geometry is a biometric trait that deserves special attention and for which a number of devices are sold and commonly adopted. However, different factors, such as a relative ease of spoofing in 2D systems, a capture process that is not particularly comfortable due to the required hand positioning, and hygienic issues because the physical contact of the hand with the capture device is required, limit its spread. Passing from 2D to 3D (see Figure 15.4) solves some of these problems. But one has to face a technical problem. A 3D contactless device must handle the additional factor of hand position, since the latter is less constrained during the capture process.

At least five different biometric traits can be extracted from the hand: fingerprints, palm prints, hand geometry, vein pattern, and finger shape.

Fingerprints are the most widespread biometrics, thanks to their massive use in forensics. They have become an identification mean that is particularly familiar in the collective imagination, with a discrete acceptability level. On the other hand, they are often associated with use by security forces to find criminals. This factor, together with the hygienic problem of contact with the sensor, contributes to lowering the viability of 2D image-based fingerprint recognition.

Palm prints present many similarities with fingerprints: They are also characterized by lines, contours, and minutiae, though they are generally captured at a lower resolution (commonly about 100 dpi), with the advantage of a possible contactless capture. The true limit of this kind of biometrics is

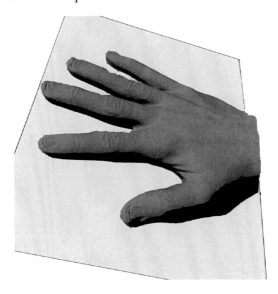

FIGURE 15.4
Three-dimensional pose-invariant hand recognition.

in the capture devices' dimensions, which make their integration in small systems, like mobile devices, difficult if not impossible. Furthermore, they are very easy to deceive and therefore prone to spoofing attacks.

These kinds of problems are common with systems based on hand geometry recognition, which are easily deceivable and attackable by impostors. However, this biometric trait, which is based on the measurement of some ratios between distances of hand elements, has conquered a large chunk of the market thanks to the speed of extraction and the matching process. However, such speed is paid for in terms of accuracy. As a matter of fact, hand geometry does not guarantee the same accuracy levels typical of fingerprints and palm prints. Therefore, it is not used in high-security applications. Finally, individual hand geometry features are not descriptive enough for identification and are not thought to be as unique as, for example, fingerprints.

Blood vessels within the hand make up an intricate reticulum, the pattern of which is certainly characteristic of the single subject. This offers a number of advantages compared with other hand characteristics, apart from its uniqueness. First of all, it is possible to capture it contactless, therefore eliminating potential problems related to sensor cleanness. Furthermore, the extracted features are particularly stable in time and not affected by particular external conditions if captured through near infrared (NIR) or thermal (IR) technology, and the obtained pattern is difficult to replicate in a spoofing attack.

The analysis of fingers' shapes and textures often joins the previous traits, both for the back and for the fingertip. In particular, lines on the knuckles seem to present a highly distinctive property.

Research on these biometrics has produced a high number of solutions based on 2D images that often present common problems such as pose and illumination variations, pattern deformation or bending, and spoofing vulnerability. In general, a 2D imaging system for capturing hand features includes a charge-coupled device (CCD) camera, an oriented light source, and a flat support surface. When the system requires hand contact with the surface, pegs may guide hand positioning; in some cases, the equipment is integrated with a mirror, making the hand side projected on the CCD so as to have a side image of the hand also. In a few cases the third dimension has been exploited and therefore the input sample is captured through a 3D scanner, producing a range image.

Woodard and Flynn [24] used a 3D scanner with laser light to capture a color image and a range image of the hand. In Malassiotis, Aifanti, and Strintzis [25], the sensor was expressly created through a color camera and a standard video projector to capture range images. With regard to hand geometry, the two methods cited differ in the way the range image is segmented to extract matching features. In the first case, a combination of skin and contour detection is exploited to separate the hand from a uniform background, while a convex hull algorithm run on the hand contour allows extracting the index, middle, and ring fingers from the range image. The discriminating features are represented by the curvature indices computed for each point in the extracted regions. Malassiotis et al. exploit the only range image for segmentation by modeling the hand in it through a mixture of Gaussian curvatures. Matching features include the 3D width and average curvature of each of the four considered fingers. This last method was tested on a relatively small archive, and therefore the evaluations of the provided performances are not truly representative for large-scale use.

Woodard and Flynn [26,27] presented the first 3D approach for finger recognition. The method analyzes the back of the hand—in particular of the index, middle, and ring fingers. For each pixel, minimum and maximum curvatures are computed in order to compute shape index. In this way a shape index image is obtained for each finger. These images then represent the input for the matching phase.

In other cases, such as in Raskar et al. [1], the capture technique is the true original element with respect to other techniques in the literature. In the specific case, 3D ultrasonic images are captured from the internal region of the palm of the hand. The capture system is composed of an ultrasound imaging machine with a high-frequency linear array (12 MHz).

Kanhangad, Kumar, and Zhang [28] deepen the aspects related to fusing information from both a 2D image and a 3D model of the hand, captured at the same time, for a hand-geometry-based recognition. The method works without hand position constraints and identifies position and orientation for the four considered fingers (excluding the thumb) based on knuckles (tips) and bendings (valleys). For each point on the fingers' surface, the average curvature and the unit normal vector are computed as features that are then used for matching according to ad hoc defined metrics. Kanhangad, Kumar, and Zhang [29] presented a further development of this strategy using a method that estimates the hand pose starting from information from range image (3D) and texture image (2D) that is aimed at normalizing texture and extracting palm print and hand geometry. Three-dimensional features include the surface code, based on curvature, and the shape index; both are evaluated in the points of the palm of the hand. Information from the 2D image and the 3D range image is fused together through a weighted sum of scores.

Tsalakanidou, Malassiotis, and Strintzis [30] experimented with fusing face biometrics with hand geometry. Samples were captured by a real-time, low-cost sensor that produced a color 2D image (texture) and a 3D model (range image). The user is asked to position the hand in front of the face with the back toward the sensor, which captures a number of samples while the user is asked for different finger positions (closed, open, slanted). The face is detected from the range image according to a priori knowledge of the searched object. An interactive procedure guides the user toward a correct face position in front of the sensor, while captured images are normalized and passed to the classifier. The latter is constituted by a probabilistic matching algorithm [31] applied to both 2D and 3D images. Assuming that the hand is not still, it is extracted by exploiting information from face positioning and segmented through a blob detector. Signature functions are extracted from finger geometry and then passed to the classifier. Fusion is performed at score level after score normalization. The authors experimented with different normalization functions (min–max, z-score, median–MAD, tanh, quadratic-line-quadratic [QLQ]) and different fusion rules (simple sum, product, max-score, weighted sum). The best performance in terms of equal error rate (EER) came from using QLQ and weighted sum.

15.5 Signature

Electronic signature from the capture of a handwritten one is designed for contract, document, receipt, and register signing. The dynamic aspect of a handwritten signature is very important, not only because it is the perfect way to authenticate a voluntary act, but also because it allows identifying the author (i.e., to associate each electronic signature to a person). Electronic signature includes static and dynamic data. Static data are revealed by the signature's two-dimensional outline and may indicate some unique traits to a graphologist.

Dynamic data of an electronic signature are much more difficult to analyze but are more precise. Only dynamic data, pressure, direction, speed, and speed variations during the signature can offer high security for the identification of a signature. Anybody, after a suitable training period, can imitate the outline of the handwritten signature of another person, and it is not difficult to obtain a sample of a signature from any person. Nevertheless, the impostor cannot know, and in any case could not imitate, the signature's dynamic elements. Moreover, the acceptability level of signature recognition-based systems is high, since users usually sign and do not notice differences between the traditional and the biometric methods. System vulnerability is deemed quite low, and capture devices are reasonably cheap (from 100 to 1000 euros). The main disadvantage is the stability of the sample, since the way of signing may vary over time.

As observed before, traditional signature verification methods are subject to spoofing when they operate in a 2D setting. However, a number of works in the literature exploit the third dimension, intended as the degree of pressure during signing, in order to increase the degree of security of this biometric. In particular, Rubesh, Bajpai, and Bhaskar [32] proposed a system with a special pad, the aim of which was to measure the pressure so as to be able to use any kind of pen (see Figure 15.5). The presented method exploits further additional information, such as speed, acceleration, direction, interruption points (pen ups/downs), the total time, and the length. Further parameters are derived through curve fitting, surface fitting, and solid-angle computation methods. This approach was designed to be used in the context of financial document and transaction signing.

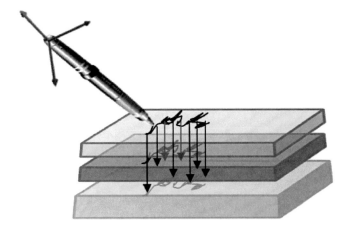

FIGURE 15.5
A nonlinearly spaced pad for the 3D acquisition of a signature.

Haskell, Hanna, and Sickle [33] characterize the signature through curvature moments, computed in a 3D space including spatial coordinates and signing speed. For each signature, the average vector and the associated covariance matrix are computed, while matching is performed through Mahalanobis distance.

15.6 Gait

In the biometrics field, the term *gait* indicates each person's particular way of walking. Therefore, gait recognition is the verification/recognition of the identity of a subject based on the particular way of walking. Gait recognition is a relatively recent biometric, though the architecture of a gait-based system is not substantially different from more traditional biometrics; as a matter of fact, such a system captures input from a moving subject, extracts unique characteristics and translates them into a biometric signature, and then exploits the signature during the matching phase. According to the claim by Gafurov [34], gait recognition systems based on 2D imaging can be classified into three main groups based on: machine vision, floor sensors, and wearable sensors. Most methods in the first group are based on silhouette extraction [35,36], though some exceptions can be found [37,38] where stride and cadence of a person, or some static parameters of the body (e.g., maximum distance between head and pelvis, distance between feet), are exploited.

An important advantage of this approach is that input samples can be captured at a distance, when other biometrics cannot be used. It is worth noting that, in such a context, sensor-based approaches could not be adopted. As a matter of fact, they require mounting a number of sensors in the layer below the floor. When a person treads upon this kind of flooring, they capture data that are input to the rest of the processing pipeline. Features typically used in these kinds of approaches are the maximum heel strike time or the maximum heel strike width. Though nonintrusive, such systems present the great disadvantage of requiring the installation of specific sensors.

As for wearable sensor-based systems, sensors are worn by the subject. Though there is no specific constraint and therefore they might be positioned at any body part, experiments in the literature put them on the pelvis [39] or on the legs [40]. Even in this case, the obtained error rates demonstrate that, at present, gait recognition can be mainly exploited to support other kinds of biometrics. Moreover, classical approaches to gait recognition are limited by internal factors related to the subject (e.g., pathologies, accidents, aging, weight variations) as well as by external factors (e.g., viewing angles, illumination conditions, type of flooring). Given the intrinsic nature of the former, it is highly unlikely that possible technological solutions, though advanced, can address them. On the contrary, since the second group is tightly related to the exploited imaging type and to the method, it seems more viable to devise enhancing solutions in terms of both usability and accuracy. The third dimension represents, in this case, a valid approach to 2D imaging.

In 3D-based gait analysis systems, the subject is typically captured by two or more view angles, which allow one to reconstruct a three-dimensional model. Static and dynamic features of the model are then analyzed.

Using more cameras and, consequently, more viewpoints allows reconstructing a 3D volumetric space of the subject; to this aim, visual hulls, introduced by Laurentini [41], are one of the most commonly used techniques. Among the first to use this gait recognition technique were Shakhnarovich, Lee, and Darrell [42], who reconstructed the visual hull starting from the silhouettes extracted from video clips captured from four cameras. This allows identification of the subject's trajectory and synthesis of a silhouette in canonical pose by positioning a virtual camera within the reconstructed volumetric space. However, matching exploits purely 2D techniques, so the approach performs 2D gait analysis within a 3D reconstructed scenario.

Dockstader and Tekalp [43] introduced a relatively structured method, based on strong and weak kinematic constraints, through which they extracted specific gait patterns to be matched. However, the work is more focused on the model construction than on actual gait analysis. In the same way, Bhanu and Han [47] exploited a kinematic model to extract static and kinematic features from monocular video sequences. This is done by performing the fitting of the 3D kinematic model to the 2D silhouette of the subject, which is extracted from the same sequences. The system implements different fusion strategies (product, sum rule, min rule, and max rule) of static and dynamic model parameters.

Urtasun and Fua [44] exploited a 3D temporal model in a similar setting. Using a motion model based on PCA and exploiting a robust human tracking algorithm, the authors reformulated the problem of people tracking in terms of minimization of a differentiable objective function, the variables of which are represented by the PCA weights. Such a model allows using deterministic optimization methods without resorting to probabilistic ones, which are computationally more expensive. In this way the system is robust to changes in viewpoint or to partial occlusions.

A further proposal along this line was presented by Guoying et al. [45], where the subject was videotaped by more video cameras and therefore from different viewpoints. In this way it is possible to create a three-dimensional model of the subject. A local optimization algorithm is used to divide the model into key segments, the length of which represents a set of model static features. Lower limb trajectories represent a dynamic feature of the subject; after being normalized with respect to time through a linear model, they are used for matching. Such an approach allows significant improvement in system performance when the surface upon which the subject moves is nonoptimal in terms of capture conditions.

The method proposed by Shakhnarovich, Lee, and Darrell [46] represents one of the first attempts to fuse face and gait in a multibiometric setting. Along this line, research was further extended by Seely et al. [48,49], who focused on fusing gait, face, and ear in a particular capture system (multibiometric tunnel) implemented at Southampton University. This tunnel facilitates the capture of biometric data for a large number of persons, so the university was able to collect a suitable biometric database. Moreover, this tunnel allowed studies addressed to the modalities of gait change when the subject carries an object.

15.7 Conclusions

Biometric systems based on three dimensions have high potential and can significantly improve, in many cases, the accuracy that can be obtained with techniques solely based on 2D imaging. Such high potential is the base for noticeable research efforts in the last years on different fronts. First of all, there is a continuous search for new technologies to overcome the limits of current, mostly diffused 3D capture systems. Such limits include, among others, the fact that the capture process with structured light may be uncomfortable for the user or that some devices require too long for capture or 3D model reconstruction, and occlusions are still an open problem.

Present research is moving along this line by studying alternative approaches such as imperceptible scanning. Graphics processing unit programming may be a potential solution to capture 3D modeling times. The field of view and the depth of field of present technologies must be further increased to allow capturing larger objects in wider settings, such as the full subject body for gait recognition. Second, research must still work on improving the accuracy of the model reconstruction and segmentation process to facilitate the tasks that follow in the pipeline of recognition systems. The third dimension significantly increases problem complexity and present segmentation techniques are not yet able to handle the huge amount of data coming from a 3D capture efficiently and effectively.

Finally, systems accuracy is not yet at optimal levels, since the used classifiers do not completely exploit the 3D potential. Most techniques used for 3D have been directly inherited from the 2D setting. In other words, 3D data have often been mapped on a new kind of 2D image (e.g., depth map images), to which standard methods such as PCA have been applied. Only in a following phase have ad hoc techniques been especially devised for 3D models—for example, shape indices and curvature histograms. Even more recently, attempts have been made to combine 3D with 2D in a specific way or the different 3D techniques through suitable protocols and fusion rules; the results seem to encourage proceeding along this direction.

References

1. Raskar, R., Welch, G., Cutts, M., Lake A., Stesin, L., and Fuchs, H. The office of the future: A unified approach to image-based modeling and spatially immersive displays. *Proceedings of the 25th Annual Conference on Computer Graphics and interactive Techniques (SIGGRAPH '98).* ACM, New York, pp. 179–188, 1998.
2. Iula, A., Savoia, A., Longo, C., Caliano, G., Caronti, A., and Pappalardo, M. 3D ultrasonic imaging of the human hand for biometric purposes. *Ultrasonics Symposium (IUS), 2010 IEEE,* pp. 37–40, Oct. 11–14, 2010.
3. Phillips, P. J., Grother, P., Michaels, R. J., Blackburn, D. M., Tabassi, E., and Bone, J. FRVT 2002: Overview and summary. Available at www.frvt.org (March 2003).
4. Xu, C., Wang, Y., Tan, T., and Long, Q. A new attempt to face recognition using 3D eigenfaces. *6th Asian Conference on Computer Vision (ACCV)* 2:884–889, 2004.
5. http://www.sic.rma.ac.be/~beumier/DB/3d_rma.html (October 25 2011).
6. Gordon, G. G. Face recognition based on depth and curvature features. *Proceedings of IEEE Computer Society Conference on Computer Vision & Pattern Recognition* 808–810, June 1992.
7. Achermann, B., Jiang, X., and Bunke, H. Face recognition using range images. *Proceedings of the International Conference on the Virtual Systems and Multimedia (VSMM97),* pp. 129–136, September 1997.
8. Blanz, V. and Vetter, T. Face recognition based on fitting a 3D morphable model. *IEEE Transactions on Pattern Analysis and Machine Intelligence* 25 (9):1063–1074, 2003.

9. Ansari, A. and Abdel-Mottaleb, M. 3D face modeling using two orthogonal views and a generic face model. *Proceedings of the International Conference on Multimedia and Expo (ICME '03)* 3:289–292, 2003.

10. Wang, Y., Chua, C., and Ho, Y. Facial feature detection and face recognition from 2D and 3D images. *Pattern Recognition Letters* 3(23):1191–1202, 2002.

11. Ping, Y. and Bowyer, K. W. Ear biometrics using 2D and 3D images. *IEEE Computer Society Conference on Computer Vision and Pattern Recognition* 1, 121–128, 2005.

12. Heng, L., Yan, J., and Zhang, D. 3D ear reconstruction attempts: Using multi-view. *International Conference on Intelligent Computing* 578–583, 2006.

13. Zeng, H., Mu, Z-C., Wang, K., and Sun, C. Automatic 3D ear reconstruction based on binocular stereo vision. *IEEE International Conference on Systems, Man and Cybernetics* 5205–5208, Oct. 11–14, 2009.

14. Chen, H. and Bhanu, B. Human ear detection from side face range images. *Proceedings International Conference Image Processing* 3, 574–577, 2004.

15. Yan, P. and Bowyer, K. Multi-biometrics 2D and 3D ear recognition, audio- and video-based biometric person authentication. *Lecture Notes in Computer Science* 3546:459–474, 2005.

16. Jones, M. J. and Rehg, J. M. Statistical color models with application to skin detection. *International Journal Computer Vision* 46 (1):81–96, 2002

17. Chen, C. S., Hung, Y. P., and Cheng, J. B. RANSAC-based DARCES: A new approach to fast automatic registration of partially overlapping range images. *IEEE Transactions Pattern Analysis and Machine Intelligence* 21 (11):1229–1234, Nov. 1999.

18. Cadavid, S. and Abdel-Mottaleb, M. 3D ear modeling and recognition from video sequences using shape from shading. *19th International Conference on Pattern Recognition* 1–4, Dec. 8–11, 2008.

19. Zhou, J., Cadavid, S., and Abdel-Mottaleb, M. Histograms of categorized shapes for 3D ear detection. 2010 *Fourth IEEE International Conference on Biometrics: Theory Applications and Systems* (BTAS) 1–6, Sept. 27–29, 2010.

20. Cadavid, S. and Abdel-Mottaleb, M. Human identification based on 3D ear models. *First IEEE International Conference on Biometrics: Theory, Applications, and Systems* 1–6, Sept. 2007.

21. Lucas, B. D. and Kanade, T. An iterative image registration technique with an application to stereo vision. *International Joint Conference on Artificial Intelligence* 674–679, 1981.

22. Chen, H. and Bhanu, B. Human ear recognition in 3D. *IEEE Transactions on Pattern Analysis and Machine Intelligence* 29 (4):718–737, 2007.

23. Dalal, N. and Triggs, B. Histograms of oriented gradients for human detection. *CVPR 2005* 886–893.

24. Woodard, D. L. and Flynn, P. J. Finger surface as a biometric identifier. *Computer Vision and Image Understanding* 100 (3):357–384, 2005.

25. Malassiotis, S., Aifanti, N., and Strintzis, M. G. Personal authentication using 3-D finger geometry. *IEEE Transactions on Information Forensics and Security* 1 (1):12–21, 2006.

26. Woodard, D. L. and Flynn, P. J. 3D finger biometrics. *BioAW 2004*, LNCS 3087, 238–247, 2004.

27. Woodard, D. L. and Flynn, P. J. Finger surface as a biometric identifier. *Computer Vision and Image Understanding* 100 (3):357–384, 2005.

28. Kanhangad, V., Kumar, A., and Zhang, D. Combining 2D and 3D Hand geometry features for biometric verification. *CVPR 2009.*

29. Kanhangad, V., Kumar, A., and Zhang, D. Human hand identification with 3D hand pose variations. Computer Vision and Pattern Recognition Workshop, 17–21, 2010.

30. Tsalakanidou, F., Malassiotis, S., and Strintzis, M. G. A 3D face and hand biometric system for robust user-friendly authentication. *Pattern Recognition Letters* 28:2238–2249, 2007.

31. Moghaddam, B., Wahid, W., and Pentland, A. Beyond eigenfaces: Probabilistic matching for face recognition. *Proceedings International Conference on Automatic Face and Gesture Recognition* (FGR '98) 30–35, Nara, Japan, 1998.

32. Rubesh, P. M., Bajpai, G., and Bhaskar, V. Online multi-parameter 3D signature verification through curve fitting. *IJCSNS International Journal of Computer Science and Network Security* 9 (5): 38–44, May 2009.

33. Haskell, R. E., Hanna, D. M., and Sickle, K. V. 3D signature biometrics using curvature moments. *International Conference on Artificial Intelligence* 718–721, Las Vegas, Nevada, 2006.

34. Gafurov, D. A survey of biometric gait recognition: Approaches, security and challenges. This paper was presented at the NIK-2007 conference; see http://www.nik.no/

35. Liu, Z. and Sarkar, S. Simplest representation yet for gait recognition: Averaged silhouette. *International Conference on Pattern Recognition* 211–214, 2004.

36. Liu, Z., Malave, L., and Sarkar, S. Studies on silhouette quality and gait recognition. *Computer Vision and Pattern Recognition* 704–711, 2004.

37. BenAbdelkader, C., Cutler, R., and Davis, L. Stride and cadence as a biometric in automatic person identification and verification. *Fifth IEEE International Conference on Automatic Face and Gesture Recognition* 357–362, May 2002.

38. Johnson, A. Y. and Bobick, A. F. A multi-view method for gait recognition using static body parameters. *Third International Conference on Audio- and Video-Based Biometric Person Authentication* 301–311, June 2001.

39. Gafurov, D., Snekkenes, E., and Buvarp, T. E. Robustness of biometric gait authentication against impersonation attack. *First International Workshop on Information Security (IS'06),* OnTheMove Federated Conferences (OTM'06), 479–488, Montpellier, France, Oct. 30–Nov. 1 2006. Springer LNCS 4277.

40. Gafurov, D., Helkala, K., and Sondrol, T. Gait recognition using acceleration from MEMS. *1st IEEE International Conference on Availability, Reliability and Security (ARES)* 432–437, Vienna, Austria, April 2006.

41. Laurentini, A. The visual hull concept for silhouette-based image understanding. *IEEE Transactions on Pattern Analysis and Machine Intelligence* 16 (2): 150–162, 1994.

42. Shakhnarovich, G., Lee, L., and Darrell, T. Integrated face and gait recognition from multiple views. *CVPR* 1:439–446, 2001.

43. Dockstader, S. L. and Tekalp, A. M. A kinematic model for human motion and gait analysis. *Proceedings of the Workshop on Statistical Methods in Video Processing (ECCV)*, 49–54, Copenhagen, Denmark, June 2002.

44. Urtasun, R. and Fua, P. 3D tracking for gait characterization and recognition. *Proceedings 6th International Conference Automatic Face and Gesture Recognition* 17–22, 2004.

45. Guoying, Z., Guoyi, L., Hua, L., and Pietikäinen, M. 3D gait recognition using multiple cameras. *Proceedings of the 7th International Conference on Automatic Face and Gesture Recognition (FGR'06)*, 529–534, April 2006.

46. Shakhnarovich, G., Lee, L., and Darrell, T. Integrated face and gait recognition from multiple views. *CVPR* 1:439–446, 2001.

47. Bhanu, B. and J. Han. Human recognition on combining kinematic and stationary features. *Proceedings International Conference Audio, Video-Based Biometric Person Authentication* 2688:600–608, Guildford, UK, 2003.

48. Seely, R. D., Goffredo, M., Carter, J. N., and Nixon, M. S. View invariant gait recognition. *Handbook of Remote Biometrics Advances in Pattern Recognition* Part I, 61–81, 2009.

49. Seely, R., Carter, J., and Nixon, M. Spatio-temporal 3D gait recognition. In *3D video—analysis, display and applications*. London: The Royal Academy of Engineering, 2008. ePrint ID:267082, http://eprints.soton.ac.uk/id/eprint/267082

Index

Printed and bound by CPI Group (UK) Ltd, Croydon, CR0 4YY

21/10/2024

01777112-0007